OXFORD LOGIC GUIDES

Series Editors

A. J. MACINTYRE   D. S. SCOTT

Emeritus Editors

D. M. Gabbay
John Shepherdson

OXFORD LOGIC GUIDES

For a full list of titles please visit
http://www.oup.co.uk/academic/science/maths/series/OLG/

21. C. McLarty: *Elementary Categories, Elementary Toposes*
22. R. M. Smullyan: *Recursion Theory for Metamathematics*
23. Peter Clote and Jan Krajícek: *Arithmetic, Proof Theory, and Computational Complexity*
24. A. Tarski: *Introduction to Logic and to the Methodology of Deductive Sciences*
25. G. Malinowski: *Many Valued Logics*
26. Alexandre Borovik and Ali Nesin: *Groups of Finite Morley Rank*
27. R. M. Smullyan: *Diagonalization and Self-Reference*
28. Dov M. Gabbay, Ian Hodkinson, and Mark Reynolds: *Temporal Logic: Mathematical Foundations and Computational Aspects, Volume 1*
29. Saharon Shelah: *Cardinal Arithmetic*
30. Erik Sandewall: *Features and Fluents,* Volume I: *A Systematic Approach to the Representation of Knowledge about Dynamical Systems*
31. T. E. Forster: *Set Theory with a Universal Set: Exploring an Untyped Universe,* second edition
32. Anand Pillay: *Geometric Stability Theory*
33. Dov M. Gabbay: *Labelled Deductive Systems*
34. R. M. Smullyan and M. Fitting: *Set Theory and the Continuum Problem*
35. Alexander Chagrov and Michael Zakharyaschev: *Modal Logic*
36. G. Sambin and J. Smith: *Twenty-Five Years of Martin-Löf Constructive Type Theory*
37. María Manzano: *Model Theory*
38. Dov M. Gabbay: *Fibring Logics*
39. Michael Dummett: *Elements of Intuitionism,* second edition
40. D. M. Gabbay, M. A. Reynolds, and M. Finger: *Temporal Logic: Mathematical Foundations and Computational Aspects,* Volume 2
41. J. M. Dunn and G. Hardegree: *Algebraic Methods in Philosophical Logic*
42. H. Rott: *Change, Choice and Inference: A Study of Belief Revision and Nonmonotoic Reasoning*
43. Peter T. Johnstone: *Sketches of an Elephant: A Topos Theory Compendium,* Volume 1
44. Peter T. Johnstone: *Sketches of an Elephant: A Topos Theory Compendium,* Volume 2
45. David J. Pym and Eike Ritter: *Reductive Logic and Proof Search: Proof Theory, Semantics and Control*
46. D. M. Gabbay and L. Maksimova: *Interpolation and Definability: Modal and Intuitionistic Logics*
47. John L. Bell: *Set Theory: Boolean-Valued Models and Independence Proofs,* third edition
48. Laura Crosilla and Peter Schuster: *From Sets And Types to Topology and Analysis: Towards Practicable Foundations for Constructive Mathematics*
49. Steve Awodey: *Category Theory*
50. Roman Kossak and James Schmerl: *The Structure of Models of Peano Arithmetic*
51. André Nies: *Computability and Randomness*
52. Steve Awodey: *Category Theory,* second edition
53. Byunghan Kim: *Simplicity Theory*

Simplicity Theory

BYUNGHAN KIM

Yonsei University, Seoul, Republic of Korea

OXFORD
UNIVERSITY PRESS

OXFORD
UNIVERSITY PRESS

Great Clarendon Street, Oxford, OX2 6DP,
United Kingdom

Oxford University Press is a department of the University of Oxford. It furthers the University's objective of excellence in research, scholarship, and education by publishing worldwide. Oxford is a registered trade mark of Oxford University Press in the UK and in certain other countries

First Edition published in 2014

Published in the United States of America by Oxford University Press
198 Madison Avenue, New York, NY 10016, United States of America

British Library Cataloguing in Publication Data
Data available

Library of Congress Control Number: 2013938289

ISBN 978-0-19-856738-7

PREFACE

This book is about simple first-order theories. The class of simple theories was introduced by S. Shelah in the early 1980s [112]. Then several specific algebraic structures having simple theories were studied by leading researchers, notably E. Hrushovski [23], [26], [55], [58], [62]. In the mid-90s the author established in his thesis the symmetry and transitivity of non-forking in simple theories [65], [66] and, with A. Pillay, type-amalgamation over models and Lascar strong types [74]. Since then a great deal of research work has been produced on *simplicity theory*, the study of simple theories and structures.

The class of simple theories properly contains that of stable theories. Since the 1960s and 1970s, Shelah's stability theory (= the study of stable theories/structures) [113] has become a main subject of model theory. In particular, geometric stability theory [99], studying the geometrical nature of forking for stable structures, has turned out to be a major technical bridge connecting model theory and its applications to number theory and algebraic geometry. Hrushovski's full resolution of the function-field version of the Mordell–Lang conjecture in number theory [53] using Zilber's principle [59] is a spectacular example.

Simplicity theory has been developed by generalizing the results in stability theory, investigating the more complicated nature of simple structures, and revealing the interrelationship between stability theory and simple structures. The initial developments in simplicity theory, mostly focused on general foundational issues, are surveyed in [75], and an expository book by F. O. Wagner was published in 2000 [120].

During the past several years, the research trends in simplicity theory have expanded to include geometric simplicity theory, which primarily concerns the combinatorial geometric or algebraic sides of simplicity theory, parallel to geometric stability theory. One of the main goals of this new book, in addition to being a guide to graduate students studying model theory, is to introduce the new results on geometric simplicity theory.

Throughout this book, some familiarity with basic model theory and stability theory summarized in Chapter 1 is assumed. In Chapter 2 basic notions such as dividing, forking, simplicity, and the tree property are introduced. The fundamental theorem of forking is proved, and several rank notions are presented. The non-finiteness (except 1) of the number of a countable supersimple theory is shown, and some examples of simple theories are given.

Chapter 3 is devoted to Lascar strong types, and type amalgamation over Lascar strong types is proved. One goal is to show that basic independence relations together with type amalgamation characterize simplicity and forking.

In Chapter 4 we begin to work with hyperimaginaries. A hyperimaginary is an equivalence class of a type-definable equivalence relation. These are necessary as the canonical base for a Lascar type in a simple theory exists in the form of a hyperimaginary. Forking makes sense with respect to hyperimaginaries and basic non-forking independence relations and type amalgamation can be restated in terms of hyperimaginaries. Deeper forking relationships between types are introduced and studied in the last section.

Chapter 5 addresses the issue of elimination of hyperimaginaries. The Lascar topological group of automorphisms and its quotient groups are investigated as well. A theory has elimination of hyperimaginaries if on each complete type, every type-definable equivalence relation is a conjunction of definable equivalence relations. Equivalently, any hyperimaginary is interdefinable with a sequence of imaginaries. If a simple theory has elimination of hyperimaginaries then Lascar types are strong types and canonical bases exist as imaginaries. That low (small, resp.) simple theories eliminate bounded (finitary, resp.) hyperimaginaries is shown. The main result by Buechler, Pillay, and Wagner is that any supersimple theory eliminates hyperimaginaries. The result has consequences in regard to the definability and canonical bases of Lascar types.

In Chapter 6, several methods of constructing simple structures, including the Hrushovski construction, are introduced. These methods are used in finding a counter-example, or producing a canonical unstable simple structure (preserving non-forking independence) from a given stable structure, or used as a technical tool for obtaining deeper results.

Chapter 7 develops the theory of groups in simple structures, most of which were established by F. O. Wagner following A. Pillay's initial work [98]. The group can either be type-definable or hyperdefinable. Commensurativity, a generalization of the group theoretic notion commensurability, and local connectivity play crucial roles in developing hyperdefinable group theory. Important results on 1-based groups and supersimple groups are stated. In particular, the canonical vector space structure found in a minimal connected 1-based group will be used in Chapter 9, together with the procedure for obtaining a hyperdefinable group and its hyperdefinable action from generically given data.

Chapter 8 deals with geometric simplicity theory. Investigating and extending Zilber's principle (for stability theory) into simplicity theory is a motivation for this chapter. Although its full resolution, as in stability theory, is far beyond the reach of current knowledge, some initial results and partial answers are provided for ω-categorical context.

Chapter 9 continues Chapter 8 in an arbitrary context. The group configuration theorem is established under generalized amalgamation. Then by combining results from Chapter 7, a canonical vector space over the division ring of endogenies is hyperdefinably recovered from any non-trivial modular structure. Generalized amalgamation notions themselves are studied as well. Its evolvement toward homology theory is an interesting topic but is not dealt with in this book. Current topics of research, some are even non-first order, going beyond simplicity (but analogous to it) are not dealt with either.

It should be noted that there is absolutely no intention here to make an encyclopaedic presentation of simplicity theory, which nowadays is an almost impossible task as too much has been developed to completely absorb (or even present) it all. The results in this book were selected (within the author's narrow knowledge) mainly to bring the reader up to speed on the current works on (geometric) simplicity theory.

In March 2012, during his visit to Bogotá, Colombia, the author's computer was stolen, in which part of the LaTeX file for this book written until then had been stored. No other saved LaTeX files were found; however, a pdf file was stored online. Special thanks are given to Hyeung-Joon Kim, SunYoung Kim, and Junguk Lee, who helped re-type a compilable LaTeX file from the pdf file. Part of this book was written during the author's visit to MIT in 2012 and thanks are given for the use of its facilities. The author expresses his deep thanks to his family for their love and patience. This book was supported by NRF of Korea Grant 2011-812-A00042.

CONTENTS

1

Introduction

We assume that the reader has some familiarity with model theory and basic stability theory described in [47], [88], [114] or [113], [5], [97], [99], [119]. We use standard notations and terminologies common in modern model theory. Ordinals are denoted by $\alpha, \beta, \ldots$ and cardinals by $\kappa, \lambda, \ldots$. As usual $\omega = \omega_0 = \aleph_0$. For a set S, $|S|$ or $\mathrm{Card}(S)$ denotes its cardinality. A set being "countable" means, depending on context, either "at most countable" or "countably infinite." A symbol ":=" abbreviates "is defined as." For ordinals, we take the Cantor normal form convention. Namely, when we write $\alpha = \sum_{i \leq k} \omega^{\alpha_i}.n_i$, it means $\alpha = \omega^{\alpha_0}.n_0 + \cdots + \omega^{\alpha_k}.n_k$, where ordinals $\alpha_0 > \ldots > \alpha_k$ and $n_i < \omega$. Any nonzero ordinal is uniquely expressed as such (called the *Cantor normal form*) if we require all the summands $n_i > 0$. For $\beta = \sum_{i \leq k} \omega^{\alpha_i}.m_i$, we define the *commutative sum* $\alpha \oplus \beta := \sum_{i \leq k} \omega^{\alpha_i}.(n_i + m_i)$. Note that $\alpha + \beta \leq \alpha \oplus \beta = \beta \oplus \alpha$, and $(\alpha + 1) \oplus \beta = (\alpha \oplus \beta) + 1$. When both are finite, clearly $\alpha + \beta = \alpha \oplus \beta$. For $\bar{u} \in \omega^n$, a sequence of numbers of length n, and $i \in \omega$, the concatenation $\bar{u}^\frown \langle i \rangle \in \omega^{n+1}$ can loosely be written as $\bar{u}^\frown i$ or even $\bar{u}i$. When $i \leq n$, of course $\bar{u} \lceil i$ means the restriction of $\bar{u}$ to the domain $i = \{0, \ldots, i-1\}$. We use Pr_i to denote a projection function sending $(a_1, \ldots, a_n)$ to a_i.

Throughout this book T is a complete first-order theory in a language $\mathcal{L}$. We assume $|T|$ is always infinite (by adding sentences implied from T), so $|T| = |\mathcal{L}| + \omega$. By abusing notation we write $\varphi \in \mathcal{L}$ to mean that φ is an $\mathcal{L}$-formula. We work with its fixed saturated model $\mathcal{M}$, so $\mathrm{Th}(\mathcal{M}) := \{\sigma \in \mathcal{L} \mid \sigma \text{ sentence and } \mathcal{M} \models \sigma\} = \{\sigma \in \mathcal{L} \mid \sigma \text{ sentence and } T \models \sigma\}$. We also fix $\bar{\kappa}$ the size (= cardinality) of (the universe of) $\mathcal{M}$. Throughout this book, *boundedly many* means 'some κ-many' with $\kappa < \bar{\kappa}$. We also use $\mathcal{M}$ to denote its universe. Unless said otherwise, sets $A, B, C, \ldots$ are *small* subsets of $\mathcal{M}$ (i.e., whose sizes are $< \bar{\kappa}$), and models $M, N, \ldots$ are small elementary submodels of $\mathcal{M}$. But occasionally we do deal with subsets of $\mathcal{M}$ whose size is $\bar{\kappa}$, particularly when the set is the solution set of a type. Tuples $\bar{a}, \bar{b}, \ldots$ are from $\mathcal{M}$, i.e., $\bar{a}, \bar{b}, \ldots \in \mathcal{M}^\alpha$ (often α is omitted) with $|\alpha| < \bar{\kappa}$. We may even simply use $a, b, c, \ldots$ to denote (possibly infinite) tuples, so $a \in A$ indeed means $a \in A^\alpha$, i.e., a is a tuple from A with its domain some α. It will be clear up to context whether they are singletons or finite or possibly infinite tuples. Similarly $\bar{x}, \bar{y}, \ldots, x, y, z, \ldots$ denote sequences of variables whose lengths will be clear up to context. For example, for a formula $\varphi(x, y) \in \mathcal{L}$ and its *instance* $\varphi(x, a)$, the variables x, y and the parameter tuple a are finite. We may loosely say $\bar{x} = (x_i | i \in I)$ a variable (instead of a variable sequence) or variables by referring to x_i's. (But from Chapter 4 when we start to deal with hyperimaginaries, those letters used for tuples and sets will be used for hyperimaginaries.) As usual, AB denotes $A \cup B$ and $A\bar{b}$ denotes $A \cup \{b_i | i \in I\}$, where $\bar{b} = (b_i | i \in I)$. Similarly, up to context, $\{\bar{a}_i | i \in I\}$ may mean $\{a_i^j | i \in I, j \in I_i\}$, where $\bar{a}_i = (a_i^j | j \in I_i)$. In this sense we may occasionally confuse $\in$ and $\subseteq$, since $\bar{c} = (c_i | i \in I) \in A$ can mean $\{c_i | i \in I\} \subseteq A$. For a set A, T_A or $T(A)$ means $\mathrm{Th}(\mathcal{M}_A)$, where, as usual, $\mathcal{M}_A$ is $\mathcal{M}$ with the named A.

A formula is said to be *over* a (parameter) set A if it is an $\mathcal{L}(A)$-formula. Whether a given formula contains a parameter will be clear up to context. A loose convention is that if a formula is written as $\varphi(x)$ then it possibly contains some parameter; and if written $\psi(x, y)$ then it is an $\mathcal{L}$-formula, where x is an object variable and y is a parameter variable, so $\psi(x, a)$ is an instance of $\psi(x, y)$. A *type* (of a variable $\bar{x}$) over A (in T) is a collection of formulas over A having free variables among $\bar{x}$. A type is sometimes called a *partial type*. A single formula φ (instead of $\{\varphi\}$) may also be called a type by identifying those two. Mostly we use $p, q, r, \ldots$ to denote types. Unless said otherwise, a type $p(\bar{x})$ is consistent (with T); $\mathrm{dom}(p) :=$ the set of parameters of p; and its *solution set* means the set of realizations, i.e., $\{\bar{a} \in \mathcal{M} | \mathcal{M} \models p(\bar{x})[\bar{a}]\}$. For a type $q(\bar{x})$ we write $p \equiv q$ if the solution sets of p, q are equal; and $p(x) \vee q(x) := \{\varphi(x) \vee \psi(x) | \varphi \in p, \psi \in q\}$; $p(x) \wedge q(x) := p \cup q$. The type $p(\bar{x})$ is called *isolated* if there are finitely many formulas in $p(\bar{x})$ whose conjunction implies every formula in the type under T; it is said to be *algebraic* if its solution set is finite (so isolated by some algebraic formula). The type $p(\bar{x})$ is said to be a *complete type* over A if it is over A and is maximal consistent (in $\bar{x}$ over A with T), so then $A = \mathrm{dom}(p)$; if in addition $|\bar{x}| = n < \omega$, it is called a complete n-type. For a tuple $\bar{a}$ and a set A, $\mathrm{tp}(\bar{a}/A)$ (in a variable $\bar{x}$ with a bijection $\bar{a} \to \bar{x}$) is the collection of formulas $\varphi(\bar{x}')$ over A where $\bar{x}'$ is a finite subvariable of $\bar{x}$ such that $\mathcal{M} \models \varphi(\bar{x}')[\bar{a}']$ with the corresponding finite $\bar{a}' \subseteq \bar{a}$. Mostly the bijection is implicit. So when we write $\mathrm{tp}(B/A)$, an enumeration of the set B is assumed. Clearly $\mathrm{tp}(\bar{a}/A)$ is a complete type and any complete type is of this form.

The set of automorphisms of $\mathcal{M}$ is denoted by $\mathrm{Aut}(\mathcal{M})$, and

$$\mathrm{Aut}_A(\mathcal{M}) := \{f \in \mathrm{Aut}(\mathcal{M}) | f \text{ fixes } A \text{ pointwise}\}.$$

An automorphism $f \in \mathrm{Aut}_A(\mathcal{M})$ is said be an *A-automorphism*. A (possibly big) set X is said to be *A-invariant* (or invariant over A) if X is setwise fixed by any A-automorphism. It is said to be *type-definable* over A (or A-type-definable) if it is the solution set of some type $\pi(x)$ over A. It is *definable* over A (A-definable) if X is the solution set of some formula over A (so then $X \subseteq \mathcal{M}^n$ for some finite n). Due to $\mathcal{M}$ being saturated, $\mathrm{tp}(\bar{a}/A) = \mathrm{tp}(\bar{b}/A)$ iff there is an A-automorphism f such that $f(\bar{a}) = \bar{b}$. We often write $\bar{a} \equiv_A \bar{b}$ to mean $\mathrm{tp}(\bar{a}/A) = \mathrm{tp}(\bar{b}/A)$. Given a set A (with an implicit enumeration of distinct 1-variables $x_1, x_2, \ldots$) $S_n(A)$ is the collection of complete n-types over A (in $x_1 \ldots x_n$), and $S_{<\omega}(A) := \bigcup_{n<\omega} S_n(A)$, $S_n(T) := S_n(\emptyset)$. Note that $S_n(A)$ forms a Stone topological space where the basic open set is of the form $\{p(x) \in S_n(A) | \phi(x) \in p(x)\}$ for each $\phi(x) \in \mathcal{L}$. We reserve $S(A)$ for a larger set. Namely $p(\bar{x}) \in S(A)$ means that $p(\bar{x})$ is a complete type over A having an arbitrary variable $\bar{x}$.

1.1 Model Theory

We point out first the following folklore on invariant or type-definable sets which will be used freely throughout.

Fact 1.1.1

(1) *If a set X is A-invariant then X is a union of boundedly many sets X_i where each X_i is the solution set of a complete type over A.*

(2) *If X is type-definable and at the same time A-invariant, then X is A-type-definable.*

Now we recall $\mathcal{M}^{eq}$. Given an $\emptyset$-definable equivalence relation E on $\mathcal{M}^n$ (so on M^n too for any $M \prec \mathcal{M}$), we often want to deal with its equivalence classes as objects in definable manner. More precisely it would be desirable if there is an $\emptyset$-definable function f such that for $\bar{a}, \bar{b} \in \mathcal{M}^n$, $E(\bar{a}, \bar{b})$ holds iff $f(\bar{a}) = f(\bar{b})$. Then we can consider the definable set $\operatorname{ran} f$ as M^n/E. Indeed seemingly a weaker condition called *elimination of imaginaries* is equivalent to this.

Definition 1.1.2 T is said to have *elimination of imaginaries* (in short *e.i.*) if for any $n \geq 1$ and any definable set X on $\mathcal{M}^n$ over a parameter, there is some finite tuple c from $\mathcal{M}$ such that for any $\sigma \in \operatorname{Aut}(\mathcal{M})$, σ fixes X setwise iff it fixes c pointwise. Such c is called a *canonical parameter* or a *name*, or a *code* of the definable set X.

Fact 1.1.3 *The following are equivalent.*

(1) *T has elimination of imaginaries.*

(2) *For each $n \geq 1$, and every $\emptyset$-definable equivalence relation E on $\mathcal{M}^n$ with $a \in \mathcal{M}^n$, there is a finite tuple c from $\mathcal{M}$ such that for any automorphism σ of $\mathcal{M}$, σ fixes the equivalence class a/E setwise iff it fixes c pointwise.*

(3) *For each $n \geq 1$, and every $\emptyset$-definable equivalence relation E on $\mathcal{M}^n$, $\mathcal{M}^n/E$ is partitioned into finitely many $\emptyset$-definable sets $Y_0, \ldots, Y_k$ of $\mathcal{M}^n$, and there are $\emptyset$-definable functions $f_i : Y_i \to \mathcal{M}^{k_i}$ $(i \leq k)$ such that for $a, b \in Y_i$, $E(a, b)$ holds iff $f_i(a) = f_i(b)$.*

In particular, T has elimination of imaginaries with two $\emptyset$-definable m-tuples if and only if, for each $n \geq 1$, and every $\emptyset$-definable equivalence relation E on $\mathcal{M}^n$, there is an $\emptyset$-definable function $f : \mathcal{M}^n \to \mathcal{M}^k$ such that for $a, b \in \mathcal{M}^n$, $E(a, b)$ holds iff $f(a) = f(b)$. Unfortunately, not every theory has elimination of imaginaries. But there is a standard way to naturally impose this condition to the original structure without varying the original definable relations, and the resulting extended/expanded structure is called $\mathcal{M}^{eq}$.

There are a couple of different approaches to construct $\mathcal{M}^{eq}$, although both are eventually the same. In this book we prefer the following approach. For each $n \geq 1$ and each $\emptyset$-definable equivalence relation on $\mathcal{M}^n$ defined by say $E(\bar{x}, \bar{y})$, we add a unary predicate S_E and an n-ary function symbol f_E to $\mathcal{L}$, which constitute $\mathcal{L}^{eq}$. In particular there is $S_=$ in $\mathcal{L}^{eq}$ where $=$ is an equivalence relation on $\mathcal{M}^1$ defined by equality $x = y$. By putting indexes if necessary, we can assume $\mathcal{M}^n/E$ (= the set of E-equivalence classes) and $\mathcal{M}^n/F$ are disjoint for distinct equivalence relations (this assumption is for the sake of convenience, not essential though). The universe of $\mathcal{M}^{\mathrm{eq}}$ is the disjoint union of all of $\mathcal{M}^n/E$. The interpretations of S_E are $\mathcal{M}^n/E$ called S_E*-sort*, and f_E is the projection map $\mathcal{M}^n \to \mathcal{M}^n/E$ (while fixing all other n-tuples from $\mathcal{M}^{\mathrm{eq}}$). We identify $\mathcal{M}$ as $\mathcal{M}/=$, which is called the *home-sort*. Now T^{eq} is the theory of $\mathcal{M}^{\mathrm{eq}}$. It can be easily seen that T^{eq} does not depend on the choice of $\mathcal{M}$. T^{eq} is axiomatized by $T_=$, which is the same as T but each quantified variable is relativized in $S_=$, together with sentences saying that all sorts are disjoint, and $f_E : (S_=)^n \to S_E$ is a surjection such that for $\bar{a}, \bar{b} \in (S_=)^n$, $E(\bar{a}, \bar{b}) \leftrightarrow f_E(\bar{a}) = f_E(\bar{b})$, (while f_E fixes all other n-tuples). When we say $\varphi(x_0, \ldots, x_n)$ is an $\mathcal{L}^{\mathrm{eq}}$-formula with the variable x_i in the sort S_{E_i} $(i \leq n)$, it means it is an abbreviation of $\bigwedge_{i \leq n} S_{E_i}(x_i) \wedge \varphi(x_0, \ldots, x_n)$.

Fact 1.1.4

(1) *T^{eq} has elimination of imaginaries.*

(2) *For finite tuples $a, b \in \mathcal{M}$, $\mathrm{tp}_{\mathcal{M}}(a) = \mathrm{tp}_{\mathcal{M}}(b)$ iff $\mathrm{tp}_{\mathcal{M}^{\mathrm{eq}}}(a) = \mathrm{tp}_{\mathcal{M}^{\mathrm{eq}}}(b)$.*

(3) *Let $\varphi(x_1, \ldots, x_k)$ be an $\mathcal{L}^{\mathrm{eq}}$-formula with $x_i \in S_{E_i}$. Then there is an $\mathcal{L}$-formula $\psi(y_1, \ldots, y_k)$ (so y_i in home-sort) such that*

$$T^{\mathrm{eq}} \models \forall y_1 \ldots y_k (\psi(y_1, \ldots, y_k) \leftrightarrow \varphi(f_{E_1}(y_1), \ldots, f_{E_k}(y_n))$$

holds.

Note that a model of T^{eq} can contain an element which does not belong to any sort. $\mathcal{M}^{\mathrm{eq}}$ is simply an elementary substructure of a saturated model of T^{eq}, but in which these superfluous elements carrying no relations are omitted. In so far as dealing with elements and sets from the sorts, $\mathcal{M}^{\mathrm{eq}}$ is a saturated model of T^{eq}. For $M \prec \mathcal{M}$ the same construction gives rise to $M^{\mathrm{eq}} \models T^{\mathrm{eq}}$. Any isomorphism between two models M and N of T is uniquely extended to an isomorphism between M^{eq} and N^{eq}, and if $M \prec N$ then $M^{\mathrm{eq}} \prec N^{\mathrm{eq}}$. Any $\sigma \in \mathrm{Aut}(M)$ is uniquely extended to an automorphism of $\mathrm{Aut}(M^{\mathrm{eq}})$, so there is a canonical bijection between $\mathrm{Aut}(M)$ and $\mathrm{Aut}(M^{\mathrm{eq}})$ and we can identify those two. A tuple from $\mathcal{M}$ is called *real*, and a tuple from $\mathcal{M}^{\mathrm{eq}}$ is *imaginary*.

Another standard way to see $\mathcal{M}^{\mathrm{eq}}$ is considering it as a many-sorted structure, so each S_{E_i} is a separate sort and each variable is designated to a specific sort. As a many-sorted structure, $\mathcal{M}^{\mathrm{eq}}$ does not have superfluous elements. In Chapter 4, we shall introduce another canonical way of dealing with quotient spaces of $\mathcal{M}$, in terms of *hyperimaginaries*. Until then we stay with our construction of $\mathcal{M}^{\mathrm{eq}}$.

Now given a set $X \subset \mathcal{M}^{\mathrm{eq}}$ (X can be small or big),

$$\mathrm{acl}(X) := \{\text{singleton } c \in \mathcal{M}^{\mathrm{eq}} |\ \mathrm{tp}(c/A) \text{ is algebraic for some small } A \subseteq X\}$$

is called the *algebraic closure* of X and

$$\mathrm{dcl}(X) := \{\text{singleton } c \in \mathcal{M}^{\mathrm{eq}} |\ c \text{ is the unique realization of } \mathrm{tp}(c/A) \text{ with small } A \subseteq X\}$$

is the *definable closure* of X. Occasionally we write $\mathrm{acl}^{\mathrm{eq}}(X), \mathrm{dcl}^{\mathrm{eq}}(X)$ to emphasize the closures are taken from $\mathcal{M}^{\mathrm{eq}}$. Given an A-invariant set X,

$$X^{\mathrm{eq}} := \mathrm{dcl}^{\mathrm{eq}}(X \cup A).$$

That a structure is *interpretable* in X means it is definable in X^{eq}. An element in $\mathrm{acl}(A)$ is said to be *algebraic* over A, and that in $\mathrm{dcl}(A)$ is *definable* over A. The set A is *algebraically* (*definably*, resp.) *closed* if $A = \mathrm{acl}(A)$ ($= \mathrm{dcl}(A)$, resp.). If $\mathrm{acl}(a) = \mathrm{acl}(b)$ ($\mathrm{dcl}(a) = \mathrm{dcl}(b)$, resp.) we say a, b are *interalgebraic* (*interdefinable*, resp.). For example, any two canonical parameters of a definable set are interdefinable; T has elimination of imaginaries iff any imaginary element is interdefinable with some finite real tuple. In a similar manner we say T has *weak* (or *geometric*) *elimination of imaginaries* if for any imaginary element $e \in \mathcal{M}^{\mathrm{eq}}$, there is $c \in \mathcal{M}^n$ such that $e \in \mathrm{dcl}(c)$ and $c \in \mathrm{acl}(e)$.

Throughout this book, we shall freely pass from T to T^{eq}, and work with $\mathcal{M}^{\mathrm{eq}}$. In other words, we will not bother with mentioning precisely when working with T will suffice and when working with T^{eq} is necessary, although the distinction will be clear up to context. Therefore it will be safe to assume that, unless obviously meant otherwise, our underlying theory is T^{eq} (even if written T) and our $\mathcal{M}$ indeed is $\mathcal{M}^{\mathrm{eq}}$. In this regard, an *imaginary* tuple is a *real* tuple too. But of course when needed, we will clearly distinguish $\mathcal{M}^{\mathrm{eq}}$ and its home-sort $\mathcal{M}$.

Recall that an equivalence relation is *finite* if it has finitely many equivalence classes. We say finite tuples a, b have the same *strong* (or *Shelah*) *type* over A, written $a \equiv^s_A b$, if for any A-definable finite equivalence relation E, they are in the same E class. A well-known fact is $a \equiv^s_A b$ iff $a \equiv_{\mathrm{acl}(A)} b$. Hence we take $\mathrm{stp}(a/A)$, the strong type of a over A, as $\mathrm{tp}(a/\mathrm{acl}(A))$. Now then the strong type of an infinite tuple over A makes sense too.

Indiscernible sequences will play an important role in the development of forking theory throughout the book. A sequence $I = \langle a_i | i \in \Lambda \rangle$ of tuples indexed by a linearly ordered set Λ is said to be *A-indiscernible* if for any $i_0 < \ldots < i_n \in \Lambda$ and $j_0 < \ldots < j_n \in \Lambda$, $a_{i_0} \ldots a_{i_n} \equiv_A a_{j_0} \ldots a_{j_n}$. We may call I an A-indiscernible sequence, or simply an A-indiscernible by taking 'indiscernible' as a noun. We say I is an *indiscernible in* (or *of*) $p(x) \in S(A)$ if $p(x) = \mathrm{tp}(a_i/A)$. Ramsey-style arguments are used in finding/extending indiscernible sequences. Recall that for cardinals $\kappa, \mu \leq \lambda$ and $n < \omega$, the notation $\lambda \to (\mu)^n_\kappa$ means that every function $f : [\lambda]^n \to \kappa$ is constant on some μ subset of λ, where $[\lambda]^n = \{F \subseteq \lambda \,|\, |F| = n\}$. Ramsey's theorem says $\omega \to (\omega)^n_m$ hold for all positive integers n, m. The Erdös–Rado theorem is the following: For any infinite λ, and $n \in \omega$, $\beth_n(\lambda)^+ \to (\lambda^+)^{n+1}_\lambda$. In particular, $(2^\omega)^+ \to (\omega_1)^2_\omega$.

The following are consequences of compactness applied to Ramsey's theorem: For any linear order extension Λ' of Λ above, I can be extended to an A-indiscernible I' indexed by Λ'. There is a model $M \supseteq A$ such that I is M-indiscernible. Moreover, if $\Gamma(x_0, x_1, \ldots)$ is a type over A such that $\Gamma_{0\ldots n}(x_0, \ldots, x_n) = \Gamma_{i_0\ldots i_n}(x_0, \ldots, x_n)$ for any $i_0 < \ldots < i_n \in \omega$ and $\Gamma_{i_0\ldots i_n}(x_{i_0}, \ldots, x_{i_n}) := \Gamma \lceil x_{i_0} \ldots x_{i_n}$, then Γ is satisfied by some A-indiscernible. In other words, if a relation on $(x_0, x_1, \ldots)$ is *uniformly type-describable* over A then the relation is satisfied by an A-indiscernible sequence.

Another useful application is as follows.

Lemma 1.1.5 *For any cardinal κ, there is a cardinal λ such that for any set A of size $\leq \kappa$ and for any sequence $I_\lambda = \langle \bar{a}_i : i < \lambda \rangle$ of tuples of size $\leq \kappa + |T|$, there is an A-indiscernible $I = \langle \bar{b}_i : i < \omega \rangle$ so that for each $n < \omega$,* $\mathrm{tp}(\bar{b}_0, \ldots, \bar{b}_n/A) = \mathrm{tp}(\bar{a}_{i_0}, \ldots, \bar{a}_{i_n}/A)$ *for some $i_0 < \ldots < i_n < \lambda$.*

Proof Let $\delta = 2^{\kappa+|T|}$ and $\lambda = \beth_{\delta^+}$.

Claim There is a family $\{\tau_n | n \in \omega\}$ of subsets of δ^+ with $|\tau_n| = \delta^+$ and $\tau_{n+1} \subseteq \tau_n$, satisfying the following: For each τ_n, there is a family $\{X^n_\gamma | \gamma \in \tau_n\}$ of subsets of λ such that

(i) $X^{n+1}_\gamma \subseteq X^n_\gamma$ for $\gamma \in \tau_{n+1}$,

(ii) $|X^n_\gamma| > \beth_{\delta+\alpha}$ where γ is the αth element of τ_n,

(iii) There is a family of types $\{p_n | n \in \omega\}$ such that $p_n \subseteq p_{n+1}$ and $p_n = \mathrm{tp}(\bar{a}_{i_1} \ldots \bar{a}_{i_n}/A)$ for any $i_1 < \ldots < i_n \in X^n_\gamma$ and any $\gamma \in \tau_n$.

Proof of Claim We define τ_n inductively. For $n = 0$, let $p_0 = \emptyset$, $\tau_0 = \delta^+$, and $X^0_\gamma = \lambda$. Now let us suppose that we have found $\tau_0, \ldots, \tau_n$ satisfying the conditions. Let $\nu_n = \{\gamma \in \tau_n | \; |(\gamma \setminus \gamma') \cap \tau_n| = n \text{ for some } \gamma' \in \tau_n\}$. Note that $|\nu_n| = \delta^+$, and γ is the $(\alpha + n)$th element of τ_n if and only if γ is the αth element of ν_n. Then given γ, an αth element of $\nu_n \subseteq \tau_n$, we have:

(1) $|X^n_\gamma| > \beth_{\delta+\alpha+n} = \beth_n(\beth_{\delta+\alpha})$.

We shall define X^{n+1}_γ. Assign a complete type $\mathrm{tp}(\bar{a}_{i_1}, \ldots, \bar{a}_{i_{n+1}}/A)$ to each $i_1 < \ldots < i_{n+1} \in X^n_\gamma$. Since there are $\leq \delta$ many such types, the Erdös–Rado theorem with (1) enables us to find a subset, say X^{n+1}_γ, of X^n_γ such that

(2) $|X^{n+1}_\gamma| > \beth_{\delta+\alpha}$ (γ is the αth element of ν_n),

and $\mathrm{tp}(\bar{a}_{i_1} \ldots \bar{a}_{i_{n+1}}/A)$ are all the same for each $i_1 < \ldots < i_{n+1} \in X^{n+1}_\gamma$. Let us denote the fixed type as p^γ_{n+1}, for $\gamma \in \nu_n$. Since $|\nu_n| = \delta^+$, there is a subset τ_{n+1} of ν_n such that $|\tau_{n+1}| = \delta^+$ and $p_{n+1} = \mathrm{tp}(\bar{a}_{i_1} \ldots \bar{a}_{i_{n+1}}/A)$ for all $i_1 < \ldots < i_{n+1} \in X^{n+1}_\gamma$ and all $\gamma \in \tau_{n+1}$. Moreover, since $X^{n+1}_\gamma \subseteq X^n_\gamma$, we have $p_n \subseteq p_{n+1}$. Finally, since $\tau_{n+1} \subseteq \nu_n \subseteq \tau_n$, (2) says (ii) is true for $n + 1$. Hence, the claim follows.

Now by the conditions in the claim, any sequence realizing $\bigcup_{n<\omega} p_n$ is a desired A-indiscernible sequence. ■

Now let us recall the notion of coheir extension. Assume that $p(x)$ is a type over A which is *finitely satisfiable* in A; i.e., any finite subset of $p(x)$ is realized by a tuple from A. Then by compactness for any $B \supseteq A$, there is an extension $p'(x) \in S(B)$ of p which is also finitely satisfiable in A. $p'(x)$ is called a *coheir* (*extension*) of $p(x)$ over B. Note that any type over a model M is finitely satisfiable. Hence for any $q(x) \in S(M)$ and $B \supseteq M$, there is a coheir of q over B. By duality, q too has an *heir* (*extension*) $q'(x) \in S(B)$ such that for any $\varphi(x, b) \in q'$ with $\varphi(x, y) \in \mathcal{L}(M)$ and $b \in B$, there is $m \in M$ such that $\varphi(x, m) \in q$. Additional facts on coheirs are as follows.

Fact 1.1.6

(1) *Let $A \subseteq B \subseteq C$ and $p \in S(A)$. If $q \in S(B)$ is a coheir of p, then q has an extension r in $S(C)$ which is a coheir of p. Moreover if r in $S(C)$ is a coheir of p, then $r \upharpoonright B$ is a coheir of p.*

(2) *Let $A \subseteq B \subseteq C$ and $p \in S(C)$ be a coheir of $p\lceil A$. If $b_1, b_2 \subseteq C$ and $\text{tp}(b_1/B) = \text{tp}(b_2/B)$, then for any $c \models p$, $\text{tp}(cb_1/B) = \text{tp}(cb_2/B)$.*

We say $I = \langle a_i | i < \omega \rangle$ is a *coheir sequence* of $q \in S(A)$ if $\text{tp}(a_i/A \cup \{a_j | j < i\})$ is a coheir of q and a subset of $\text{tp}(a_{i+1}/A \cup \{a_j | j < i+1\})$ for each $i < \omega$. According to Fact 1.1.6, for any T, a coheir sequence of given finitely satisfiable $q \in S(A)$ (so any $q \in S(M)$) exists and any coheir sequence of q is an indiscernible in q.

Recall that T is said to be *κ-categorical* for an infinite $\kappa \geq \text{Card}(\mathcal{L})$ if any two models of T having the same size κ are isomorphic. Clearly T is κ-categorical iff each sort in T^{eq} is κ-categorical. Morley's categoricity theorem [93] says if a countable T is categorical in some uncountable cardinal then it is categorical in every uncountable cardinal. As is known, T is ω-categorical (so countable) iff for each $n \geq 1$, there are only finitely many $\emptyset$-definable subsets of $\mathcal{M}^n$. In this regard, we say an invariant set X *ω-categorical* if it is definable with its canonical parameter a, and for each $n \geq 1$, there are only finitely many a-definable subsets of X^n. We say T is *small* if it is countable, and for each finite n, $S_n(T)$ is countable. Hence if T is ω-categorical then it is small. As is known a countable T is small iff every $\mathcal{L}$-formula attains an ordinal-valued Cantor–Bendixson (CB) rank (defined in Section 1.2). A model $(M, <, \ldots)$ is *o-minimal* if $<$ ($\in \mathcal{L}$) linearly orders M, and every definable subset of M^1 is a finite union of intervals. As is known if M is o-minimal then so is any elementary equivalent model. Hence we say T is o-minimal if some (every) model of T is o-minimal.

For other basic notions in model theory such as quantifier elimination (q.e.), model-completeness, model-companion, and Fraïsé construction, we refer to [88].

1.2 Stability

In this section we summarize the basics of stability theory. Recall that a formula $\varphi(x, y) \in \mathcal{L}$ is *stable* if there is no sequence $\langle a_i b_i | \ i \in \omega \rangle$ such that $\varphi(a_i, b_j)$ holds iff $i < j$. T is stable if every $\mathcal{L}$-formula is stable. As is known T is stable iff T^{eq} is stable. The formula $\varphi(x, y)$ has the *independence property* if there is $\{a_i | \ i < \omega\}$ such that for each $\tau \subseteq \omega$, $\{\varphi(x, a_i) | \ i \in \tau\} \cup \{\neg\varphi(x, a_j) | \ j \notin \tau\}$ is consistent. $\varphi(x, y)$ has the *strict order property* if there is $\{b_i | \ i < \omega\}$ such that $\models \varphi(x, b_j) \rightarrow \varphi(x, b_i)$ iff $i \leq j$. Shelah showed that T is unstable iff there is a formula having the independence property or the strict order property. In the rest of this section, T in $\mathcal{L}$ indeed is T^{eq} in $\mathcal{L}^{\text{eq}}$.

Now let us first talk about local stability, a theory on stable formulas while underlying T has no restriction. Let an arbitrary formula $\varphi(x; y) \in \mathcal{L}$ be given. Recall that for an imaginary tuple b, we say $\varphi(x; b)$ is an *instance* of φ. By a *φ-formula over A*, we mean a formula over A equivalent to a finite Boolean combination of instances of φ. For example, if $\models \delta(x, a) \leftrightarrow \varphi(x, b) \wedge \neg\varphi(x, c)$, then $\delta(x, a)$ is a φ-formula over a. Also, both $x = x$ and

$c \neq c$ are φ-formulas. By a *complete φ-type over A*, we mean a maximal consistent collection of φ-formulas over A. We shall use $S_\varphi(A)$ to denote the set of all complete φ-types over A. Now for $p(x) \in S_n(A)$, we let

$$p\lceil\varphi := \{\psi(x) \in p \mid \psi \text{ is a } \varphi\text{-formula}\}.$$

Then $p\lceil\varphi \in S_\varphi(A)$ and every complete φ-type over A has this form. Let $p(x)$ be a complete n-type over A, or a complete $\varphi(x, y)$-type over A. We say p is *definable* over B if there is a formula $\psi(y)$ over B such that for any $b \in A$, $\varphi(x, b) \in p$ iff $\psi(b)$ holds. The type p is *definable* if it is definable over A, and we call $\psi(y)$ a *φ-definition* of p.

For an infinite cardinal λ, we say φ (T resp.) is *λ-stable* if for any set $|A| \leq \lambda$, $|S_\varphi(A)| \leq \lambda$ ($|S_{<\omega}(A)| \leq \lambda$, resp.).

Fact 1.2.1 *Let $\varphi(x, y) \in \mathcal{L}$. The following are equivalent:*

(1) *$\varphi(x, y)$ is stable.*
(2) *Any complete φ-type is definable.*
(3) *For any infinite λ, φ is λ-stable.*
(4) *For some infinite λ, φ is λ-stable.*

Fact 1.2.2

(1) *T is stable.*
(2) *Any complete n-type is definable.*
(3) *T is λ-stable for some λ.*

Since any finite structure is ω-stable, unless otherwise there is no harm to assume that T has only infinite models. Now we begin to talk about dividing/forking in stability. We refer to Chapter 2 for Shelah's definitions of dividing and forking.

Fact 1.2.3 *Let $\varphi(x, y)$ be stable. The following are equivalent:*

(1) *$\varphi(x, a)$ does not divide over A.*
(2) *For some model $M \supseteq Aa$, there is $(\varphi(x, a) \in) p(x) \in S_\varphi(M)$ definable over* acl(A).
(3) *$\varphi(x, a)$ is satisfied in any model $M \supseteq A$.*

Fact 1.2.4 *Let $\delta(x, y)$ be stable. (T any.)*

(1) *Let $A \subseteq B$ and let $p(x) \in S_\delta(B)$. Then p does not divide over A iff for some model $M \supseteq B$, p has an extension $p'(x) \in S_\delta(M)$ which is definable over* acl(A).
(2) *Let $A \subseteq B \subseteq C$ and let $q(x) \in S(C)$. Then $q\lceil\delta$ does not divide over A iff $q\lceil\delta$ does not divide over B and $(q\lceil B)\lceil\delta = (q\lceil\delta)\lceil B$ does not divide over A.*
(3) *Let $A \subseteq B \subseteq C$ and let $q(x) \in S_\varphi(B)$. If q does not divide over A then there is an extension $q'(x) \in S_\varphi(C)$ of $q(x)$ which does not divide over A.*

(4) *Let* $A \subseteq B$ *with* $A = \mathrm{acl}(A)$. *If* $q_0(x), q_1(x) \in S_\varphi(B)$ *and both do not divide over* A, *then* $q_0 = q_1$.

(5) *Let* $p(x)$ *be a complete type. For any* $\varphi(x, b) \in p\lceil\delta$, *there is* $\phi(y) \in \mathrm{tp}(b)$ *such that* $\varphi(x, y) \wedge \phi(y)$ *is stable.*

Point (4) says any complete φ-type over an algebraically closed set (e.g., a model) is *stationary*; i.e., over any larger set, the type has the unique non-dividing complete φ-type extension.

Caution In general Shelah's forking is different from dividing even for an instance of a stable formula. In Example 2.1.5(1), $x = x \wedge a = a$ forks over $\emptyset$ but it does not divide over $\emptyset$. Thus in some of the classical literature dealing with local stability, the forking described there is *not* Shelah's forking and only equivalent to dividing.

Now let us turn our attention to the case where the underlying T is stable. If T is stable then forking is equal to dividing, and non-forking supplies a nice independence notion to sets from $\mathcal{M}$, as described in Fact 2.1.4. One of the initial goals of this book is to generalize the fact to the context of simple theories. We add some more facts on forking in stability theory.

Theorem 1.2.5 (Open mapping theorem) *T is stable. Let $A \subseteq B$. Then the map $W \to S_n(A)$ where*

$$W = \{q \in S_n(B) |\ q \text{ does not fork over } A\}$$

sending q to $q\lceil A$ is open (in their Stone topologies).

Corollary 1.2.6 *T stable. Assume that $q(x) \in S_n(B)$ does not fork over $A(\subseteq B)$. If q is isolated then so is $q\lceil A$.*

We now talk about the notion of canonical base for a complete type in a stable theory. Call $p(x) \in S_n(A)$ *stationary* if for any $B \supseteq A$, p has the unique non-forking extension in $S_n(B)$. For stationary p,

$$\mathrm{Cb}(p) := \{\text{the name of } \varphi(x, y)\text{-definition of } p : \varphi \in \mathcal{L}\}.$$

For an n-tuple a and a set A, $\mathrm{Cb}(a/A) := \mathrm{Cb}(\mathrm{tp}(a/\mathrm{acl}(A)))$. Two complete types $p(x), q(x)$ are *parallel* (written $p||q$) if they have a common non-forking extension.

Fact 1.2.7 *T stable. $p(x) \in S_n(A)$ stationary.*

(1) $\mathrm{Cb}(p) \subseteq \mathrm{dcl}(A)$.

(2) *p does not fork over* $\mathrm{Cb}(p)$.

(3) *p does not fork over $B(\subseteq A)$ iff* $\mathrm{Cb}(p) \subseteq \mathrm{acl}(B)$.

(4) *p does not fork over $B(\subseteq A)$ and $p\lceil B$ is stationary iff* $\mathrm{Cb}(p) \subseteq \mathrm{dcl}(B)$.

(5) *For any $\sigma \in \mathrm{Aut}(\mathcal{M})$, $p||\sigma(p)$ iff σ fixes* $\mathrm{Cb}(p)$ *pointwise.*

Recall that for a definable set X, say by $\phi(x) \in \mathcal{L}(\mathcal{M})$, we define $MR(X) = MR(\phi(x))$ the *Morley rank* of X or $\phi(x)$ as follows:

(1) $MR(X) \geq 0$ if $X \neq \emptyset$;

(2) $MR(X) \geq \alpha + 1$ if X is partitioned into infinitely many definable subsets Y_i such that $MR(Y_i) \geq \alpha$;

(3) $MR(X) \geq \delta$ for a limit ordinal δ if $MR(X) \geq \alpha$ for all $\alpha < \delta$;

(4) $MR(X) = \alpha$ if $MR(X) \geq \alpha$ but $MR(X) \not\geq \alpha + 1$.

(5) For a type $p(x)$,

$$MR(p) := \min\{MR(\phi_0 \wedge \ldots \wedge \phi_n) | \ \phi_i(x) \in p\}.$$

If $MR(X) = \alpha$, then X can be partitioned into maximally $n(< \omega)$-many Morley rank α subsets of X; and we put $n = dM(X)$, the *Morley degree* of X. For the type p, its *Morley degree* is the minimal degree of the finite conjunctions of formulas in p having the same Morley rank. The definable set X, or $\phi(x)$, is *strongly minimal* if $MR(X) = 1$ and $dM(X) = 1$; equivalently X is infinite and any definable subset of X is finite or co-finite. We say T (or M) is strongly minimal if $(M \prec)\mathcal{M}^1$ is so. A major step in the proof of Morley's categoricity theorem is to find a strongly minimal definable set which 'controls' a given uncountable categorical structure.

The *Cantor–Bendixson (CB-)rank* of a formula or type is defined by the same manner but taking only $\mathcal{L}$-formulas, i.e., dealing with $\emptyset$-definable sets. If $R(-, \ldots)$ is any one of the introduced ranks in the book, then $R(b/A, \ldots)$ means $R(\text{tp}(b/A), \ldots)$.

We now recall important subclasses of stable theories. We say T is *totally transcendental* (*t.t.*) if the Morley rank of every definable set attains an ordinal value. For a countable T, T is ω-stable iff it is t.t. T is *superstable* if it is stable and any $p(x) \in S_n(A)$ does not fork over a finite subset of A. A fact is that T is superstable iff it is λ-stable for every $\lambda \geq 2^{|T|}$. We say T is *strictly stable* if it is stable but not superstable. If T is ω-stable then it is λ-stable for every infinite λ, so it is superstable. If T is ω-categorical then it is ω-stable iff superstable. We say stable T is *1-based* or *modular* if for any finite a and a set A, $\text{Cb}(a/A) \subseteq \text{acl}(a)$.

Example 1.2.8 Subclasses of stable theories:

$$\{\text{countable } \aleph_1\text{-categorical theories}\} \subset \{\omega\text{-stable theories}\} \subset \{\text{superstable theories}\}.$$

Examples in (n) below are not in the class of (m) for $m < n$.

(1) $\aleph_1$-categorical structures: An infinite set. A vector space $(V, +, 0, \{r : V \to V\}_{r \in F})$ (each r is a scalar multiplication) over a field F. Algebraically closed fields. All are indeed strongly minimal.

(2) ω-stable structures: Differentially closed fields. An infinite set with equivalence relations E_i $(i \in \omega)$ on the set such that E_i refines each E_{i+1} class into infinitely many infinite E_i classes.

(3) Superstable structures: $(\mathbb{N}, +, 0)$. An infinite set equipped with equivalence relations E_i $(i \in \omega)$ such that E_{i+1} refines each E_i class into two infinite E_{i+1} classes.

(4) Stable structures: Separably (but not algebraically) closed fields. A set with equivalence relations E_i $(i \in \omega)$ such that E_{i+1} refines each E_i class into infinitely many infinite E_i classes.

All the field examples above are *not* 1-based, while all others are 1-based. Examples of simple theories will be presented throughout the book, particularly in Section 2.6. Non-forking independence is linear independence in vector spaces, and is algebraic independence in algebraically closed fields. More precisely for $A, B, C \subset V$ the vector space in (1), $A \mathop{\smile\!\!\!\!\!\!|}_B C$ (i.e., $\mathrm{tp}(A/BC)$ does not fork over B) iff whenever vectors $v_0, \ldots, v_n \in A$ are linearly independent over B then so are over BC. For subsets A, B, C of an algebraically closed field, $A \mathop{\smile\!\!\!\!\!\!|}_B C$ iff for any finite tuple $\bar{a}$ from A, the transcendental degree of $\bar{a}$ over B is equal to that over BC.

For the theory of groups in stable structures we refer the reader to [107], [119]. In this book we explicitly use $a.b$ to denote a product of two elements a, b (in groups or ordinal/cardinal products, or even occasionally for a composition of functions) to distinguish from ab, which just means the union or the concatenation. As usual given a subset A of a field F, $\langle A \rangle$ is a subfield generated by A and $\langle A \rangle^{\mathrm{alg}}$ is its field-theoretic algebraic closure in F. More notational conventions on groups will be explained in Chapter 7. Throughout this book, when we restrict our underlying T to (super)stable/simple, we will mention it somewhere clearly (mostly in the beginning of the chapter or section). Of course, (super)stable/simple models/structures/fields will refer to those having mentioned theories. But we save the term 'simple group' for the group theoretic notion, *not* for a group having a simple theory.

1.3 Bibliographical Remarks

The initial landmark result on stability theory is Morley's categoricity theorem [93]. Then S. Shelah developed the entire saga of stability theory [113]. Geometric stability theory comes in mainly by Hrushovski answering Zilber's trichotomy principle (see Chapter 8), which results in spectacular applications to number theory and geometry [59], [53]. Several books on stability theory are available [97], [5], including Pillay's notable exposition on geometric stability theory [99].

The technique used in the proof of Lemma 1.1.5 is according to Morley [20, Theorem 7.2.2]. Notions of heirs and coheirs are according to Lascar and Poizat [86]. Local stability on dividing is introduced by Hrushovski and Pillay [58], [99].

2

Dividing, Forking, and Simplicity

In this chapter we introduce fundamental notions of dividing and forking which place at root in developing the entire subjects of simplicity theory. The notions were originally introduced by Shelah in [113], then they were replaced by equivalent notions but which only make sense in stable theories [86], [97]. In [65], the notions are restudied as stated originally, which discovers the true meanings and roles of the notions in the broader context of simple theories.

2.1 Dividing and Forking

In specific algebraic structures, independence notions between the elements of the structures supply dimension or degree notions on the structures in terms of "independent basis." For example, linear independence in vector spaces and algebraic independence in fields give the notions of "linear basis" and "transcendental basis" to the structures, respectively.

In model theoretic context, we have a uniform notion of "non-forking independence" for arbitrary structures, which corresponds to linear independence and algebraic independence in vector spaces and fields, respectively, as exemplified in Chapter 1. Showing that this "non-forking independence" well-behaves within the class of simple theories is the main goal of this chapter. We start this section by introducing "dividing" and "forking" concepts.

Definition 2.1.1

(1) For an integer $k \geq 1$, and a formula $\varphi(x, c)$, we say $\varphi(x, c)$ *k-divides over* a set A if there is a sequence $\langle c_i | i \in \omega \rangle$ such that:

  (a) $\mathrm{tp}(c/A) = \mathrm{tp}(c_i/A)$ for all i, and

  (b) $\{\varphi(x, c_i) | i \in \omega\}$ is k-inconsistent, i.e., any finite subset of size k is inconsistent.

(2) A type $p(x')$, possibly x' infinite, *k-divides over* A if $p \vdash \varphi(x, a)$, and the formula $\varphi(x, a)$ k-divides over A. (Note that we can always assume $x \subseteq x'$.) The type p *divides over* A if it k-divides over A for some k. (Conventionally, any inconsistent formula divides over any set.)

(3) *p forks over A* if there are formulas $\varphi_1(x, a_1), \ldots, \varphi_n(x, a_n)$ such that:
  (a) $p \vdash \bigvee_{1 \le i \le n} \varphi_i(x, a_i)$, and
  (b) $\varphi_i(x, a_i)$ divides over A for each i.

(4) For $A \subseteq B$, if a complete type p over B does not fork (divide, resp.) over A, then we say p is a *non-forking* (*non-dividing*, resp.) *extension* of $p \lceil A$, and conversely the latter an *unforked* (*undivided*, resp.) *restriction* of the former. For $p_i(x) \in S(A_i)$ $(i = 0, 1)$, we say they are *parallel* (written $p_0 || p_1$) if there is $q(x) \in S(A)$, which is a non-forking extension of both p_0 and p_1.

From the definitions of dividing and forking, we easily obtain the following, which will be used freely throughout the book.

Remark 2.1.2

(1) Assume a type p k-divides over A. Then p k'-divides over A for $k' \ge k$. Moreover, it forks over A.

(2) If both $\phi(x, a)$ and $\psi(x, b)$ fork over A, then $\phi(x, a) \vee \psi(x, b)$ forks over A.

(3) (Monotonicity) Let p, q be types and $p \vdash q$. If q divides (forks, resp.) over A, then p divides (forks, resp.) over any subset of A.

(4) For an $\mathcal{L}$-formula $\varphi(x, y)$ and k, there is a partial type $\Sigma^k_\varphi(y, \bar{z})$ over $\emptyset$ so that $\Sigma^k_\varphi(b, \bar{c})$ holds iff $\varphi(x, b)$ k-divides over $\bar{c}$.

(5) $\{\varphi(x, c)\}$ k-divides over A iff $\varphi(x, c)$ k-divides over A iff for all finite $a \subseteq A$, $\varphi(x, c)$ k-divides over a.

(6) The following are equivalent.
  (a) A type p divides over A.
  (b) There are a formula $\varphi(x, a)$ which is a conjunction of members of p, and an infinite A-indiscernible sequence I containing a such that $\{\varphi(x, a') | \, a' \in I\}$ is inconsistent.
  (c) There is $k \ge 2$ such that for each cardinal λ, we have k-inconsistent $\{p_i | \, i < \lambda\}$, where each p_i is an A-automorphic image of p.

(7) (Existence) If p is over $\mathrm{acl}(A)$, then p does not divide over A.

(8) (Finite character) Let $A \subseteq B$. Then $\mathrm{tp}(C/B)$ does not *divide* (*fork*, resp.) over A if and only if $\mathrm{tp}(c/Ab)$ does not divide (fork, resp.) over A for every pair of finite tuples $b \in B$, $c \in C$.

(9) (Partial transitivity) Let $A \subseteq B \subseteq C$. If $\mathrm{tp}(a/C)$ does not divide (fork, resp.) over A, then $\mathrm{tp}(a/B)$ does not divide (fork, resp.) over A, and $\mathrm{tp}(a/C)$ does not divide (fork) over B.

(10) (Anti-reflexivity) If $a \notin \mathrm{acl}(A)$, then $\mathrm{tp}(a/Aa)$ divides over A.

Proof (Sketch) For (5), use (4) and compactness.

(6) (a)$\Leftrightarrow$(b) comes by the standard argument using Ramsey's theorem. Then (b)$\Rightarrow$(c) follows by compactness. For (c)$\Rightarrow$(a), let p' be the closure of p under finite

conjunctions. If we take λ sufficiently large, then by the Erdös–Rado theorem, we can fix $q(\subseteq p')$ with $|q| \leq k$ whose A-automorphic images are k-inconsistent. Hence q divides over A, so does p.

For (10), use the formula $x = a$. All others are easy consequences from the definitions. ■

The prototypical example of dividing appears in the theory of equivalence relation E having infinitely many infinite equivalence classes. There, xEa 2-divides over the empty set. We have examples of proper k-dividing for $k > 2$ too. Consider $(\mathbb{Q}^2, R)$ where, for any $a = (a_1, a_2), b = (b_1, b_2) \in \mathbb{Q}^2$, $R(a, b)$ iff $(a_1 = b_1$ or $a_2 = b_2)$. Here xRb 3-divides (but does not 2-divide) over $\emptyset$.

Intuitively, dividing is the right notion of dependence: For type-definable non-disjoint sets $X, Y \subseteq \mathcal{M}^\alpha$, we may temporarily say that Y *k-divides* X if there are arbitrary many automorphisms f_i fixing X setwise such that $\{Y_i = f_i(Y)\}_i$ is k-inconsistent. Then $X_i = X \cap Y_i \neq \emptyset$ as well. It intuitively means that the set Y chops up X to unboundedly many pieces X_i by conjugates Y_i. It follows from Remark 2.1.2(6) that $\mathrm{tp}(c/Ab)$ divides over A iff the set Y of realizations of $\mathrm{tp}(c/Ab)$ *divides* X, that of $\mathrm{tp}(c/A)$. This, in a certain sense, means that c has a more satisfying relation with (or more dependent on) Ab than it has with A.

Why, then, is forking introduced over dividing? With forking, we have the following extension axiom which is quite useful in developing further arguments in Section 2.2.

Fact 2.1.3 (Extension) *Let $p(x)$ be a type over C which does not fork over A. Then there is a complete type $q(x)$ over C extending p, which does not fork over A.*

Proof Let

$$\Gamma = \{\psi(x', a) \mid \psi(x', y) \in \mathcal{L},\ x' \subseteq x,\ a \in C,\ \psi(x', a) \text{ forks over } A\},$$

and let $q' = p \cup \{\neg\psi(x', a) \mid \psi(x', a) \in \Gamma\}$. First we claim that q' is consistent: Suppose not; then for some $n < \omega$,

$$p \cup \{\neg\psi_0(x', a_0), \ldots, \neg\psi_n(x', a_n)\},$$

where $\psi_i \in \Gamma$, is inconsistent. Hence, $p \vdash \bigvee_{i \leq n} \psi_i(x', a_i)$. As each ψ_i forks over A, p forks over A, a contradiction. Hence, the claim follows.

Now we take q as any completion of q' over C. It remains to show q does not fork over A: If it does, then some $\varphi(x', a)(\in q)$ forks over A: Hence, $\neg\varphi(x', a) \in q' \subseteq q$. This certainly is a contradiction. ■

The following important fact gives a motivation for our work. It basically says that non-forking supplies a well-behaved independence relation for any stable structure. We shall generalize the result to broader context by showing that non-forking behaves well in all simple theories too (Theorem 3.3.1).

Fact 2.1.4 *If T is a stable theory, then non-forking has the following properties:*

(1) *Let p be a type. Then, p does not fork over A if and only if p does not divide over A.*

(2) (a) (Symmetry) $\mathrm{tp}(c/Ab)$ *does not fork over A if and only if* $\mathrm{tp}(b/Ac)$ *does not fork over A.*

(b) (Transitivity) *Let $A \subseteq B \subseteq C$. Then* $\mathrm{tp}(a/C)$ *does not fork over A if and only if* $\mathrm{tp}(a/B)$ *does not fork over A and* $\mathrm{tp}(a/C)$ *does not fork over B.*

(c) (Extension) *Let $A \subseteq B \subseteq C$. If $p(x) \in S(B)$ does not fork over A, then there is an extension $q(x) \in S(C)$ of p, which does not fork over A either.*

(d) (Local character) *For any $p(x) \in S(A)$ with finite x, there is a subset A_0 of A with $|A_0| \leq |T|$ such that p does not fork over A_0.*

(e) (Finite character) *Let $A \subseteq B$. Then* $\mathrm{tp}(C/B)$ *does not fork over A if and only if* $\mathrm{tp}(c/Ab)$ *does not fork over A for every pair of finite tuples $b \in B, c \in C$.*

(f) (Uniqueness over a model) *Any complete type over a model is stationary; i.e., if $p(x) \in S_n(M)$ and $M \subseteq A$, then p has the unique non-forking extension $q(x) \in S_n(A)$.*

(g) (Uniqueness for strong types) *If $a_0 \equiv^s_A a_1$ and $B \supseteq A$, $a_i \mathop{\smile\!\!\!\!\!\!\;|}_A B$, then $a_0 \equiv^s_B a_1$. In other words, any complete type over a set $A = \mathrm{acl}^{\mathrm{eq}}(A)$ is stationary.*

Moreover if, in a theory T, an automorphism-invariant relation between complete types (of finite tuples) and sets satisfies the preceding axioms (a) to (f) in (2), then the theory is stable and the relation must be non-forking.

For example, as said in Chapter 1, non-forking coincides linear independence and algebraic independence, respectively, in vector spaces and algebraically closed fields, and both structures are stable.

The notions of forking and dividing need not be equivalent if T is unstable.

Example 2.1.5

(1) Let T be the theory of a ternary relation $R(x, y, z)$ defined on a circle C as follows: $R(x, y, z)$ holds if and only if x, z are non-antipodal points on C and y lies on the shorter arc between x and z. It is straightforward to show that T is ω-categorical, and any two non-antipodal distinct points have the same type. Now we can choose $a_i \in C$ $(i < 3)$ such that

$$\models x = x \leftrightarrow R(a_0, x, a_1) \vee R(a_1, x, a_2) \vee R(a_2, x, a_0)$$

and $R(a_i, x, a_j)$ divides over $\emptyset$. Hence $x = x$ forks, but does not divide over $\emptyset$. Moreover, $\mathrm{tp}(a_0)$ has no non-dividing extension over C.

(2) Consider $(\mathbb{Q}, <)$. Quantifier elimination shows that forking is dividing. However, symmetry fails. Choose rational numbers $a < b < c$. Then $\mathrm{tp}(b/ac)$ divides over $\emptyset$ witnessed by $a < x < c$, whereas $\mathrm{tp}(ac/b)$ does not divide over $\emptyset$.

Let us investigate further on *dividing* for arbitrary T.

Lemma 2.1.6 *The following are equivalent:*

(1) $\text{tp}(c/Ab)$ *does not divide over* A.

(2) *For any* A*-indiscernible sequence* I *with* $b \in I$*, there is a tuple* c' *realizing* $\text{tp}(c/Ab)$ *such that* I *is* Ac'*-indiscernible.*

(3) *If* I *is an* A*-indiscernible sequence with* $b \in I$*, then some* J*, an* Ab*-automorphic image of* I*, is* Ac*-indiscernible.*

Proof (1)⇒(2) Suppose $p(x, b) = \text{tp}(c/Ab)$ does not divide over A. Let an A-indiscernible sequence I containing b be given. We claim first that $q(x) = \bigcup\{p(x, b') | b' \in I\}$ is consistent: If not, then by compactness, we have $\varphi(x', a, b)$ in $p(x, b)$ $(a \in A)$ such that

$$\varphi(x', a, b_1) \wedge \ldots \wedge \varphi(x', a, b_n)$$

is inconsistent for a subsequence $\langle b_1, \ldots, b_n \rangle$ of I. Since I is indiscernible, it follows $p(x, b)$ n-divides over A, a contradiction. We have proved the claim.

Now, let $\Gamma(x)$ be a set of formulas over $A \cup I$ saying I is Ax-indiscernible. It follows, by the claim and Ramsey's theorem, that $q \cup \Gamma$ is consistent. Then, the desired c' is a tuple realizing $q(x) \cup \Gamma(x)$.

(2)⇒(1) Assume (2) holds. To prove (1), according to 2.1.2(6) it suffices to show that

$$q = \{\varphi(x, a, b') | \, b' \in I\}$$

is consistent, for any $\varphi(x, a, b)$ in $\text{tp}(c/Ab)$ $(a \in A)$, and any A-indiscernible sequence I with $b \in I$. Actually by (2), there is c' realizing q.

(2)⇔(3) Consider Ab-automorphisms sending c' to c, and I to J. ■

Hence, if $\text{tp}(a/Ac)$ does not divide over A and $c \in \text{acl}(Aa)$, then $c \in \text{acl}(A)$.

Lemma 2.1.7 (Left transitivity) *Assume for each* $i \leq n$*,* $\text{tp}(a_i/Ba_0 \ldots a_{i-1})$ *does not divide over* $Aa_0 \ldots a_{i-1}$*, where* $A \subseteq B$*. Then* $\text{tp}(a_0 \ldots a_n/B)$ *does not divide over* A*.*

Proof Write $B = Ab$. Assume an A-indiscernible sequence I with $b \in I$ is given. According to Lemma 2.1.6, it suffices to find tuples c_i $(i \leq n)$ such that

(i) $c_0 \ldots c_i$ realizes $\text{tp}(a_0 \ldots a_i/Ab)$ (i.e., $c_0 \ldots c_i \equiv_{Ab} a_0 \ldots a_i$),

(ii) I is $Ac_0 \ldots c_i$-indiscernible.

Suppose that $c_0 \ldots c_{i-1}$ have been appropriately chosen. Now there is c such that

$$(*) \quad c_0 \ldots c_{i-1}c \equiv_{Ab} a_0 \ldots a_{i-1}a_i.$$

Hence $p = \text{tp}(c/Abc_0 \ldots c_{i-1})$ does not divide over $Ac_0 \ldots c_{i-1}$. Therefore again by 2.1.6, there is c_i realizing p such that I is $Ac_0 \ldots c_i$-indiscernible. Thus,

(ii) follows, and according to (*), $c_0 \ldots c_{i-1}c_i \equiv_{Ab} c_0 \ldots c_{i-1}c \equiv_{Ab} a_0 \ldots a_{i-1}a_i$. Hence, (i) holds too. ■

Corollary 2.1.8 *In any T, if dividing*[1] *satisfies symmetry (2.1.4), then it satisfies transitivity as well.*

Proof One direction of transitivity already holds (2.1.2(9)). Now, assume that $\text{tp}(b/Aa_0)$ does not divide over A and $\text{tp}(b/Aa_0a_1)$ does not divide over Aa_0. Then by symmetry, $\text{tp}(a_0/Ab)$ does not divide over A and $\text{tp}(a_1/Aba_0)$ does not divide over Aa_0. Hence, by the case $n = 1$ of Lemma 2.1.7, $\text{tp}(a_0a_1/Ab)$ does not divide over A. Then again by symmetry, $\text{tp}(b/Aa_0a_1)$ does not divide over A. We have proved the transitivity of dividing. ■

2.2 Simplicity and Forking

By now we introduce a class of "simple" theories properly containing that of stable theories. Studying the class of simple theories is the main theme of this book.

Definition 2.2.1 T is said to be *simple* if non-forking satisfies the local character axiom; i.e., for any complete type $p \in S_n(B)$, there is a subset A of B such that $|A| \leq |T|$ and p does not fork over A.

Remark & Example 2.2.2

(1) According to Fact 2.1.4, every stable theory is simple.

The prototypical examples of simple unstable structures are the random graph and its variations. The random graph is a countable graph $(V; R)$ (so R is an irreflexive, symmetric relation on the countable set V) such that for any distinct $a_1, \ldots, a_n, b_1, \ldots, b_n \in V$ there is $c \in V$ such that a_iRc but $\neg b_iRc$ $(i = 1, \ldots, n)$ hold. The theory of the random graph is the Fraïsé limit of all finite graphs, so is ω-categorical having quantifier elimination.

The bipartite random graph is a countable structure $(V; R, P_0, P_1)$, where $(V; R)$ is a graph, and unary predicates P_0, P_1 partition V such that $xRy \to x \in P_i \wedge y \in P_j$ $(\{i, j\} = \{0, 1\})$, and for any distinct $a_1, \ldots, a_n, b_1, \ldots, b_n \in P_i$ there is $c \in P_j$ such that a_iRc but $\neg b_iRc$ hold. Again the theory of the bipartite random graph is ω-categorical and has quantifier elimination. The reasons for the (bipartite) random graph being simple will be explained in 2.3.8.

More examples will be introduced in Section 2.6. Of course, the reader can jump to the section to see those.

(2) If T is simple, then due to finite character and partial transitivity (Remark 2.1.2), it satisfies the *existence axiom of forking*; i.e., for any $p \in S(A)$, p does not fork over A.

---

1. From now, saying that dividing (forking, resp.) satisfies an axiom is the same as saying non-dividing (non-forking, resp.) satisfies it.

(3) One may wonder whether simplicity of T is independent from the choice of a saturated model. Because of finite character and invariance of non-forking, one easily sees that if local character holds in a $|T|^+$-saturated model then it holds in *every* $|T|^+$- saturated model. In this sense simplicity only depends on T. Section 2.3 will give more obvious reasons for this.

Now we introduce the notion of Morley sequences of complete types which will be frequently used in future developments.

Definition 2.2.3 Let $A \subseteq B$ and $p \in S(B)$. By a *Morley sequence of p over A*, we mean a B-indiscernible sequence $I = \langle a_i : i \in \omega \rangle$ in p such that for every $i \in \omega$, $\mathrm{tp}(a_i/Ba_0 \dots a_{i-1})$ does not fork over A. When $A = B$, we omit mentioning "over A."

Proposition 2.2.4 *If $p(x) \in S(A)$ does not fork over $A_0(\subseteq A)$, then there exists a Morley sequence of p over A_0. Thus for simple T, a Morley sequence of a complete type exists.*

Proof The second statement automatically follows from the previous 2.2.2, if the first statement is true. Let us prove this. For any cardinal λ, the extension axiom (2.1.3) guarantees by induction the existence of a sequence $I_\lambda = \langle a_\alpha | \alpha < \lambda \rangle$ such that for every $\alpha < \lambda$, $\mathrm{tp}(a_\alpha/A \cup \{a_\beta | \beta < \alpha\})$ extends $p(x)$ and does not fork over A_0. We now let $\kappa = |A| + |x|$, and obtain an A-indiscernible I from I_λ where λ and I are as in Lemma 2.1.5. Clearly I is a desired Morley sequence. ■

The following lemma plays a crucial role in proving the main proposition 2.2.6.

Lemma 2.2.5 *Let $I = \langle a_i : i \in \omega \rangle$ be a Morley sequence of $\mathrm{tp}(a_0/A)$. Suppose that for some $n < \omega$, $\langle a_0^j \dots a_n^j | j < \omega \rangle$ is an A-indiscernible sequence of $\mathrm{tp}(a_0 \dots a_n/A)$. Then there is a sequence J, an A-automorphic image of I, such that $\langle a_0^j, \dots, a_n^j \rangle ^\frown J$ is an A-indiscernible sequence for each $j < \omega$.*

Proof We are going to select tuples $b_1, \dots, b_m$ by induction on m satisfying the following:

(i) $\mathrm{tp}(a_0^j \dots a_n^j b_1 \dots b_m/A) = \mathrm{tp}(a_0 \dots a_n a_{n+1} \dots a_{n+m}/A)$ for every $j < \omega$.

(ii) $\langle a_0^j \dots a_n^j b_1 \dots b_m | j < \omega \rangle$ is an A-indiscernible.

Eventually, $\langle b_m | 1 \leq m < \omega \rangle$ will form the desired sequence J, since for each $j < \omega$, $\mathrm{tp}(a_0^j \dots a_n^j J/A) = \mathrm{tp}(I/A)$ by (i). Now for $m = 0$, nothing is to be proved. Hence, we may assume that $b_1, \dots, b_m$ satisfying (i), (ii) have already been chosen. Let us find b_{m+1} so that (i), (ii) hold for $m + 1$: Since I is a Morley sequence, $\mathrm{tp}(a_{n+m+1}/Aa_0 \dots a_{n+m})$ does not divide over A. Hence by (i), $p(a_0^0, \dots, a_n^0, b_1, \dots, b_m, x_{n+m+1})$ does not divide over A where

$$p(x_0, \dots, x_n, x_{n+1}, \dots, x_{n+m}, x_{n+m+1}) = \mathrm{tp}(a_0, \dots, a_n, a_{n+1}, \dots, a_{n+m}, a_{n+m+1}/A).$$

Thus, by applying Lemma 2.1.6 with (ii), we can find a tuple b_{m+1} realizing $p(a_0^0, \ldots, a_n^0, b_1, \ldots, b_m, x_{n+m+1})$ such that $\langle a_0^j \ldots a_n^j b_1 \ldots b_m | j < \omega \rangle$ is an Ab_{m+1}-indiscernible. Therefore (i), (ii) hold for $m+1$ with $b_1, \ldots, b_{m+1}$. ■

The next proposition is the key property of simplicity which leads to show the equivalence of forking and dividing (Proposition 2.2.7), forking symmetry and transitivity (Theorem 2.2.8). The reader will see many more basic results are derived from it.

Proposition 2.2.6 *If T is simple, then the following are equivalent:*

(1) *$\varphi(x, c)$ divides over A.*

(2) *For any Morley sequence I of $\mathrm{tp}(c/A)$, $\{\varphi(x, c') | c' \in I\}$ is inconsistent.*

(3) *For some Morley sequence I of $\mathrm{tp}(c/A)$, $\{\varphi(x, c') | c' \in I\}$ is inconsistent.*

Proof (1)⇒(2) Assume that a Morley sequence $I = \langle c_n : n \in \omega \rangle$ of $\mathrm{tp}(c/A)$ is given. Now $\varphi(x, c_0)$ k-divides over A for some k. We claim that $\varphi(x, c_0)$ k-divides over $A \cup \{c_m | 0 < m\}$ too: By 2.1.2(6), there is an A-indiscernible sequence $\langle c^j : j < \omega \rangle$ with $c^0 = c_0$ such that $\{\varphi(x, c^j) | j < \omega\}$ is k-inconsistent. Now from the previous Lemma 2.2.5, there is an A-automorphic image I' of I such that $c^j I' \equiv_A I$ for all $j < \omega$. Thus $c^0 \equiv_{AI'} c^j$ for all $j < \omega$ too. This says, from the definition of dividing, that $\varphi(x, c_0)$ k-divides over AI', so it does over $A \cup \{c_m | 0 < m\}$ since $c_0 I' \equiv_A I$. We have proved the claim.

Now to lead a contradiction, suppose that $\{\varphi(x, c_n) | n < \omega\}$ is consistent. We can find that (by reversing the order of I) an A-indiscernible $J = \langle c'_\alpha | \alpha < |T|^+ \rangle$ such that $\mathrm{tp}(c'_{\alpha_0} \ldots c'_{\alpha_n}/A) = \mathrm{tp}(c_n \ldots c_0/A)$ for $\alpha_0 < \ldots < \alpha_n < |T|^+$. By compactness, $\{\varphi(x, c'_\alpha) | \alpha < |T|^+\}$ is consistent and hence realized by, say, b. Moreover, according to the preceding claim with 2.1.2(5), we have that

$$(*)\quad \varphi(x, c'_\alpha)\ k\text{-divides over } A \cup \{c'_\beta | \beta < \alpha\} \text{ for each } \alpha < |T|^+.$$

Note now, since T is simple, there is some $J_0 \subseteq J$ of cardinality $\leq |T|$ such that $\mathrm{tp}(b/J)$ does not fork (hence, neither divide) over J_0. Then there must exist an ordinal $\alpha_0 < |T|^+$ such that $\alpha_0 > \beta$ whenever $c'_\beta \in J_0$. From (*), $\varphi(x, c'_{\alpha_0})$ divides over J_0, and then since $\varphi(x, c'_{\alpha_0}) \in \mathrm{tp}(b/J)$, $\mathrm{tp}(b/J)$ divides over J_0, a contradiction. Therefore, $\{\varphi(x, c_n) | n < \omega\}$ cannot be consistent.

(2) ⇒ (3) Obvious from 2.2.4.

(3) ⇒ (1) According to the definition of dividing. ■

Proposition 2.2.7 *If T is simple, then the following are equivalent:*

(1) *$\varphi(x, c)$ divides over A.*

(2) *$\varphi(x, c)$ forks over A.*

Proof (1)⇒(2) Obvious.

(2)⇒(1) Since $\varphi(x,c)$ forks over A, there are $\psi_i(x,b_i)$ $(0\leq i\leq r)$ such that $\models \varphi(x,c)\rightarrow\bigvee_{i\leq r}\psi_i(x,b_i)$ and for each i, $\psi_i(x,b_i)$ divides over A. Let $b^0 = b_0,\dots,b_r$ and $I = \langle c_n b^n | n\in\omega\rangle$ be a Morley sequence of $\mathrm{tp}(cb/A)$ with $c_0b^0 = cb$ and $b^n = b^n_0,\dots,b^n_r$. Then, clearly each of $\langle c_n|n\in\omega\rangle$, $\langle b^n_i|n<\omega\rangle$ is a Morley sequence, and

$$(^{**})\quad \models\varphi(x,c_n)\rightarrow\bigvee_{i\leq r}\psi_i(x,b^n_i).$$

Now, it suffices to show that $\{\varphi(x,c_n)|n<\omega\}$ is inconsistent. Suppose not; say it is realized by d. Then from (**), for each n, there is $i_n\leq r$ such that $d\models\psi_{i_n}(x,b^n_{i_n})$. Hence, by the pigeon-hole principle, there must exist an $i_0\leq r$ and some infinite $\tau\subseteq\omega$ such that $\psi_{i_0}(d,b^n_{i_0})$ for all $n\in\tau$. Then since $\langle b^n_{i_0}|n\in\tau\rangle$ is Morley as well, from Proposition 2.2.6, $\psi_{i_0}(x,b_{i_0})$ does not divide over A. This is a contradiction, and the proof is completed. ■

Theorem 2.2.8 *If T is simple, then non-forking satisfies symmetry and transitivity.*

Proof (Symmetry) Suppose that $\mathrm{tp}(b/Ac)$ does not fork over A. We want to prove that $\mathrm{tp}(c/Ab)$ does not fork over A. According to finite character, it suffices to show when b is finite. Now, there is a Morley sequence $I = \langle b_i|i\in\omega\rangle$ of $\mathrm{tp}(b/Ac)$ over A where $b_0 = b$ (by 2.2.4). Clearly, I is a Morley sequence of $\mathrm{tp}(b/A)$ as well, so for $a\in A$, $\langle ab_i|i<\omega\rangle$ is a Morley sequence of $\mathrm{tp}(ab/A)$ too. Hence, by 2.2.6, it suffices to show that for given $\varphi(x,a,b)$ in $\mathrm{tp}(c/Ab)$ with finite $a\in A$, $p:=\{\varphi(x,a,b_i)|i\in\omega\}$ is consistent. But by Ac-indiscernibility of I, c realizes p.

(Transitivity) It follows from Corollary 2.1.8 and symmetry. ■

For simple T, we call a family of sets $\{C_i|i\in I\}$ *independent over A* (or equivalently *A-independent*) if for every $i\in I$,

$$\mathrm{tp}\left(C_i/A\cup\{C_j|\,j(\neq i)\in I\}\right)$$

does not fork over A. If $\{B,C\}$ is A-independent, i.e., $tp(B/AC)$ does not fork over A, then we say B *is independent from (or with) C over A* and write $B\mathrel{\perp\!\!\!\!\smile}_A C$.

From now on, because of symmetry and transitivity, basic forking calculation is possible for simple theories. For example, if $\{C_i|\ i\in I\}$ is independent over A, and $\{C_i|\ i\in I\}\mathrel{\perp\!\!\!\!\smile}_A B$, then $\{C_i|\ i\in I\}$ is independent over AB. Also $\{C_i|\ i<\omega\}$ is A-independent iff for each $i>0$, we have $C_i\mathrel{\perp\!\!\!\!\smile}_A C_0\dots C_{i-1}$. In particular, for a sequence I of realizations of $p\in S(B)$, I is a Morley sequence of p over $A(\subseteq B)$ iff I is independent and indiscernible over B, and p does not fork over A. We also have $A\mathrel{\perp\!\!\!\!\smile}_B C$ iff $\mathrm{acl}(A)\mathrel{\perp\!\!\!\!\smile}_{\mathrm{acl}(B)}\mathrm{acl}(C)$; Moreover, in simple T there does not exist finite a and $B=\{b_i|i<|T|^+\}$ such that B is A-independent while $a\not\mathrel{\perp\!\!\!\!\smile}_A b_i$ for all i, since, otherwise, it easily follows that $a\not\mathrel{\perp\!\!\!\!\smile}_{A\cup\{b_j|j<i\}} a_i$ for each $i<|T|^+$ and thus $\mathrm{tp}(a/B)$ does not satisfy local character, a contradiction. We will use these kinds of forking computations freely.

The following cases, non-forking independence is type-describable. This will be used frequently in future arguments.

Remark 2.2.9 Assume T is simple.

(1) Let $p_i \in S(A)$ $(i = 1, \ldots, n)$. There is a partial type $\Sigma(x_0, \ldots, x_n)$ over A such that for any $a_i \models p_i$ and b with $|b| = |x_0|$, we have $\{b, a_1, \ldots, a_n\}$ is A-independent iff $(b, a_1, \ldots, a_n) \models \Sigma(x_0, \ldots, x_n)$.

(2) Assume $p(x) \in S(B)$ does not fork over A. Then there is a partial type $\Gamma(x_0, x_1, \ldots)$ over B of infinite variables with $|x| = |x_i|$ such that for any $I = \langle a_0, a_1, \ldots \rangle$,

$$I \text{ is a Morley sequence of } p \text{ over } A \text{ iff } I \models \Gamma(x_0, x_1, \ldots).$$

Proof

(1) Let $\Sigma(x_0) = \{x_0 = x_0\}$. For $n \geq 1$, fix $a'_n \models p(x_n)$ and recursively put

$$\Sigma(x_0, \ldots, x_{n-1}, x_n) := \Sigma(x_0, \ldots, x_{n-1}) \cup \Sigma_n(x_0, \ldots, x_n),$$

where

$$\Sigma_n(x_0, \ldots, x_n) := \big\{\neg\varphi(x_0, \ldots, x_{n-1}, x_n) \in \mathcal{L}(A) \mid \varphi(x_0, \ldots, x_{n-1}, a'_n) \text{ forks over } A\big\}.$$

Now assume inductively for $a_i \models p_i$, that $\{b, a_1, \ldots, a_{n-1}\}$ is independent over A iff $(b, a_1, \ldots, a_{n-1}) \models \Sigma(x_0, \ldots, x_{n-1})$. Then, by symmetry and transitivity, $\{b, a_1, \ldots, a_n\}$ is independent over A iff $(b, a_1, \ldots, a_{n-1}) \models \Sigma(x_0, \ldots, x_{n-1})$ and $ba_1 \ldots a_{n-1} \perp_A a_n$. By considering contrapositive, it is easy to see that $ba_1 \ldots a_{n-1} \perp_A a_n$ holds iff $ba_1 \ldots a_n \models \Sigma_n$. We have proved (1).

(2) follows from (1).

■

2.3 $D_{\varphi,k}$-Ranks and the Tree Property

Rank notions are very important in further developments of simplicity theory. We shall see various kinds of ranks as we proceed. In this section we introduce the $D(-, \varphi, k)$-rank for each φ, k, which characterizes simplicity (2.3.7) and reflects forking (2.3.9). We shall also introduce a combinatorial property called "tree property" characterizing the negation of simplicity. The study of the ranks and the tree property give rise to the converse of 2.2.8.

Definition 2.3.1 Let $\varphi(x; y)$ be an $\mathcal{L}$-formula, and let k be an integer ≥ 2. For any type $p(x')$ with $x \subseteq x'$, $D(p(x'), \varphi(x, y), k) = D_{\varphi,k}(p)$ (either a natural number or ∞) is recursively defined as follows:

(1) $D(p,\varphi,k) \geq 0$ for any consistent type p.

(2) $D(p,\varphi,k) \geq n+1$ if there are tuples a_i $(i < \omega)$ such that $D(p(x') \cup \{\varphi(x;a_i)\}, \varphi, k) \geq n$ for each $i < \omega$, and $\{\varphi(x;a_i) | i < \omega\}$ is k-inconsistent.

(3) $D(p,\varphi,k) = n$ if $D(p,\varphi,k) \geq n$ and $D(p,\varphi,k) \not\geq n+1$.

(4) $D(p,\varphi,k) = \infty$ if $D(p,\varphi,k) \geq n$ for all $n \in \omega$.

Lemma 2.3.2 *Given $p(x')$, $\varphi(x,y) \in \mathcal{L}$, $k \geq 2$, $n \geq 1$, the following are equivalent:*

(1) $D_{\varphi,k}(p) \geq n$.

(2) There is a set of finite tuples $\{a_{\bar{v}} | \bar{v} \in \omega^{\leq n}\}$ such that:

(a) for each $\bar{v} \in \omega^{<n}$, $\{\varphi(x, a_{\bar{v}\frown i}) | i \in \omega\}$ is k-inconsistent,

(b) for every $\bar{u} \in \omega^n$, $p(x') \cup \{\varphi(x, a_{\bar{u}\lceil i}) |\, 0 < i \leq n\}$ is consistent.

(3) For variables $x_{\bar{u}} \subseteq x'_{\bar{u}}$ and $y_{\bar{v}}$ where $\bar{u} \in \omega^n, \bar{v} \in \omega^{\leq n}$,

$$\bigcup \{p(x'_{\bar{u}}) | \bar{u} \in \omega^n\} \cup \{\varphi(x_{\bar{u}}, y_{\bar{u}\lceil i}) |\, 0 < i \leq n, \bar{u} \in \omega^n\}$$

$$\cup \left\{ \neg\exists z \bigwedge_{i\in\tau} \varphi(z, y_{\bar{v}\frown i}) | \bar{v} \in \omega^{<n}, \tau \subseteq \omega, |\tau| = k \right\}$$

is consistent.

Proof (1)⇔(2) If $n = 1$, then (1) and (2) are obviously equivalent. Hence, let us show that (1) and (2) are equivalent for $n+1$ with the induction hypothesis for n: If $D(p,\varphi,k) \geq n+1$, then there exist tuples a_i $(i < \omega)$ such that $D(p \cup \{\varphi(x;a_i)\}, \varphi, k) \geq n$ for each $i < \omega$ and $\{\varphi(x;a_i) | i < \omega\}$ is k-inconsistent. By the induction hypothesis, there are tuples $a_{i\bar{v}}$ $(\bar{v} \in \omega^{\leq n}, i < \omega)$ such that for each $\bar{v} \in \omega^{<n}$, $i \in \omega$, $\{\varphi(x, a_{i\bar{v}\frown j}) | j \in \omega\}$ is k-inconsistent and for every $\bar{u} \in \omega^n$ and $i \in \omega$, $p \cup \{\varphi(x,a_i)\} \cup \{\varphi(x, a_{i\bar{u}\lceil j}) | 0 < j \leq n\}$ is consistent. Hence (2) holds for $n+1$, witnessed by $\{a_i | i < \omega\} \cup \{a_{i\bar{v}} | \bar{v} \in \omega^{\leq n}\}$. Conversely, if (2) holds for $n+1$, witnessed by $\{a_{\bar{v}} | \bar{v} \in \omega^{\leq n+1}\}$, then firstly $\{\varphi(x;a_i) | i < \omega\}$ is k-inconsistent and secondly for each i, tuples $a_{i\bar{v}}(\bar{v} \in \omega^{\leq n})$ witness $D(p \cup \{\varphi(x;\bar{a}_i)\}, \varphi, k) \geq n$, by the induction hypothesis. Therefore, $D(p,\varphi,k) \geq n+1$.

(2)⇔(3) (3) simply says (2) is true. ■

Remark 2.3.3

(1) From Lemma 2.3.2 with compactness and Ramsey's theorem, it routinely follows that $D(p,\varphi,k) \geq n+1$ iff there is c such that $D(p \cup \{\varphi(x;c)\}, \varphi, k) \geq n$ and $\varphi(x;c)$ k-divides over $\mathrm{dom}(p)$. To visualize this, in particular when x' is infinite, we may consider that the solution set X of $p(x')$ in $\mathcal{M}^{|x'|}$ is placed in $(x, y = x' \setminus x)$-coordinate. Then the solution set of $p \cup \varphi(x,c)$ is $X \cap Y_c$ where Y_c is that of $\varphi(x,c)(\wedge\, y = y)$. Moreover, Y_c k-divides X (see the remark above 2.1.3).

(2) Given $\psi_i(x, y_i)$ $(i \leq r)$, there is a formula $\psi(x, z, y)$ where $y = y_0, \ldots, y_r$, called a *code formula for* the formulas ψ_i, such that for each $i \leq r$, there is c_i with $\models \psi(x, c_i, y) \leftrightarrow \psi_i(x, y_i)$. Let $\{\varphi_i(z) | i \leq r\}$ be an arbitrary set of $\mathcal{L}$-formulas forming a partition of the universe. For example, $z = z_0, \ldots, z_r$ and

$$\varphi_i(z) \equiv \left(\bigwedge_{j<i} \neg (z_j = z_{j+1}) \right) \wedge z_i = z_{i+1}$$

for $i < r$, and $\varphi_r(z) \equiv \bigwedge_{j<r} \neg(z_j = z_{j+1})$. Now define

$$\psi(x, z, y) \equiv \bigwedge_{i \leq r} (\varphi_i(z) \rightarrow \psi_i(x; y_i)).$$

Clearly it fulfills the condition.

We are ready to state the basic properties of the $D_{\varphi,k}$-rank.

Lemma 2.3.4

(1) $D(p_1, \varphi, k_1) \leq D(p_2, \varphi, k_2)$ if $p_1 \vdash p_2$ and $k_1 \leq k_2$.

(2) For every $p(x')$, $\varphi(x, y)$, k, there is a finite subset $p_0(x)$ of p such that $D(p, \varphi, k) = D(p_0, \varphi, k)$. In particular,

$$D(p, \varphi, k) = \min\{D(q, \varphi, k) | \ q \subseteq p, \ q \text{ finite}\}.$$

(3) $D(p \cup \bigvee_{i<n} q_i, \psi, k) = \max\{D(p \cup q_i, \psi, k) | \ i < n\}$, where

$$\bigvee_{i<n} q_i := \left\{ \bigvee_{i<n} \phi_i | \ \phi_i \in q_i \right\}.$$

(4) Let $\varphi(x, y)$, k be given. For a type $p(x')$ over A, there is a completion $p'(x') \in S(A)$ of p such that $D(p, \psi, k) = D(p', \psi, k)$.

(5) Assume that a partial type $p(x')$ over B forks over $A(\subseteq B)$ witnessed by $\psi_i(x, b_i)$ with k_i $(i \leq r)$. Let $\psi(x; z, y)$ $(y = y_0 \ldots y_r)$ be a code formula for $\psi_i(x, y_i)$, and let $k = \max\{k_0, \ldots, k_r\}$. Then $D(p, \psi, k) < \omega$ implies that $D(p, \psi, k) < D(p \lceil A, \psi, k)$.

Proof (1) Let us show that

$$(*) \quad D(p_1, \varphi, k_1) \geq n \text{ implies that } D(p_2, \varphi, k_2) \geq n$$

whenever $p_1 \vdash p_2$, and $k_1 \leq k_2$. If $n = 0$, then (*) is clearly true. Suppose that (*) is true for n. If finite tuples a_i $(i < \omega)$ witness $D(p_1, \varphi, k_1) \geq n + 1$, then by the induction hypothesis $D(p_2 \cup \{\varphi(x; a_i)\}, \varphi, k_2) \geq n$. Since $\{\varphi(x; a_i) | i < \omega\}$ is k_2-inconsistent as well, $D(p_2, \varphi, k_2) \geq n + 1$.

(2) Lemma 2.3.2 with compactness says that $D(p, \varphi(x, y), k) \geq n$ whenever $D(q, \varphi(x, y), k) \geq n$ for each finite $q \subseteq p$. Hence, if $D(p, \varphi, k) = n$, then for some finite $p_0 \subseteq p$, $D(p_0, \varphi, k) \leq n$. Since $p \vdash p_0$, we have $D(p_0, \varphi, k) = n$.

(3) Since $p \cup q_i \vdash p \cup \bigvee_{i<n} q_i$,

$$D\Big(p \cup \bigvee_{i<n} q_i, \psi, k\Big) \geq \max\{D(p \cup q_i, \psi, k) \mid i < n\}.$$

Hence, it suffices to show that $D(p \cup \bigvee_{i<n} q_i, \psi, k) \geq s+1$ implies that $\max\{D(p \cup q_i, \psi, k) \mid i < n\} \geq s+1$, using the induction hypothesis on s. (If $s = 0$, then the statement is true.): Now there are tuples b_j $(j < \omega)$ such that $D(p \cup \bigvee_{i<n} q_i \cup \{\psi(x, b_j)\}, \psi, k) \geq s$ for each $j < \omega$ and $\{\psi(x, b_j) \mid j < \omega\}$ is k-inconsistent. By the induction hypothesis, $\max\{D(p \cup q_i \cup \{\psi(x, b_j)\}, \psi, k) \mid i < n\} \geq s$ for each $j < \omega$. Thus, there are an infinite subset τ of ω, and a number $r < n$ such that $D(p \cup q_r \cup \{\psi(x, b_j)\}, \psi, k) \geq s$ for each $j \in \tau$, and $\{\psi(x, b_j) \mid j \in \tau\}$ is k-inconsistent. Therefore, $D(p \cup q_r, \psi, k) \geq s+1$, and so is

$$\max\Big\{D(p \cup q_i, \psi, k) \mid i < n\Big\}.$$

(4) Suppose that $\{\varphi_i \mid i < \lambda\}$ is a well-ordering of all the formulas over A having free variables from x'. Given $\alpha < \lambda$, assume inductively that for each $\beta < \alpha$, there is a type p_β such that

(i) $p \subseteq p_\beta$ and $D(p, \psi, k) = D(p_\beta, \psi, k)$,

(ii) p_β contains either φ_β or $\neg\varphi_\beta$,

(iii) $p_\gamma \subseteq p_\beta$ for $\gamma < \beta$.

Now, let $p'_\alpha = \bigcup\{p_\beta \mid \beta < \alpha\}$. Then for each finite $q \subseteq p'_\alpha$, $D(q, \psi, k) \geq D(p, \psi, k)$, as $q \subseteq p_\beta$ for some $\beta < \alpha$. Hence, by (1)(2), $D(p, \psi, k) = D(p'_\alpha, \psi, k)$. Also, (3) implies that $D(p'_\alpha \cup \{\varphi'_\alpha\}, \psi, k) = D(p'_\alpha, \psi, k)$, where either $\varphi'_\alpha \equiv \varphi_\alpha$ or $\varphi'_\alpha \equiv \neg\varphi_\alpha$. Let us define $p_\alpha = p'_\alpha \cup \{\varphi'_\alpha\}$. Then p_α satisfies the induction hypothesis for α. We set $p' = \bigcup\{p_\alpha \mid \alpha < \lambda\}$.

(5) Assume $D(p, \psi, k) = s$. Since $p \vdash \bigvee_{i \leq r} \psi_i(x, b_i)$, by (3), $D(p\lceil A \cup \{\psi_i(x, b_i)\}, \psi, k) \geq s$ for some $i \leq r$. Now, since $\psi_i(x, b_i)(\leftrightarrow \psi(x, c_i, d_i)$ for some $c_i d_i)$ k-divides over A, there are tuples b_i^j $(j < \omega)$ which witness it. Then clearly by taking an A-automorphism, we find $c_i^j d_i^j \models \mathrm{tp}(c_i d_i/A)$ such that $\psi_i(x, b_i^j) \leftrightarrow \psi(x, c_i^j, d_i^j)$. As the rank is invariant under the automorphism, $D(p\lceil A \cup \{\psi_i(x, b_i^j)\}, \psi, k) = D(p\lceil A \cup \{\psi(x, c_i^j, d_i^j)\}, \psi, k) \geq s$ for each $j < \omega$. Moreover, $\{\psi(x, c_i^j d_i^j) \mid j < \omega\}$ is k-inconsistent, so $D(p\lceil A, \psi, k) \geq s+1$. ∎

As promised we now introduce the tree property which is closely related to the $D_{\varphi,k}$-rank and characterizes non-simplicity. The investigation of the tree property will lead to show the fundamental theorem of forking (Theorem 2.4.3).

Definition 2.3.5 An $\mathcal{L}$-formula $\varphi(x, y)$ has the *k-tree property* for $k(\geq 2)$, if there exist tuples a_α with $\alpha \in \omega^{<\omega}$ such that:

(1) For each $\alpha \in \omega^{<\omega}$, $\{\varphi(x, a_{\alpha^\frown n}) | n \in \omega\}$ is k-inconsistent, and
(2) For each $\beta \in \omega^{\omega}$, $\{\varphi(x, a_{\beta\lceil i}) | i \in \omega\}$ is consistent.

We say T has the *tree property* if there is a formula having the k-tree property for some k.

For example, it is easy to see that $(\mathbb{Q}, <)$ has the 2-tree property via the formula $y_1 < x < y_2$, where x, y_1, y_2 are single variables, by iteratively finding countably many pairwise disjoint open subintervals of the nth level interval to form $(n+1)$th level of the tree.

Remark 2.3.6

(1) The k-tree property is a type-definable condition. Namely, a formula $\varphi(x, y)$ has the k-tree property iff for variables x_β, y_α where $\alpha \in \omega^{<\omega}, \beta \in \omega^{\omega}$,

$$\left\{\neg\exists z \bigwedge_{i\in\tau} \varphi(z, y_{\alpha i}) \,|\, \alpha \in \omega^{<\omega}, \tau \subseteq \omega, |\tau| = k\right\} \cup \{\varphi(x_\beta, y_{\beta\lceil n}) \,|\, n \in \omega, \beta \in \omega^{\omega}\}$$

is consistent.

(2) Suppose that $\varphi(x, y)$ has the k-tree property. Then according to (1) with compactness, the height and the branch number of the tree can increase arbitrary large. Namely, for any infinite cardinals κ, λ, there is a set of tuples $\{a_\alpha | \alpha \in \lambda^{<\kappa}\}$ such that:
   (a) For each $\alpha \in \lambda^{<\kappa}$, $\{\varphi(x, a_{\alpha^\frown i}) | i \in \lambda\}$ is k-inconsistent,
   (b) For every $\beta \in \lambda^{\kappa}$, $\{\varphi(x, a_{\beta\lceil\alpha}) | \alpha < \kappa\}$ is consistent.

A sequence of complete types $\langle p_i(x) | i \in \kappa\rangle$ in finite x is called a *forking (dividing, resp.) chain* of length κ if for $i < j \in \kappa$, p_j is a forking (dividing, resp.) extension of p_i.

Proposition 2.3.7 *The following are all equivalent:*

(1) T is simple, i.e., forking satisfies local character.
(2) There does not exist a forking chain of length $|T|^+$.
(3) (Local character of dividing) For any $p(x) \in S_n(A)$, there is $A_0(\subseteq A)$ with $|A_0| \leq |T|$ such that p does not divide over A_0.
(4) There is no dividing chain of length $|T|^+$.
(5) T does not have the tree property.
(6) For any p, φ, k, $D(p, \varphi, k)$ is finite.
(7) For $p(x), \varphi(x, y)$ with a single variable x, and k, $D(p, \varphi, k)$ is finite.

Proof (1)⇒(2) We prove the contraposition. Suppose that there is a forking chain $\langle p_i | i \in |T|^+\rangle$, where $p_i \in S_n(A_i)$. Let $p = \bigcup\{p_i | i \in |T|^+\} \in S_n(B)$, where $B = \bigcup\{A_i | i \in |T|^+\}$. Then for any $A = \{a_i | i < |T|\} \subseteq B$, there is $i_0 \in |T|^+$ such that $A \subseteq A_{i_0}$. Now since $p\lceil A_{i_0+1}$ forks over A_{i_0}, p forks over A. Therefore T is not simple.

(2)⇒(1) Suppose that T is not simple, so that there is $p \in S_n(B)$ such that for any subset A of B with $|A| \leq |T|$, p forks over A. Thus, for finite $A_0 \subseteq B$, some $\psi(x, a_0)$

in p forks over A_0. Let $A_1 = A_0 \cup \{a_0\}$. Then $p\lceil A_1$ forks over A_0. By iteration, we can construct a forking chain $\langle p\lceil A_i |\ i \in |T|^+\rangle$, where $A_i \subseteq B$ and $A_{i+1} \smallsetminus A_i$ is finite (possibly $\emptyset$).

(2)$\Rightarrow$(4) Obvious.

(3)$\Leftrightarrow$(4) Similar to the proof for (1)$\Leftrightarrow$(2).

(4)$\Rightarrow$(5) Suppose that $\varphi(x,y)$ and $\{c_\alpha | \alpha \in \lambda^{<\kappa}\}$, where $\kappa = |T|^+$ witness the k-tree property. Since there are boundedly many types over a given set, by taking λ sufficiently large, we can find $\beta \in \lambda^\kappa$ such that for each $\alpha < \kappa$, $\varphi(x, c_{\beta\lceil\alpha+1})$ divides over $\bigcup\{c_{\beta\lceil\gamma} |\ \gamma \leq \alpha\} = A_\alpha$. Now let p be some completion of $\{\varphi(x, c_{\beta\lceil\gamma})|\ \gamma < \kappa\}$. Then clearly $\langle p\lceil A_\alpha |\ \alpha \in |T|^+\rangle$ is a dividing chain.

(5)$\Leftrightarrow$(6) comes from Lemma 2.3.2.

(6)$\Rightarrow$(2) Assume (6). Suppose that there is a forking chain $\langle p_i | i \in |T|^+\rangle$, where $p_i \in S(A_i)$. As there are $|T|$-many formulas, from 2.3.4(5), we can find a fixed (ψ, k) such that $D(p_{i+1}, \psi, k) < D(p_i, \psi, k)$ for $|T|^+$-many i. But since each of $D(p_i, \psi, k)$ is finite, this leads a contradiction.

(6)$\Rightarrow$(7) Obvious.

(7)$\Rightarrow$(4) Assume (7). Then by the similar reason of (6)$\Rightarrow$(2), there is no dividing chain of length $|T|^+$ in a single variable. Now to lead a contradiction, suppose that there are a finite sequence $\bar{a} = a_1 \ldots a_{m+1}$, and an increasing sequence of sets A_i ($i \in |T|^+$) such that $\text{tp}(\bar{a}/A_{i+1})$ divides over A_i. By 2.1.7, either $\text{tp}(a_1 \ldots a_m/A_{i+1})$ divides over A_i or $\text{tp}(a_{m+1}/A_{i+1}a_1 \ldots a_m)$ divides over $A_i a_1 \ldots a_m$. One of the cases happens $|T|^+$ times. In the first case we continue the argument until we get a 1-tuple to reach a contradiction. In the second case, we are done. ■

Remark & Example 2.3.8 Proposition 2.3.7 enables us to make the following observations on simplicity:

(1) T is simple if and only if T^{eq} is simple. Assume T^{eq} has the k-tree property via $\varphi(z, a_\alpha)$ where z is a variable in a sort S_E. Then it is straightforward to see that the formulas

$$\exists z(f_E(x) = z \wedge \varphi(z, a_\alpha))$$

also witness the k-tree property. Those define sets in the home-sort which are also definable by the instances of an $\mathcal{L}$-formula. Hence, T has the tree property too.

(2) The proof of the proposition says, to check simplicity it suffices to verify the local character axiom for complete 1-types. Because of this we now see the simplicity of the theories of the random graph, and the bipartite random graph. Each theory is ω-categorical having quantifier elimination. Then by the axioms of the graphs it is easy to see that any complete 1-type does not divide over a finite set (in fact, over the empty set if p is not algebraic).

(3) If $\varphi(x,y)$ has the tree property, then there are uncountably many φ-types over a countable set. (This can be seen more directly from the fact that some conjunction of the formula has the 2-tree property. See the remark after Proposition 2.3.10.) Hence the formula is unstable. Of course though there is an

unstable formula which does not have the tree property. The binary edge relation in the random graph serves it.

(4) If T has the strict order property via $\varphi(x,y)$, then T is *not* simple since, because of compactness, $\varphi(x,y_0) \wedge \neg\varphi(x,y_1)$ supplies the 2-tree property. In particular, the real field and more generally o-minimal structures are *not* simple. Thus by Shelah's result pointed out in Section 1.2, any unstable formula in a simple theory must have the independence property.

(5) There is a non-simple theory not having the strict order property: (The random parametrized equivalence relations) Let T_0 be a theory with two sorts (P, E) and a ternary relation $\sim$ on $P \times P \times E$ saying that, for each $e \in E$, $x \sim_e y$ forms an equivalence relation on P. Let T be a Fraïssé limit theory of the class of finite models of T_0. Then T is ω-categorical having quantifier elimination. The formula $x \sim_z y$ with the parameter variable yz witnesses the 2-tree property, and quantifier elimination shows that any formula $\varphi(u,v)$ with a single variable u does not have the strict order property.

We can now show that the rank $D_{\varphi,k}$ reflects forking in simple T.

Proposition 2.3.9 *Let T be simple. Let $p \in S(B)$ with $A \subseteq B$. Then p does not fork over A if and only if $D(p, \psi(x,y), k) = D(p\lceil A, \psi(x,y), k)$ for any ψ, k.*

Proof ($\Leftarrow$) From 2.3.7(6) and 2.3.4(5).

($\Rightarrow$) Let $p = \mathrm{tp}(a/B)$, $q = p\lceil A$. Clearly $D(q, \psi(x,y), k) \geq D(p, \psi(x,y), k)$. Now let us suppose $D(q, \psi(x,y), k) \geq n+1$. We shall show that $D(p, \psi(x,y), k) \geq n+1$ assuming the induction hypothesis for n: As $D(q, \psi, k) \geq n+1$, from 1.3.3(1), there is c'' such that:

(i) $\psi(x; c'')$ k-divides over A, and $D(q \cup \{\psi(x; c'')\}, \psi, k) \geq n$.

Now by 2.3.4,

(ii) There is $\mathrm{tp}(a'/Ac'')$ containing $q \cup \{\psi(x; c'')\}$ having the same $D_{\psi,k}$-rank.

Then since $a' \equiv_A a$, we find c', an A-automorphic image of c'', such that $c'a \equiv_A c''a'$. Thus, (i)(ii) still hold with ac'. Now, by symmetry we have $B \mathop{\smile\!\!\!\!\!\!|}_A a$, and by extension there is $B' \equiv_{Aa} B$ such that $B' \mathop{\smile\!\!\!\!\!\!|}_{Aa} c'$. Then, by taking an Aa-automorphism, we have c such that $Bc \equiv_{Aa} B'c'$. Hence, again (i)(ii) hold with ac and $B \mathop{\smile\!\!\!\!\!\!|}_{Aa} c$. In particular,

(iii) $D_{\psi,k}(a/Ac) \geq n$; $\psi(x,c) \in \mathrm{tp}(a/Ac)$; and $\psi(x;c)$ k-divides over A.

Now by transitivity and symmetry, we have $a \mathop{\smile\!\!\!\!\!\!|}_{Ac} B$. Then (iii) with the induction hypothesis implies $D(\mathrm{tp}(a/Bc), \psi, k) \geq n$, and $D(\mathrm{tp}(a/B) \cup \{\psi(x,c)\}, \psi, k) \geq n$. Moreover there is an A-indiscernible sequence I containing c which witnesses k-dividing of $\psi(x;c)$ over A. Then since $B \mathop{\smile\!\!\!\!\!\!|}_A c$, from 2.1.6, we can assume that I is B-indiscernible. This says that $\psi(x;c)$ k-divides over B. Thus, again according to 2.3.3, as desired, $D_{\psi,k}(a/B) \geq n+1$. ■

We finish this section by giving some type-counting criteria for (non)simplicity.

Proposition 2.3.10 *The following are equivalent. Here, types are in finite variables.*

(1) *T has the tree property; i.e., there is a formula $\varphi(x, y)$ which has the k-tree property for some k. (Equivalently, T is not simple.)*

(2) *There is a formula $\psi(x, y)$ having the 2-tree property.*

(3) *For any infinite cardinals λ, κ with $\lambda^{<\kappa} = \lambda$, there is a family F of pairwise inconsistent types over a set A such that:*
   (a) *$|A| = \lambda$, $|F| = \lambda^{\kappa}$, and*
   (b) *$|p| = \kappa$ for each $p \in F$.*

(4) *There are an infinite cardinal κ and a family F of types over A such that*
   (a) *$|F| > |A|^{|T|} + 2^{|T|+\kappa}$,*
   (b) *$|p| \leq \kappa$ for each $p \in F$,*
   (c) *whenever $F' \subseteq F$ and $|F'| > \kappa$, then $\bigcup F'$ is inconsistent.*

Proof (2)$\Rightarrow$(1) Obvious.

(1)$\Rightarrow$(4) Let $\delta = |T|^+$ and $\kappa = 2^{\delta}$. Suppose that $\varphi(x, y)$ has the k-tree property. Because of compactness there is a k-tree of $\varphi(x, y)$ such that its height is δ and each node has $\lambda := \beth_{\delta}(2^{\kappa})$ many branches. Then by considering each path as a type, we obtain λ^{δ} many types of size δ over a set A of size $\lambda^{<\delta}$. Since the cofinality of λ is δ, and $\lambda^{<\delta} = \lambda$ [61, 5.20],

$$\lambda^{\delta} > |A|^{|T|} + 2^{|T|+\kappa} = \max(\lambda, 2^{\kappa}) = \lambda.$$

Finally among them, at most $(k-1)^{\delta} \leq \kappa$ many types can have a common realization.

(4)$\Rightarrow$(2) Assume (4). Because of (a), there is no harm to suppose that $\kappa \leq |A|$, and hence by adding tautologies we can assume $|p| = \kappa$ for all $p \in F$. In order to show (2), we need the following two claims:

Claim 1 Let $\lambda = |A|^{|T|} + 2^{|T|+\kappa}$. Then, there is a family $S = \{q_i | i < \lambda^+\}$ of pairwise inconsistent types of size κ over A such that $q_i = \{\varphi_\alpha(x, a^i_\alpha) | \alpha < \kappa\}$ for each $i < \lambda^+$, and $\text{tp}(A^i) = \text{tp}(A^0)$ for every $i < \lambda^+$, where $A^i = \langle a^i_\alpha | \alpha < \kappa \rangle$.

Proof of Claim 1 First, we take each q_i as the union of a maximal consistent subset of F. Then $|q_i| = \kappa \cdot \kappa = \kappa$, and i runs through $< \lambda^+$. Let us write $q_i = \{\varphi^i_\alpha(x, a^i_\alpha) | \alpha < \kappa\}$. As the number of sequences $\langle \varphi^i_\alpha(x, y^i_\alpha) | \alpha < \kappa \rangle$ is $\leq |T|^{\kappa} \leq 2^{|T|+\kappa} \leq \lambda$, by the pigeon hole principle we can choose the family S consisting of the types $q_i = \{\varphi_\alpha(x, a^i_\alpha) | \alpha < \kappa\}$. Moreover as there are at most $2^{|T|+\kappa}$ many types of the form $\text{tp}(A^i)$, we can assume again $\text{tp}(A^i)$ $(i < \lambda^+)$ is fixed.

Claim 2 In Claim 1, $|T|^+ \leq \kappa$ and there is a family $\{h_j | j < |T|^+\}$ of pairwise disjoint subsets of κ such that

(i) each h_j is finite,

(ii) for each $j < |T|^+$, $\bigwedge_{\alpha \in h_j} \varphi_\alpha(x, a^0_\alpha)$ 2-divides over $\{a^0_\beta | \beta \in H_j\}$ where $H_j = \bigcup_{i<j} h_i$.

Proof of Claim 2 First we note that $|S| = \lambda^+$ cannot be greater than $|A|^\kappa$. (If $|S| > |A|^\kappa$, then $q_i = q_j$ for some $i \neq j$, a contradiction.) Hence, $|A|^{|T|} < |S| \leq |A|^\kappa$, so $|T|^+ \leq \kappa$. Now, define h_j inductively. Assume that given $j < |T|^+$, h_i $(i < j)$ satisfying (i),(ii) are defined. Since $|H_j| \leq |T|$ and $|A|^{|T|} < |S|$, by the pigeon hole principle there is $\tau \subseteq \lambda^+$ with $|\tau| = \lambda^+$ such that $a^u_\beta = a^v_\beta$ for every $\beta \in H_j$ and each $u \neq v \in \tau$. As $q_u \cup q_v$ is still inconsistent, there is a finite subset $h(u,v)$ of κ $(H_j \cap h(u,v) = \emptyset)$ witnessing the inconsistency of $q_u \cup q_v$. By the Erdös–Rado theorem $((2^\kappa)^+ \to (\kappa^+)^2_\kappa$, hence $|\tau| \to (\kappa^+)^2_\kappa)$, there are a fixed finite $h_j \subseteq \kappa$ and $\tau' \subseteq \tau$ with $|\tau'| = \kappa^+$ such that $h(u,v) = h_j$ for every $u \neq v \in \tau'$. Finally, since $\mathrm{tp}(A^u) = \mathrm{tp}(A^0)$, (ii) follows.

Now we can finish the proof of (4)⇒(2). Let us consider $q = \{\bigwedge_{\alpha \in h_j} \varphi_\alpha(x, a^0_\alpha) \mid j \in |T|^+\}$. Note that q is consistent and there are $|T|$-many $\mathcal{L}$-formulas. Then it easily follows from Claim 2(ii) that for some j, $\bigwedge_{\alpha \in h_j} \varphi_\alpha(x, y_\alpha)$ witnesses the 2-tree property.

(2)⇒(3) Compactness.

(3)⇒(4) Let λ $(> 2^{|T|+\kappa})$ be a strong limit cardinal having cofinality $\kappa = |T|^+$. Then $\lambda^{<\kappa} = \lambda$ and $|F| = \lambda^\kappa > |A|^{|T|} + 2^{|T|+\kappa} = \lambda$. ■

The proof of the previous proposition also says that if $\varphi(x,y)$ has the k-tree property, then for some n, $\varphi(x,y_1) \wedge \ldots \wedge \varphi(x,y_n)$ has the 2-tree property.

2.4 Fundamental Theorem of Forking

In this section we prove the fundamental theorem of forking, which says the symmetry, transitivity, and local character of dividing or forking are all equivalent. In other words, the converse of Theorem 2.2.8 holds. This is a somewhat surprising fact of dividing and forking. The proof uses a further investigation of the tree property. Throughout this section, T is arbitrary.

Lemma 2.4.1 *The following are equivalent:*

(1) *A formula $\varphi(x,y)$ has the k-tree property.*

(2) *There is an indiscernible $\langle c_i b_i | i < \omega\rangle$ such that for every i, $\varphi(c_i, b_0)$ holds and $\varphi(x, b_i)$ k-divides over $\{c_j b_j | j < i\}$.*

(3) *There are a tuple c, and some c-indiscernible sequence $\langle b_i | i \leq \omega\rangle$ such that $\varphi(c, b_0)$ and $\varphi(x, b_\omega)$ k-divides over $\{b_i | i < \omega\}$.*

Proof (1)⇒(2) Suppose that $\{\varphi(x, a_\alpha) | \alpha \in \lambda^{<\lambda}\}$ witnesses the k-tree property $(\lambda > 2^{|T|})$. We write $0(n)$ to denote the sequence $00\cdots00$ of length n.

Claim We can assume that there are tuples c_n $(n < \omega)$ such that

$$c_n \models \varphi(x, a_0) \wedge \varphi(x, a_{00}) \wedge \cdots \wedge \varphi(x, a_{0(n+1)}),$$

and $\varphi(x, a_{0(n+1)})$ k-divides over $\{c_l a_{0(l+1)} | l < n\}$: Let c_0 be a tuple realizing $\varphi(x, a_0)$. Assume inductively that we have chosen $c_0, \ldots, c_{n-1}$ to satisfy the claim for $n-1$.

Now since $\lambda > 2^{|T|}$($\geq$ number of types over a finite set), there is infinite $\tau \subseteq \lambda$ such that for each $i, j \in \tau$, $\text{tp}(a_{0(n)i}/\{c_l a_{0(l+1)} \mid l < n\}) = \text{tp}(a_{0(n)j}/\{c_l a_{0(l+1)} \mid l < n\})$. Without loss of generality, we can assume $0 \in \tau$. Hence, $\varphi(x, a_{0(n+1)})$ k-divides over $\{c_l a_{0(l+1)} | l < n\}$. Then choose a tuple c_n realizing $\varphi(x, a_0) \wedge \varphi(x, a_{00}) \wedge \cdots \wedge \varphi(x, a_{0(n+1)})$. Therefore, the claim is proved.

Now note that the condition in the claim can be expressed by a type with variables for c_n and $a_{0(n)}$. Thus, we can apply Ramsey and compactness to further assume that the sequence $\langle c_i a_{0(i+1)} | i < \omega \rangle$ is a desired indiscernible.

(2)$\Rightarrow$(3) In (2), we extend the indiscernible to $\langle c_i b_i | i \leq \omega + 1 \rangle$. Then simply take $c = c_{\omega+1}$.

(3)$\Rightarrow$(1) According to 2.3.2, it suffices to show $D(\text{tp}(c), \varphi(x,y), k) = \infty$: Suppose not; say $D(\text{tp}(c), \varphi(x,y), k) = n$. Note that for each i, $\varphi(c, b_i)$ holds and $\varphi(x, b_i)$ k-divides over $b_0 \ldots b_{i-1}$. Since $\varphi(x, b_0) \in \text{tp}(c/b_0)$ and $\varphi(x, b_0)$ divides over $\emptyset$, from 2.3.4(5) we have $D(\text{tp}(c/b_0), \varphi, k) = n_0 < n$. Similarly, we see that $D(\text{tp}(c/b_0 \ldots b_i), \varphi, k) = n_i < n_{i-1} \ldots < n$. Since i can be arbitrary large, it leads a contradiction. ■

The following simple fact held in any T will be frequently used.

Lemma 2.4.2 *Let $J = \langle b_i | i \leq \omega \rangle$ be A-indiscernible, and let $I = \langle b_i | i < \omega \rangle$. Then $\text{tp}(b_\omega / I \cup A)$ does not fork over I.*

Proof Suppose not; say there is a formula $\psi(x, b_n \ldots b_0 a) \in \text{tp}(b_\omega/IA)$ $(a \in A)$ forking over I. Hence, $\psi(x, b_n \ldots b_0 a)$ implies a finite disjunction of formulas $\phi_j(x, d_j)$ $(j = 0, \ldots, m)$, each of which divides over I. Now, by A-indiscernibility, b_{n+1} realizes $\psi(x, b_n \ldots b_0 a)$. Hence, b_{n+1} realizes $\phi_i(x, d_i)$ for some $i(\leq m)$. Then $\phi_i(x, d_i)$ cannot divide over I, as $b_{n+1} \in I$. From this contradiction, the lemma is deduced. ■

We are ready to state and prove the fundamental theorem of forking. We have already seen in Proposition 2.3.7 that forking satisfies local character iff so does dividing. The same is true for symmetry and transitivity. Moreover, those are all equivalent.

Theorem 2.4.3 *(Fundamental theorem of forking) Let T be arbitrary. The following are all equivalent:*

(1) Forking (Dividing) satisfies local character (i.e., T is simple).

(2) Forking (Dividing) satisfies symmetry.

(3) Forking (Dividing) satisfies transitivity.

(4) A formula $\varphi(x, a)$ divides (forks) over a set A if and only if for any Morley sequence I of $\text{tp}(a/A)$, $\{\varphi(x, a') | a' \in I\}$ is inconsistent.

Proof (1)$\Rightarrow$(2),(3),(4). 2.2.6, 2.2.7, 2.2.8 with 2.3.7.

(2)$\Rightarrow$(1). Suppose that T is not simple. Then by Lemma 2.4.1, there are a tuple c and some c-indiscernible sequence $\langle a_i | i \leq \omega \rangle$ such that $\text{tp}(c/\{a_i | i \leq \omega\})$ divides, and hence forks, over $I = \{a_i | i < \omega\}$. On the other hand, by 2.4.2 $\text{tp}(a_\omega/Ic)$ does not fork, and hence does not divide, over I. This shows the non-symmetry of forking and dividing at the same time, and finishes the proof.

(3)⇒(1). Again if T is not simple, then by 2.4.1, there are a formula $\varphi(x,y)$ and an indiscernible sequence $\langle c_i a_i | i \in \Lambda \rangle$, where $\Lambda = \langle -1, -1/2, -1/3, \ldots, 0, 1, \ldots, 4/3, 3/2, 2 \rangle$, such that $\varphi(c_i, a_{-1})$ for all i and $\varphi(x, a_j)$ divides over $\{c_i a_i | i < j\}$. Let $I = \{c_i | i \in \Lambda, i < 0\}$, and $J = \{c_i | i \in \Lambda, i > 1\}$. Now by (the proof of) 2.4.2, $\text{tp}(c_1/IJ)$ does not fork over I. Similarly, $\text{tp}(c_1/IJa_0)$ does not fork over J, and hence does not fork over IJ either. But, since $\varphi(c_1, a_0)$ holds, we have that $\text{tp}(c_1/IJa_0)$ divides over I. Hence, the transitivity of forking and dividing fails.

(4)⇒(1). Assume that T is not simple. Now there are a formula $\varphi(x,y)$, a tuple c, and some c-indiscernible sequence $\langle a_i | i \in \Lambda \rangle$ with $\Lambda = \langle -1, -1/2, -1/3, \ldots, 1/3, 1/2, 1 \rangle$, such that for $j > 0$, $\varphi(x, a_j)$ divides over $I = \{a_i | i < 0\}$ and $\models \varphi(c, a_j)$ for all j. But, according to 2.4.2, $\langle a_1, a_{1/2}, a_{1/3}, \ldots \rangle$ is a Morley sequence of $\text{tp}(a_1/I)$. Hence, either of (4) fails. ■

2.5 Ranks and Supersimple Theories

As we have seen the rank $D_{\varphi,k}$ characterizes simplicity, and reflects forking. We shall introduce more rank notions to categorize various subclasses of simple theories. The class of supersimple theories is possibly the most representative subclass. A theory is called *supersimple* if any complete type of a finite tuple does not fork over some finite subset of the domain (2.5.10). Supersimplicity can be defined in terms of the D-rank or the SU-rank. The SU-rank reflects forking in the class, too. The class is a natural extension of that of superstable theories sharing many analogous facts. Extending results and topics held in superstable theories to the supersimple context is a common concern in the early development of simplicity theory. Many are extended positively while some negatively, but either case reveals the more primitive and hidden nature of the subjects. Still a lot remain as conjectural forms. We shall study some typical results in this section.

2.5.1 *D_φ-Rank and Low Theories*

We begin with the D_φ-rank by which we introduce the class of *low* theories. The rank D_φ for each formula φ is similarly defined as the $D_{\varphi,k}$-rank while only φ is fixed.

Definition 2.5.1 Let $\varphi(x; y)$ be an $\mathcal{L}$-formula. For any type $p(x')$ with $x \subseteq x'$, $D(p(x'), \varphi(x,y)) = D_\varphi(p)$ (either an ordinal or ∞) is recursively defined as follows:

(1) $D(p, \varphi) \geq 0$ for any consistent type p.

(2) $D(p, \varphi) \geq \alpha + 1$ if there is a such that $\varphi(x, a)$ divides over $\text{dom}(p)$ and $D(p \cup \{\varphi(x, a)\}, \varphi) \geq \alpha$.

For limit δ, $D(p, \varphi) \geq \delta$ if $D(p, \varphi) \geq \alpha$ for all $\alpha < \delta$.

(3) $D(p,\varphi) = \alpha$ if $D(p,\varphi) \geq \alpha$ and $D(p,\varphi) \not\geq \alpha + 1$.
(4) $D(p,\varphi) = \infty$ if $D(p,\varphi) \geq \alpha$ for all ordinal α.

The rank D_φ does not behave as well as $D_{\varphi,k}$. For example, differently from 2.3.2, the D_φ-rank need not be type-definable in general. Still the following properties hold:

Proposition 2.5.2

(1) *If $p \vdash q$, then $D_\varphi(p) \leq D_\varphi(q)$.*
(2) *If there is $\langle a_i |\ i < \alpha\rangle$ such that $p \cup \{\varphi(x,a_i)|\ i < \alpha\}$ is consistent, and for $i < \alpha$, $\varphi(x,a_i)$ divides over* $\mathrm{dom}(p) \cup \{a_j|\ i < j < \alpha\}$, *then $D(p,\varphi) \geq \alpha$. If $\alpha < \omega$, then the converse holds too.*
(3) *For any k, we have $D_{\varphi,k}(p) \leq D_\varphi(p)$.*
(4) *T is simple if $D(p,\varphi) < \infty$ for any p, φ.*

Proof

(1) comes from an induction.
(2) can be proved by the induction on α.
(3) also comes by an induction when $D_{\varphi,k}(p) < \omega$. If $D_{\varphi,k}(p) = \infty$, then by compactness (see 2.3.2 and 2.4.1) with (2), it follows that $D_\varphi(p) = \infty$ too.
(4) By (3), if $D(p,\varphi) < \infty$, then $D(p,\varphi,k) < \omega$. Hence, T is simple.

■

The converse of 2.5.2(4) is not true. But we shall see in 2.5.8 that if T is supersimple then $D_\varphi(x = x) < \infty$ for any $\varphi(x,y)$. Casanovas and Wagner [19] construct an example of simple structure where for some ψ, $D(x = x, \psi) = \infty$. We are mostly interested in the following class of theories.

Definition 2.5.3 We say T is *low* if $D(p,\varphi) < \omega$ for any p, φ.

Proposition 2.5.4 *The following are equivalent:*

(1) *T is low.*
(2) *T is simple, and for any $\varphi(x,y) \in \mathcal{L}$, there is k such that given an indiscernible I, if $\{\varphi(x,a')|\ a' \in I\}$ is inconsistent, then it is k-inconsistent.*
(3) *T is simple, and for any $\varphi(x,y) \in \mathcal{L}$, there is k such that given a, if $\varphi(x,a)$ divides over the empty set, then it k-divides.*
(4) *T is simple, and given $\varphi(x,y) \in \mathcal{L}$, there is k such that for any p and $k' \geq k$, we have $D(p,\varphi) = D(p,\varphi,k')$.*

Proof (1)⇒(2) By 2.5.2, a low theory is simple. For $\varphi(x,y)$, put $k := D_\varphi(x = x) + 1$. Assume an indiscernible I is given such that $q := \{\varphi(x,a')|\ a' \in I\}$ is inconsistent. If q is not k-inconsistent, then by 2.5.2(2), we have $D_\varphi(x = x) \geq k$, a contradiction.

(2)$\Rightarrow$(4) Assume $\varphi(x, y)$ and k are given as in (2). Then, by 2.3.3(1), given p and any $k' \geq k$, we have $D(p, \varphi, k) = D(p, \varphi, k')$, say n. It suffices to see $D(p, \varphi) = n$. If $D(p, \varphi) \geq n + 1$, then applying (2) to 2.5.2(2) and 2.3.3(1), we see that $D(p, \varphi, k) \geq n + 1$, a contradiction.

(4)$\Rightarrow$(1), (2)$\Rightarrow$(3) Obvious.

(3)$\Rightarrow$(1) Similarly to (2)$\Rightarrow$(4), it follows $n = D_{\varphi,k}(x = x) \geq D_{\varphi,k'}(x = x)$ for any k' where $\varphi(x, y), k$ are as in (3). If $D_\varphi(x = x) \geq n + 1$, then again by 2.5.2(2), there are $a_0, \ldots, a_n$ such that $\{\varphi(x, a_i) | i \leq n\}$ is consistent, and each $\varphi(x, a_i)$ k_i-divides over $a_{i+1} \ldots a_n$. If we let $k'' = \max\{k_i | \ i \leq n\}$, then we have $D_{\varphi,k''}(x = x) \geq n + 1$, a contradiction. Hence, T is low. ■

Examples of non-low (super)simple theories are a bit complicated and can be found in [14], [16]. Proposition 2.5.4 says all stable theories are low: If T is simple non-low, then according to 2.5.4 we have $\psi(x, y)$ such that for any k, there are a_k and $k' \geq k$ so that $\psi(x, a_k)$ k'-divides, but does not k-divides over $\emptyset$. Then by compactness it easily follows $\psi(x, y)$ has the order property.

Moreover, the proposition together with 2.2.6 and 2.2.9 yields, for low theories, the type-definability of the dividing parameter set of a given formula. This property will play an important role in showing strong type-amalgamation for low theories in Section 5.2.

Corollary 2.5.5 *Assume T is low. Given $\psi(x, y) \in \mathcal{L}$ and a set A, there is a set of formulas $\Sigma(y)$ over A such that $\models \Sigma(a)$ holds iff $\psi(x, a)$ divides over A.*

2.5.2 *D-Rank and SU-Rank*

Both the D-rank and the SU-rank (or U-rank) are originally introduced for (super)stable theories [113], [80], [97]. But those make perfect (or more) sense in the context of simple theories, and will characterize *supersimple* theories. Indeed, T is superstable if and only if it is stable and supersimple.

In the rest of Section 2.5, formulas may have parameters. The definition of the D-rank for formulas shares the same pattern as those of the $D_{\varphi,k}$ and the D_φ-ranks, but φ, k are not fixed.

Definition 2.5.6 For a formula $\psi(x)$ and an ordinal α, we define $D(\psi(x)) \geq \alpha + 1$ if there is an $\mathcal{L}$-formula $\phi(x; y)$, and for every λ there are c_i for $i < \lambda$ such that $\{\phi(x; c_i) : i < \lambda\}$ is k-inconsistent for some $k \geq 2$, and $D(\psi(x) \wedge \phi(x; c_i)) \geq \alpha$ for all i.

The other parts of the definition (such as $D(\psi(x)) \geq 0$; $D(\psi(x)) \geq \delta$ for limit δ; $D(\psi(x)) = \alpha$ or ∞) are the same as defined in 2.5.1.

If $p(x)$ is a type in finite x, then we define

$$D(p(x)) := \min\{D(\psi_0(x) \wedge \ldots \wedge \psi_n(x)) | \psi_i \in p, \ i \leq n \in \omega\}.$$

Hence, the D-rank of a complete type in finite x is represented by a formula in there. (Exercise) If there is a definable bijection between (the solution sets of) types $p(x)$ and $q(x)$, then $D(p) = D(q)$. (It is clear when p, q are formulas. Then use compactness.) We freely use these characters in future arguments. If x is infinite in $p(x)$, different from $D_{\varphi(x',y),k}$ where the rank is measured in $\mathcal{M}^{|x'|}$, the above definition does not make much sense.

Lemma 2.5.7 *For a formula $\psi(x)$ over A, and α, the following are equivalent:*

(1) $D(\psi(x)) \geq \alpha + 1$.

(2) *There is $\theta_0(x)$ such that $\theta_0(x)$ divides over A, and $D(\psi(x) \wedge \theta_0(x)) \geq \alpha$.*

(3) *There is $\theta(x)$ such that $\models \theta(x) \rightarrow \psi(x)$, $\theta(x)$ divides over A, and $D(\theta(x)) \geq \alpha$.*

Proof (1)⇒(2) From (1), we have an $\mathcal{L}$-formula $\phi(x; y)$, and c_i for $i < \lambda$ for sufficiently large λ, such that $\{\phi(x; c_i) : i < \lambda\}$ is k-inconsistent, for some $k < \omega$, and $D(\psi(x) \wedge \phi(x; c_i)) \geq \alpha$ for all i. Since λ is sufficiently large, we can assume $\text{tp}(c_0/A) = \text{tp}(c_i/A)$ for all i, and so $\phi(x; c_0)$ divides over A. Put $\theta_0(x) := \phi(x; c_0)$.

(2)⇒(3) Put $\theta(x) := \psi(x) \wedge \theta_0(x)$.

(3)⇒(1) Use that $\theta(x) \leftrightarrow \theta(x) \wedge \psi(x)$ and $\theta(x)$ divides over A. ■

Observations analogous to 2.5.2 follow.

Proposition 2.5.8

(1) *If $p(x) \vdash q(x)$, then $D(p) \leq D(q)$.*

(2) *If there is $\{\theta_i(x) | i < \alpha\}$ such that $\{\psi(x)\} \cup \{\theta_i(x) | i < \alpha\}$ is consistent, and for $i < \alpha$, $\theta_i(x)$ divides over $\text{dom}(\psi) \cup \bigcup\{\text{dom}(\theta_j) | i < j < \alpha\}$, then $D(\psi) \geq \alpha$. If $\alpha < \omega$, then the converse holds too.*

(3) $D_{\varphi,k}(\psi(x)) \leq D_{\varphi}(\psi(x)) \leq D(\psi(x))$.

Corollary 2.5.9 *For a formula $\psi(x)$ over A, the following are equivalent:*

(1) $D(\psi(x)) = \infty$.

(2) *There is $\theta(x)$ such that $\models \theta(x) \rightarrow \psi(x)$, $\theta(x)$ divides over A, and $D(\theta(x)) = \infty$.*

(3) *There are a sequence of sets $A = A_0 \subseteq A_1 \subseteq \ldots A_i \subseteq \ldots$ $(i < \omega)$ and consistent formulas $\theta_i(x)$ over A_i, such that $\theta_0 = \psi$, θ_{i+1} implies θ_i for all i, and θ_{i+1} divides over A_i for all i.*

Proof (1)⇒(2) Let $D(\psi(x)) = \infty$. For each α, there is $\theta_\alpha(x)(\models \psi(x))$ such that $\theta_\alpha(x)$ divides over A, and $D(\theta_\alpha(x)) \geq \alpha$. As there are at most $2^{|T|+|A|}$ many types over A, we can assume there are only boundedly many formulas $\theta_\alpha(x)$, by identifying the formula and its A-automorphic images. Then by the pigeon hole argument, we find the desired θ satisfying (2).

(2)⇒(3) Applying (1)⇒(2) repeatedly.

(3)⇒(1) Assume (3) holds. To get a contradiction, suppose that $D(\psi(x)) = \alpha$. By 2.5.7, $D(\theta_{i+1}(x)) < D(\theta_i(x))$. This means there is a decreasing sequence of ordinals, which is not true. ■

By strengthening the local character axiom of forking, we introduce a (sub)class of simple theories, called *supersimple*. The class is an extension of the class of *superstable* theories, or more correctly, the latter is the canonical restriction of the former. T is superstable iff T is stable and supersimple. It is well known that the D-rank characterizes superstability *among* stable theories. In fact, the D-rank characterizes supersimplicity.

Definition 2.5.10 T is said to be *supersimple* if any $p \in S_n(A)$ does not fork over some finite subset A_0 of A.

We call T *strictly simple* if it is simple but not supersimple. Supersimple theories satisfy properties analogous but stronger than what all simple theories do. For example, in supersimple T there do not exist finite a and $B = \{b_i | i < \omega\}$ such that B is A-independent while $a \not\downarrow_A b_i$ for all $i < \omega$. (See the argument stated before 2.2.9.) The following proposition analogous to 2.3.7 is another example.

Proposition 2.5.11 *The following are equivalent:*

(1) *T is supersimple.*
(2) *There does not exist a forking chain (in a finite tuple) of length ω.*
(3) *Every $p \in S_n(A)$ does not divide over some finite subset A_0 of A.*
(4) *There does not exist a dividing chain of length ω.*
(5) *$D(\psi(x)) < \infty$ for all $\psi(x)$.*
(6) *Any $p \in S_1(A)$ does not divide (fork) over some finite subset A_0 of A.*

Proof (1)⇔(2)⇒(3)⇔(4) The proof is the same as that of 2.3.7.

(4)⇔(5) Suppose that $D(\psi(x)) = \infty$. Then sets $A_0 = \mathrm{dom}(\psi) \subseteq A_1 \subseteq \ldots$. and consistent formulas $\theta_i(x)$ over A_i are given as in 2.5.9(3). Let $A := \bigcup_{i<\omega} A_i$, and let p be a complete type over A containing all $\theta_i(x)$. Then, clearly $\langle p\lceil A_i | \, i \in \omega\rangle$ is a dividing chain. Conversely, if $\langle p_i(x) | \, i \in \omega\rangle$ is a dividing chain, then we can find formulas $\varphi_{i+1} \in p_{i+1}$ such that φ_{i+1} divides over $\mathrm{dom}(p_i)$ for all i. Then that $D(x = x) = \infty$ is witnessed by $\theta_{i+1} := \bigwedge_{1 \le j \le i+1} \varphi_j$.

(4)⇒(2) Assume (4). By 2.3.7, T is simple. Thus, forking is dividing, and (2) follows.

(1)⇒(6)⇒(4) Similar to the proof (7)⇒(4) of 2.3.7. ■

According to 2.5.8(3) and 2.5.11, if T is supersimple then $D_\varphi(p) < \infty$ for any φ, p. But there is an example of a non-low supersimple theory [16].

Now we turn into the issue of forking representation of the D-rank in (super)simple theories.

Lemma 2.5.12 *Assume that T is simple. Let $p(x) \subseteq q(x)$ be complete types in finite x, and let $D(q) < \infty$. If q is a forking extension of p, then $D(q) < D(p)$.*

Proof Let $D(q) \geq \alpha$, and let $\psi(x) \in p$ be given. It suffices to show that $D(\psi(x)) \geq \alpha + 1$. Now, there is $\theta(x) \in q$ such that $\psi(x) \wedge \theta(x)$ forks over $\mathrm{dom}(p)$. As $\psi(x) \wedge \theta(x) \in q$, $D(\psi(x) \wedge \theta(x)) \geq \alpha$. Hence, by Lemma 2.5.7, $D(\psi(x)) \geq \alpha + 1$. ■

Remark 2.5.13 In (super)stable theories, the open mapping theorem implies the converse of the previous lemma. But the open mapping theorem is not true for simple unstable theories. Consider the bipartite random graph partitioned by disjoint sets U and V with the relation R between U and V. Let $A = \{a_i | i < \omega\} \subseteq U$. Choose $c \in U \setminus A$ such that $\mathrm{tp}(c/A)$ is not isolated. Also select $b \in V$ such that $\neg a_i R b$ for all i, and cRb. Then $c \mathop{\smile\!\!\!\!\!\mid}_A b$ holds, whereas $\mathrm{tp}(c/Ab)$ is isolated.

In Example 6.3.4 we will come up with a supersimple structure, where the D-rank does not represent forking.

We now introduce the SU-rank for complete types. This is defined exactly like Lascar's U-rank for stable theories.

Definition 2.5.14 Let $p(x) \in S(A)$ with possibly infinite x. We define $SU(p) \geq \alpha + 1$ if for some $B \supseteq A$, p has a forking extension $q(x) \in S(B)$ such that $SU(q) \geq \alpha$. (The other parts of the definition are as usual.) For a partial type $\pi(x)$ over A, $SU(\pi(x)) := \sup\{SU(p(x)) | p(x)(\in S(A)) \models \pi(x)\}$.

Lemma 2.5.15

(1) *If $p(x) \subseteq q(x)$ are complete types, then $SU(q) \leq SU(p)$.*
(2) *For a type $p(x) \in S(A)$, the following are equivalent:*
   (a) *$SU(p) = \infty$.*
   (b) *For some $B \supseteq A$, p has a forking extension $q \in S(B)$ such that $SU(q) = \infty$.*
   (c) *There is a forking chain $\langle p_i(x) | i \in \omega \rangle$ in x where $p_0 = p$.*

Proof (1) Use induction.

(2) The proof, similar to that of 2.5.9, is left to the reader. ■

The SU-rank characterizes the class of supersimple theories and reflects forking there.

Proposition 2.5.16

(1) *T is supersimple if and only if for any complete type p in a finite tuple, $SU(p) < \infty$.*
(2) *Let $p(x) \subseteq q(x)$ be complete types. If $SU(p) = SU(q) < \infty$ then q is a non-forking extension of p. Conversely, for simple T, if q is a non-forking extension of p then $SU(p) = SU(q)$.*

Proof (1) This follows from the preceding lemma, together with 2.5.11.

(2) The first assertion is clear by the definition of the SU-rank. For the second assertion, we use induction on α to show that if $SU(p) \geq \alpha$, then $SU(q) \geq \alpha$. This is clear

for $\alpha = 0$ or limit ordinal. Suppose $A = \mathrm{dom}(p)$, $q \in S(B)$ and $SU(p) \geq \alpha + 1$. Now there are $C \supseteq A$ and $p_1 \in S(C)$ a forking extension of p such that:

(i) $SU(p_1) \geq \alpha$.

By moving C, we can assume there is d realizing q and p_1. By moving C again, we further assume $\{C, B\}$ is independent over Ad, and hence

(ii) $\mathrm{tp}(d/C \cup B)$ is a non-forking extension of p_1.

Hence, by the induction hypothesis, $SU(\mathrm{tp}(d/C \cup B)) \geq \alpha$. Moreover, by the nature of q and p_1, $\mathrm{tp}(d/C \cup B)$ is a forking extension of q. Hence, $SU(q) \geq \alpha + 1$. ■

In simple T, the SU-rank is not greater than the D-rank.

Lemma 2.5.17 *Let T be simple. For any $p \in S_n(A)$, $SU(p) \leq D(p)$.*

Proof Let us show that $SU(p) \geq \alpha + 1$ implies that $D(q) \geq \alpha + 1$ using the induction hypothesis on α (the case $\alpha = 0$ or limit being immediate). Assume $SU(p) \geq \alpha + 1$. Then there is a forking extension $q \supseteq p$ such that $SU(q) \geq \alpha$. By the induction hypothesis, $D(q) \geq \alpha$. Now, 2.5.12 says that $D(p) \geq \alpha + 1$. ■

Example 2.5.18 Consider a model consists of infinitely many disjoint unary predicates P_i $(i < \omega)$, and equivalence relations E_i defining infinitely many infinite E_i classes on P_i. Note that $\{\neg P_i(x) |\ i < \omega\}$ axiomatizes a complete type, say $p(x)$. The model is ω-stable, and it is easy to see that $SU(p) = 1$, whereas $D(p) = 2$ and the Morley rank of p is 3. More stable examples where ranks differ can be found in [5].

Just by definition it easily follows that if $SU(p) = \alpha < \infty$ and $\beta < \alpha$, then p has an extension q such that $SU(q) = \beta$. But the above example shows that the D-rank does not have this property. (A rank having the property is said to be *connected*.)

We finish this subsection by stating the *Lascar inequalities* of the SU-rank.

Proposition 2.5.19 *T simple.*

(1) *$SU(a/bA) + SU(b/A) \leq SU(ab/A) \leq SU(a/bA) \oplus SU(b/A)$. Hence, $SU(ab/A) = SU(a/bA) + SU(b/A)$ when values are finite.*

(2) *If $a \mathop{\smile\!\!\!\!\!\mid}_A b$, then $SU(ab/A) = SU(a/A) \oplus SU(b/A)$.*

(3) *If $SU(a/bA) < \infty$ and $SU(a/A) \geq SU(a/Ab) \oplus \alpha$, then $SU(b/A) \geq SU(b/Aa) + \alpha$.*

(4) *If $SU(a/bA) < \infty$ and $SU(a/A) \geq SU(a/Ab) + \omega^\beta.n$, then $SU(b/A) \geq SU(b/Aa) + \omega^\beta.n$.*

Proof The proof is exactly the same as the U-rank case for stable theories.

(1) We first show inductively that if $SU(b/A) \geq \alpha$, then $SU(ab/A) \geq SU(a/bA) + \alpha$. For the induction step, assume $SU(b/A) \geq \alpha + 1$, so there is $(A \subseteq) B$ with $b \not\mathop{\smile\!\!\!\!\!\mid}_A B$ such that $SU(b/B) \geq \alpha$. By the extension axiom we can choose $B \mathop{\smile\!\!\!\!\!\mid}_{Ab} a$.

Then by the induction hypothesis, $SU(ab/B) \geq SU(a/bB) + \alpha = SU(a/bA) + \alpha$, whereas $SU(ab/A) \geq SU(ab/B) + 1 \geq SU(a/bA) + \alpha + 1$. It is clear when α is limit. When $\alpha = 0$, similar induction argument easily shows $SU(ab/A) \geq SU(a/bA)$.

Conversely, we show also by induction that $SU(ab/A) \geq \alpha$ implies that $SU(a/bA) \oplus SU(b/A) \geq \alpha$. It clearly holds when $\alpha = 0$ or limit. Assume now that $SU(ab/A) \geq \alpha + 1$ so that there is $B(\not\perp_A ab)$ with $A \subseteq B$ such that $SU(ab/B) \geq \alpha$, and $SU(a/bB) \oplus SU(b/B) \geq \alpha$ (the induction hypothesis). Note that by transitivity, either $a \not\perp_{Ab} B$ or $b \not\perp_A B$. For the former case, $SU(a/Ab) \geq SU(a/Bb) + 1$ so $SU(a/bA) \oplus SU(b/A) \geq \alpha + 1$ (this need not hold for usual addition). For the latter case, $SU(b/A) \geq SU(b/B) + 1$ and again $SU(a/bA) \oplus SU(b/A) \geq \alpha + 1$.

(2) Assume that $a \perp_A b$, and $SU(a/A) \geq \alpha + 1$. We will show, by induction, that $SU(ab/A) \geq (\alpha + 1) \oplus SU(b/A)$ ($\alpha = 0$ or limit case is clear). Now there is $B(\not\perp_A a)$ containing A with $SU(a/B) \geq \alpha$. By extension, we can assume that $B \perp_{Aa} b$, so $b \perp_A B$ (*). According to $ab \not\perp_A B$ and the induction hypothesis, we have

$$SU(ab/A) \geq SU(ab/B) + 1 \geq (\alpha \oplus SU(b/B)) + 1 = (\alpha + 1) \oplus SU(b/B).$$

It suffices to see $SU(b/A) = SU(b/B)$, which comes from (*).

(3) Assume (3) holds for α. We show it for $\alpha + 1$ ($\alpha = 0$, or limit case is obvious), so assume that $SU(a/A) \geq SU(a/bA) \oplus \alpha + 1$. Thus, for some $A_1 \supset A_0 = A$ with $a \not\perp_{A_0} A_1$, we have $SU(a/A_1) \geq SU(a/A_0b) \oplus \alpha \geq SU(a/A_1b) \oplus \alpha$. By extension we can suppose $b \perp_{aA_0} A_1$.

Case I of A_0A_1) $b \not\perp_{A_0} A_1$: Then, by the induction hypothesis

$$SU(b/A_0) \geq SU(b/A_1) + 1 \geq SU(b/A_1a) + \alpha + 1 = SU(b/A_0a) + \alpha + 1.$$

Case II of A_0A_1) $b \perp_{A_0} A_1$ and $a \not\perp_{bA_0} A_1$: Then, $SU(a/A_0b) \geq SU(a/A_1b) + 1$ and $SU(a/A_1) \geq SU(a/A_0b) \oplus \alpha \geq SU(a/A_1b) \oplus (\alpha + 1)$. Now by iteration we obtain $A_1 \subset A_2 \ldots \subset A_i \ldots$ such that A_iA_{i+1} play the role of A_0A_1 in the arguments above. We claim that for some finite $n > 0$, A_iA_{i+1} satisfy case II for all $i < n$, and $SU(b/A_n) \geq SU(b/A_na) + \alpha + 1$: If some first A_nA_{n+1} satisfy case I, then the claim follows. Otherwise, since $SU(a/bA) < \infty$, there must be some finite n such that $a \not\perp_{bA_{n-1}} A_n$ and $SU(a/bA_n) = 0$. Then from (1), the claim follows.

From the claim, for all $i < n$,

$$SU(b/A_i) = SU(b/A_{i+1}) \geq SU(b/A_{i+1}a) + \alpha + 1 = SU(b/A_ia) + \alpha + 1.$$

The induction step for $\alpha + 1$ is obtained.

(4) comes from (3) with the fact that for ordinals $\gamma < \omega^\beta.n$, and γ' we have $\gamma' + \omega^\beta.n > \gamma' \oplus \gamma$. ■

It follows from 2.5.19(1), in simple T if there is a definable bijection between $\pi(x)$ and $\pi'(x)$, then $SU(\pi(x)) = SU(\pi'(x))$.

2.5.3 *Counting Models of Supersimple Theories*

The aim of this subsection is to extend the following classical result of Lachlan [79] to the supersimple theory context.

Theorem 2.5.20 *Let T be a countable superstable theory. Then the number of (non-isomorphic) countable models of T is either* 1 *or* $\geq \aleph_0$.

In other words, we shall prove the following.

Theorem 2.5.21 *Let T be a countable supersimple theory. Then the number of countable models of T is either* 1 *or* $\geq \aleph_0$.

In the rest of this subsection, tuples are finite.

2.5.3.1 Forking and isolation

Let us recall Pillay's notion of semi-isolation [97]. We say $\mathrm{tp}(b/a)$ is *semi-isolated* if there is a formula $\varphi(x, a)$ in $\mathrm{tp}(b/a)$ such that $\models \varphi(x, a) \rightarrow \mathrm{tp}(b)$. This definition implies the following easy but important facts.

Remark & Example 2.5.22

(1) If $\mathrm{tp}(b/a)$ is isolated, then $\mathrm{tp}(b/a)$ is semi-isolated. If $\mathrm{tp}(c/b)$ and $\mathrm{tp}(b/a)$ are semi-isolated, then so is $\mathrm{tp}(c/a)$. As is well known, both $\mathrm{tp}(c/ba)$ and $\mathrm{tp}(b/a)$ are isolated iff $\mathrm{tp}(cb/a)$ is isolated.

(2) The notions semi-isolation and isolation are different. For example, consider the theory $\mathrm{Th}(\mathbb{Z}, S)$, where S is the successor function. If a, b are in different components of a model, then $\mathrm{tp}(b/a)$ is semi-isolated, but not isolated.

(3) Let $\mathcal{L} = \{E_i | i < \omega\}$. Let T be a theory saying that E_0 is an equivalence relation having two infinite classes, and for each $i < \omega$, the equivalence relation E_{i+1} refines every E_i class into exactly two infinite E_{i+1} classes. The theory T is superstable. Now if a and c are in the same E_i class for each i, and $\neg bE_0a$, then $\mathrm{tp}(c/b), \mathrm{tp}(b/a)$ are isolated, while $\mathrm{tp}(c/a)$ is not isolated (but is semi-isolated).

Lemma 2.5.23 *Suppose that* $\mathrm{tp}(b/a)$ *is isolated, whereas* $\mathrm{tp}(a/b)$ *is non-isolated. Then* $\mathrm{tp}(a/b)$ *is nonsemi-isolated.*

Proof Suppose that $\varphi(x, a)$ isolates $\mathrm{tp}(b/a)$. In order to deduce a contradiction, assume that $\mathrm{tp}(a/b)$ is semi-isolated witnessed by $\psi(b, y)$. Now, as $\mathrm{tp}(a/b)$ is non-isolated, there is a formula $\phi(x, y) \in \mathcal{L}$ such that $\varphi(b, y) \wedge \psi(b, y) \wedge \phi(b, y)$ and $\varphi(b, y) \wedge \psi(b, y) \wedge \neg\phi(b, y)$ are both consistent. Moreover, both formulas imply $\mathrm{tp}(a)$. Hence, $\varphi(x, a) \wedge \phi(x, a)$ and $\varphi(x, a) \wedge \neg\phi(x, a)$ are both consistent. This contradicts that $\varphi(x, a)$ is a principal formula. ■

Now we state a key proposition which describes the relationship between isolation and forking in simple theories.

Proposition 2.5.24 *Assume that T is simple, and tuples a, b with* $\text{tp}(a) = \text{tp}(b)$ *are given. If* $\text{tp}(b/a)$ *is semi-isolated, and* $\text{tp}(a/b)$ *is nonsemi-isolated, then* $\text{tp}(a/b)$ *forks over* $\emptyset$.

Proof Suppose that $\text{tp}(b/a)$ is semi-isolated witnessed by $\varphi(x, a)$. Let c be an *arbitrary* tuple such that $\text{tp}(cb) = \text{tp}(ba)$. We claim that $\varphi(c, x) \wedge \varphi(x, a)$ 2-divides over $\emptyset$, so, $\text{tp}(b/ca)$ forks over $\emptyset$: First, let $c_0 = c, b_0 = b, a_0 = a$. Now, there is a sequence of tuples $\langle c_i b_i a_i | i < \omega \rangle$ such that, for all $i < \omega$, $\text{tp}(c_i b_i a_i) = \text{tp}(cba)$ and $\text{tp}(a_{i+1} c_i) = \text{tp}(ba)$. We note that, by Remark 2.5.22, $\text{tp}(a_j/a_i)$ is semi-isolated for every $j \geq i$ (*). We shall show that $\{\varphi(c_i, x) \wedge \varphi(x, a_i) | i < \omega\}$ is 2-inconsistent. If it were not 2-inconsistent, then there is d such that $\varphi(d, a_j)$ and $\varphi(c_i, d)$ for some $j > i$. Therefore, clearly $\text{tp}(d/a_j)$, $\text{tp}(c_i/d)$ are both semi-isolated, and hence again by 2.5.22, so does $\text{tp}(c_i/a_j)$. Now, since $\text{tp}(a_j/a_{i+1})$ is semi-isolated as in (*), once more 2.5.22 implies that $\text{tp}(c_i/a_{i+1})$ is semi-isolated. But since $\text{tp}(c_i a_{i+1}) = \text{tp}(ab)$, it leads a contradiction. Hence, the claim is proved.

Now if $\{a, b\}$ is independent (over $\emptyset$), then by the extension, symmetry and transitivity of non-forking, we can find a tuple c' such that $\text{tp}(c'b) = \text{tp}(ba)$ and $\{a, b, c'\}$ is independent. This contradicts the claim above. Thus $\text{tp}(a/b)$ forks over $\emptyset$. ■

Hence, Lemma 2.5.23 leads the following.

Corollary 2.5.25 *Assume that T is simple. Let a, b realize the same type over $\emptyset$. If* $\text{tp}(b/a)$ *is isolated, and* $\text{tp}(a/b)$ *is non-isolated, then* $\text{tp}(a/b)$ *forks over* $\emptyset$.

2.5.3.2 Proof of Theorem 1.5.21

Until the end of the proof, T will be *countable*, and non $\aleph_0$-categorical.

Fact 2.5.26 *Suppose that T has only finitely many non-isomorphic countable models. Then there exist some tuple a and prime model M over a such that* $\text{tp}(a)$ *is non-isolated and every complete n-type (for all n) over $\emptyset$ is realized in M. Moreover, there is some tuple b in M such that,* $\text{tp}(b) = \text{tp}(a)$ *and* $\text{tp}(a/b)$ *is non-isolated.*

Proof Note that T must be small. Let $q_0, q_1, q_2, \ldots$ be an enumeration of all complete types of T over $\emptyset$. Suppose that $e_i \models q_i$ and $d_i = e_0 e_1 \ldots e_i$. Now there is a prime model N_i over d_i for each $i < \omega$. Thus, for some $j < \omega$, $N_j (= M)$ is isomorphic to N_i for infinitely many $i \geq j$. Therefore, the prime model M over $d_j (= a)$ realizes every complete type over $\emptyset$. As M is not atomic over $\emptyset$, $\text{tp}(a)$ is not isolated (2.5.22).

Now, since $T(a)$ is again non $\aleph_0$-categorical, some complete type $q(x, a)$ over a is non-isolated. Let $s'b (\in M)$ realize $q(x, y)$. Then as $\text{tp}(s'/b)$ is non-isolated, again by 2.5.22 $\text{tp}(a/b)$ must not be isolated. ■

We are ready to prove Theorem 2.5.21. We will continue to use the notation in Fact 2.5.26. Let T be simple, having finitely many countable models. We shall show T cannot be supersimple.

Claim Let $p = \mathrm{tp}(a)$. There are two realizations a_0 and a_1 of p such that $a_0 \mathop{\smile\!\!\!\!\!\!|}\, a_1$, and $\mathrm{tp}(a_0/a_1)$ is non-isolated: Let c be a realization of p such that $c \mathop{\smile\!\!\!\!\!\!|}\, ab$. Now, by 2.5.23, $\mathrm{tp}(a/b)$ is nonsemi-isolated. Hence, by 2.5.22, either $\mathrm{tp}(a/c)$ or $\mathrm{tp}(c/b)$ must be non-isolated. Thus, a, c or c, b are the desired two realizations of p. The claim is proved.

In the claim, we can assume a_0 and a_1 are in M. Moreover, as $\{a_0, a_1\}$ is independent, $\mathrm{tp}(a_1/a_0)$ is also non-isolated, by Corollary 2.5.25. Then, $\mathrm{tp}(a/a_0)$ and $\mathrm{tp}(a/a_1)$ are both non-isolated, since, for example, if $\mathrm{tp}(a/a_0)$ were isolated, then M is prime over a_0 and so $\mathrm{tp}(a_1/a_0)$ would be isolated, a contradiction. Therefore, again by 2.5.26, $\mathrm{tp}(a/a_0)$ and $\mathrm{tp}(a/a_1)$ both fork over $\emptyset$. Let us summarize here the relationships between three realizations $\{a, a_0, a_1\}$ of p.

(i) $\{a_0, a_1\}$ is independent.

(ii) For each $i = 0, 1$, $\mathrm{tp}(a_i/a)$ is isolated, whereas $\mathrm{tp}(a/a_i)$ is non-isolated (so nonsemi-isolated). Thus, $a \not\mathop{\smile\!\!\!\!\!\!|}\, a_i$.

Now we are able to construct a tree $\{a_\sigma | \sigma \in 2^{<\omega}\}$ such that $a_\emptyset = a$ and $\mathrm{tp}(a_\sigma a_{\sigma 0} a_{\sigma 1}) = \mathrm{tp}(a a_0 a_1)$ for each $\sigma \in 2^{<\omega}$ (**). Moreover, the basic properties of non-forking together with (i) enable us to assume that every anti-chain in the tree is independent (e.g., $\{a_{0(n)1} : n < \omega\}$ is independent where $0(n)$ is a sequence $00\cdots 0$ of length n). Now, by (ii), for each $\sigma \in 2^{<\omega}$ and each $i = 0, 1$, $\mathrm{tp}(a_{\sigma i}/a)$ is semi-isolated. But $\mathrm{tp}(a/a_{\sigma i})$ is nonsemi-isolated, since, otherwise, $\mathrm{tp}(a_\sigma/a_{\sigma i})$ would be semi-isolated, contradicting (ii) and (**). Hence, by Proposition 2.5.24, $\mathrm{tp}(a/a_{\sigma i})$ forks over $\emptyset$. Then due to the remark before 2.5.11, T is not supersimple. Therefore, Theorem 2.5.21 is proved.

The proof above also says Theorem 2.5.21 holds for countable simple T, where $\mathrm{prwt}(p)$ is *not* strictly ω for any $p \in S_n(\emptyset)$ (see 4.4.6; any supersimple T has this property).

In the rest of this subsection T is simple, and we sketch an argument that extends Theorem 2.5.21 to the theories having *finite coding*. It is exactly the same argument for stable case [49].

Definition 2.5.27

(1) We say T has *finite coding* if any $p(x) \in S_n(A)$ is parallel with some $q(x) \in S_n(b)$ with finite b.

(2) A *dense forking chain* is a sequence $\langle p_i(x) \in S_n(A_i) | i \in \mathbb{Q}\rangle$ such that for $i < j$, p_j is a forking extension of p_i.

Any supersimple or 1-based (Definition 5.5.14) theories have finite coding.

Fact 2.5.28 *Assume T is small having finite coding. Then T does not have a dense forking chain.*

Proof Suppose not; say, there is a dense forking chain $\langle \mathrm{tp}(a/A_i) | i \in \mathbb{Q}\}$. Now, for $r \in \mathbb{R}$, there is finite b_r such that $\mathrm{tp}(a/A_r)$, where $A_r := \bigcup\{A_i | \; r > i \in \mathbb{Q}\}$, and $\mathrm{tp}(a/b_r)$ are parallel. Hence, $D_{\varphi,k}(a/A_r) = D_{\varphi,k}(a/b_r)$ for any φ, k. On the other hand, for $r < s \in \mathbb{R}$, $a \not\mathop{\smile\!\!\!\!\!\!|}_{A_r} A_s$. Thus, for some φ', k', we have $D_{\varphi',k'}(a/b_s) < D_{\varphi',k'}(a/b_r)$, and so $ab_s \not\equiv ab_r$. It contradicts that T is small. ■

Using the above fact, we can extend 2.5.21 to the class of countable theories having finite coding: The same proof works. Namely, if countable non $\aleph_0$-categorical T has only finitely many countable models, then as above we have the tuple a and an independent set $\{b_i | i < \omega\}$ such that $a \not\!\downarrow b_i$. Let $f : \mathbb{Q} \to \omega$ be a bijection. Then, $\langle \text{tp}(a/\{b_{f(j)} | j \leq i\}) | i \in \mathbb{Q}\rangle$ is a dense forking chain. Hence, due to 2.5.28, T does not have finite coding.

2.6 Examples of Simple Theories

Example 2.6.1 Any stable theory is simple. More examples of stable theories are described in Chapter 1.

Example 2.6.2 The theories of the random graph and the bipartite random graph are supersimple having SU-rank 1 (2.2.2, 2.3.8).

Example 2.6.3 In [29], extending Zilber's earlier work [126], ω-categorical, ω-stable (equivalently ω-categorical superstable) structures are studied. Firstly, "rank 1" such structures are analyzed using stability theory together with the classification of finite simple groups (applied to the automorphism groups of those structures). This different approach simplifies and extends Zilber's original proof. Then the fine analysis of all those structures is made by applying deep model-theoretic arguments to the rank 1 case understanding. In particular, it is shown that the class of countable ω-categorical superstable structures is exactly that of *stable* smoothly approximated structures.

By a *smoothly approximable* (or *smoothly approximated*) structure we mean a countable ω-categorical structure M, which is the union of a chain of finite substructures $A_0 \subseteq A_1 \subseteq A_2 \ldots$ such that for any finite tuples a, b chosen from A_i, a has the same type as b in M if and only if there is an automorphism of M which fixes A setwise and takes a to b. In [62], a general study of smoothly approximable structures is undertaken, in which the permutation group-theoretic analysis is present from the beginning. The structures are shown to be built up from certain geometries, essentially variants of infinite-dimensional vector spaces over finite fields, equipped with non-degenerate bilinear forms. Later, Cherlin and Hrushovski in their book [30] introduce more model-theoretic notions reminiscent of stability theory into the matter. They prove the quasi-finite axiomatizability of smoothly approximable structures, extending the result for stable case. A rank notion and consequent notion of independence are developed, for which the independence theorem over algebraically closed sets is proved. We conclude from Section 3.3 that smoothly approximable structures are simple. Indeed they are supersimple. But different from stable case, the random graph is ω-categorical supersimple but *not* smoothly approximable.

We now speak about fields having simple theories. Recall some definitions first.

Definition 2.6.4 Let F be a field.

(1) F is said to be *pseudo-algebraically closed* if any absolutely irreducible algebraic set defined over F has an F-rational point.

(2) F is *pseudo-finite* if it is an infinite field of the theory of all finite fields.

(3) F is *bounded* if for each $n > 1$, in its algebraic closure, there are only finitely many extensions of degree n.

Fields having superstable theories are well understood.

Fact 2.6.5 (MacIntyre [87]; Cherlin and Shelah [31]) *Superstable infinite fields are algebraically closed.*

Cherlin and Shelah also show that any superstable division ring is commutative. This is extended to the supersimple context.

Fact 2.6.6 (Pillay, Scanlon, and Wagner [104]) *A division ring having a supersimple theory is a field.*

Only known strictly stable (i.e., stable but not superstable) fields are separably closed (but nonalgebraically closed) fields. It is still an open question whether any stable infinite field is separably closed. Fields of (super)simple theories are less known. A conjecture is that any supersimple infinite field is PAC. Some partial results are obtained. A stable infinite field having the weight 1 generic type is separably closed [78]. No stable infinite field has Artin–Schreier extensions while a simple field has only finitely many such extensions [63].

But if we restrict our attentions only to PAC fields, a good classification is already made.

Fact 2.6.7 *Assume a field F is PAC (so infinite).*

(1) (Chatzidakis [21]) F is simple iff F is bounded.

(2) (Hrushovski [55]; Pillay and Poizat [103]) F is supersimple iff F is perfect and bounded.

(3) (Duret [35], Wood [123]) F is stable iff F is separably closed.

Historically, the first simple unstable fields studied seriously were pseudo-finite fields by J. Ax in late 1960s [4].

Fact 2.6.8 *A field F is pseudo-finite if and only if it is perfect, PAC, and for each n it has exactly one algebraic extension of degree n.*

Hence, pseudo-finite fields are unstable supersimple. Hrushovski shows it has SU-rank 1 [55]. Chatzidakis, MacIntyre, and van den Dries continued Ax's work, studying definable sets in more detail [26].

Example 2.6.9 *ACFA* is the model companion of the theory of the class $\{(F,\sigma)|F$ is a field and σ is an automorphism of $F\}$. A deep model-theoretic study of *ACFA* and its number theoretic applications are carried out in [23], [25], [54]. Let $(\bar{F},\bar{\sigma})$ be a saturated model of *ACFA*. For a small subset A of F, let $\langle A\rangle$ denote the smallest sub- field of F containing A and closed under σ and σ^{-1}. Chatzidakis and Hrushovski introduced the following notion of independence: B is independent from C over A if $\langle A\cup B\rangle$ is algebraically independent from $\langle A\cup C\rangle$ over $\langle A\rangle$. They proved the independence theorem for complete types over $\langle A\rangle^{\mathrm{alg}}$. It follows that $\mathrm{Th}(F)$ is simple, and that this notion of independence coincides with non-forking (Theorem 3.3.1). Indeed, Th(F) is supersimple, but there are types with infinite SU-rank (such as the type of an element a such that $\{\sigma^k(a) : k \in \omega\}$ is algebraically independent).

Example 2.6.10 In [27], stable theories with "generic predicates" or "automorphisms" are studied. This means the following. One begins with a complete theory T that has quantifier elimination in a language $\mathcal{L}$. Let P be a new predicate symbol (of any arity). Then T has a model companion T_P in the language $\mathcal{L}(P) := \mathcal{L} \cup \{P\}$ iff $\exists^\infty$ is eliminable in T; i.e., for any $\varphi(x, y) \in \mathcal{L}$, there is $\psi(y) \in \mathcal{L}$ such that for any tuple c, $\models \psi(c)$ iff $\varphi(x, c)$ defines an infinite set. The basic result is that if T is stable (or just simple), then any completion of T_P (if it exists) is simple. A is independent with B over C in a model (M, P) of T_P iff it is so in the $\mathcal{L}$-reduct M.

One could try to do the same thing for automorphisms in place of predicates. Here we start with stable T again having quantifier elimination. Put $\mathcal{L}(\sigma) = \mathcal{L} \cup \{\sigma\}$ where σ is a new unary function symbol. Let T^0_σ be T together with sentences saying "σ is an $\mathcal{L}$-automorphism." The issue of whether T^0_σ has a model companion T_σ is more complicated (we have no nice necessary and sufficient condition), but if it does, then T_σ is simple. (Of course, Example 2.6.9 is a special case.)

Certain simple theories of separably closed fields with an automorphism are studied in [22], [24]. For T the theory of differentially closed fields of characteristic 0, T_σ is studied in [92] following Hrushovski's result of the existence of T_σ.

We give more detailed descriptions of these methods of adding generic predicates/automorphisms in Section 6.3.

More algebraic examples of supersimple theories are found in [38] and [6].

2.7 Bibliographical Remarks

Results in this chapter can be found in [112], [65], [66], [68], [69]. Definitions of dividing and forking, and Facts 2.1.3 and 2.1.4 are according to Shelah [113]. Lemmas 2.1.6 and 2.1.7 and Proposition 2.2.4 are also according to Shelah [112]. Lemma 2.2.5 through Theorem 2.2.8 are according to Kim [65], [66]. In [112], [113], the tree property and the $D_{\varphi,k}$-rank are introduced, and Propositions 2.3.7 and 2.3.10 are proved. Type-counting criteria for supersimple theories analogous to 2.3.10 can be found in [14]. Proposition 2.3.9 is in [66]. Theorem 2.4.3 is according to Kim [69]. The class of low theories and the D_φ-rank are introduced by Buechler [12]. The D-rank and its properties are generalization of those for (super)stable theories [113]. Similarly, the SU-rank and its properties including Lascar inequalities are generalization of those for (super)stable theories [80]. Theorem 2.5.21 is in [68].

3

Lascar Strong Types and Type Amalgamation

As we have seen in Fact 2.1.4 and Theorem 2.2.8, in a simple unstable theory forking cannot satisfy uniqueness over a model. In this chapter we shall find a substitute for it (Theorem 3.2.9) called *type amalgamation* (equivalently 3-*amalgamation* or *the independence theorem*) *over a model*. Corresponding to Fact 2.1.4 of stable theories, simple theories have type amalgamation over a model, and conversely, symmetry, transitivity, extension, finite and local characters, and type amalgamation characterize simple theories (Theorem 3.3.1).

Another axiom parallel to uniqueness over a model for stable theories is *uniqueness for strong types* (Fact 2.1.4(2)(g)). Natural expectation for simple theories to satisfy is "type amalgamation for strong types." However, the truth of the expectation is still unknown. Instead, we shall introduce *Lascar (strong) types*, a notion stronger than that of strong types, and show that "type amalgamation for Lascar types" holds in simple theories (Theorem 3.2.8).

3.1 Lascar Strong Types

Definition 3.1.1

(1) By $\mathrm{Autf}_A(\mathcal{M})$, we mean a subgroup of $\mathrm{Aut}(\mathcal{M})$ generated by

$$\left\{f \in \mathrm{Aut}(\mathcal{M}) | f \in \mathrm{Aut}_M(\mathcal{M}) \text{ for some model } M \supseteq A\right\}.$$

(2) Let a, b be tuples of the same length, possibly infinite. We say that a and b have the same *Lascar strong type* (or simply *Lascar type*) over A ($\mathrm{Ltp}(a/A) = \mathrm{Ltp}(b/A)$, or $a \equiv^L_A b$) if there is $f \in \mathrm{Autf}_A(\mathcal{M})$ such that $f(a) = b$.

Remark 3.1.2

(1) We can think of (the solution set of) Lascar strong type of a over A as an orbit of a under $\mathrm{Autf}_A(\mathcal{M})$. In fact, $\mathrm{Ltp}(a/A) = \mathrm{Ltp}(b/A)$ if and only if there are tuples

$a = a_0, a_1, \ldots, a_n = b$ and models $M_1, M_2, \ldots, M_n$, each containing A, such that $\text{tp}(a_{i-1}/M_i) = \text{tp}(a_i/M_i)$ for each $i = 1, \ldots, n$. We will use this property freely.

(2) It follows that $\text{Ltp}(a/A) = \text{Ltp}(b/A)$ implies that $\text{stp}(a/A) = \text{stp}(b/A)$ (and we know $a \equiv^s_A b$ implies that $a \equiv_A b$). For a model M, obviously $a \equiv_M b$ iff $a \equiv^s_M b$ iff $a \equiv^L_M b$.

(3) In a stable T, Ltp coincides with stp since $a \equiv^s_A b$ holds in T iff $a \equiv_M b$ for some model M containing A. It is an open question whether the coincidence still holds in all simple theories. But there is a non-simple example (3.1.7) where two notions differ.

(4) In general, $\text{stp}(c/A)$ is considered as $\text{tp}(c/\operatorname{acl}^{\text{eq}}(A))$ (Chapter 1). But it is not clear how to understand $\text{Ltp}(c/A)$ as a set of formulas, although the notion $\text{Ltp}(c/A) = \text{Ltp}(c'/A)$ is introduced. In Chapter 4, it will become clear for simple theories, as $\text{Ltp}(c/A)$ is a type of c over a set of *hyperimaginaries* (4.2.19(3)).

Recall that an equivalence relation E (on $\mathcal{M}^\alpha$, not necessarily definable) is said to be *A-invariant*, if $E(a, b)$ implies $E(f(a), f(b))$ for any A-automorphism f. It is called *bounded* if it has strictly less than $\bar{\kappa}$ many equivalence classes, or equivalently the number of E-equivalence classes has a fixed bound undertaking elementary extensions. (As any A-invariant set is some union of complete types over A, it makes sense to say E-equivalence relation on an elementary extension of $\mathcal{M}$.) It is said to be *finite* if it has finitely many classes.

Lemma 3.1.3 *Suppose E is an A-invariant bounded equivalence relation on α-tuples. Let I be an infinite A-indiscernible sequence of α-tuples. Then, $E(a, b)$ for all $a, b \in I$.*

Proof Suppose otherwise. Then by A-invariance of E and A-indiscernibility of I, we have $\neg E(a, b)$ for all $a \neq b \in I$. But for any κ, I can be extended to an A-indiscernible sequence I' of length κ. Again we have $\neg E(a, b)$ for all $a \neq b \in I'$ contradicting the boundedness of E. ■

Definition 3.1.4 Let A be a set, and let a, b be tuples such that $\text{tp}(a/A) = \text{tp}(b/A)$. We define $d_A(a, b)$ to be the least natural number $n (\geq 1)$ such that there are sequences $I_1, I_2, \ldots, I_n$ and tuples $a = a_0, a_1, \ldots, a_n = b$ such that both $a_{i-1}{}^\frown I_i$ and $a_i{}^\frown I_i$ are A-indiscernible for each $1 \leq i \leq n$. (If there is no such $n < \omega$, then we write $d_A(a, b) = \infty$.)

The following proposition says that equality of Lascar types is a conjunction of all invariant bounded equivalence relations (whereas equality of strong types is a conjunction of all definable finite equivalence relations).

Proposition 3.1.5 *The following are equivalent:*

(1) $\text{Ltp}(a/A) = \text{Ltp}(b/A)$.

(2) $d_A(a, b) < \omega$.

(3) $\models E(a, b)$ *for any A-invariant bounded equivalence relation E.*

In conclusion $x \equiv^L_A y$ is the finest A-invariant bounded equivalence relation over A.

Proof (1)⇒(2) Suppose that for some model $M \supseteq A$, $\mathrm{tp}(a/M) = \mathrm{tp}(b/M)$. It suffices to show $d_A(a,b) = 1$ as the general case proceeds by induction. We use Fact 1.1.6 freely. Let $p = \mathrm{tp}(a/M)$. Then there is a sequence $\langle a_i : i < \omega \rangle$ such that $\mathrm{tp}(a_i/Maba_0 \dots a_{i-1})$ is a coheir of p, and is realized by a_{i+1}, for $i < \omega$. Then both $\langle a, a_0, a_1, \dots . \rangle$ and $\langle b, a_0, a_1, \dots . \rangle$ are M-indiscernibles. Hence, $d_A(a,b) = 1$.

(2)⇒(3) Suppose $d_A(a,b) = 1$ (for n, use induction), witnessed by I. Then by Lemma 3.1.3, for any A-invariant bounded equivalence relation E, it holds that $E(a,c)$ and $E(c,b)$ for any $c \in I$. Hence, we have $E(a,b)$.

(3)⇒(1) Obviously $\mathrm{Ltp}(x/A) = \mathrm{Ltp}(y/A)$ is an A-invariant equivalence relation. Let $\kappa = |T| + |A|$. Now there is a model M containing A such that $|M| = \kappa$. If the number of equivalence classes of $\mathrm{Ltp}(x/A) = \mathrm{Ltp}(y/A)$ is $> 2^{\kappa+|x|}$ then, as there are at most $2^{\kappa+|x|}$ types over M, $\mathrm{tp}(c/M) = \mathrm{tp}(d/M)$ for some representatives c, d of different Ltp classes. But then $\mathrm{Ltp}(c/A) = \mathrm{Ltp}(d/A)$, a contradiction. Hence, there are at most $2^{\kappa+|x|}$-many distinct Lascar strong types over A. Therefore, (3)⇒(1) follows. ■

We reconfirm that $a \equiv^L_A b$ implies $a \equiv^s_A b$. From Proposition 3.1.5(2) and Fact 1.1.6, we observe the following.

Corollary 3.1.6 *If $a \equiv^L_A b$ and $c \equiv_{Aa} b$, then $c \equiv^L_A a$. If $a \equiv_A b$ and $\mathrm{tp}(a/A)$ is finitely satisfiable in A, then $a \equiv^L_A b$.*

Example 3.1.7 Consider a model $(C; \{U_n(x,y)\}_{0<n\in\omega})$, where C is the unit circle, and $U_n(a,b)$ holds for $a, b \in C$ iff the length of the shorter arc from a to b is $\leq n^{-1}$. Thus, $U_1(x;y_0) \wedge \neg U_1(x;y_1)$ has the strict order property, and the model is not simple. Now, let E be an equivalence relation on C type-defined by $\{U_n(x,y) | 0 < n \in \omega\}$. Note that for each n, there is a number k_n, so that whenever there are k_n distinct points on C, some two of them should realize $U_n(x,y)$. Hence, it easily follows by the Erdös–Rado theorem, the number of E classes is 2^ω. On the other hand, it is not hard to show that it has quantifier elimination up to formulas $\exists z_1 \dots z_m(U_n(x,z_1) \wedge U_n(z_1,z_2) \dots \wedge U_n(z_m,y))$, and conclude that for any $b, d \in C$, $\mathrm{stp}(b) = \mathrm{stp}(d)$. Therefore, Lascar type is different from strong type (over $\emptyset$).

3.2 Type Amalgamation

We recall Fact 2.1.4 with stable T. A complete type $p \in S(A)$ is said to be *stationary* if for any B containing A, p has the unique non-forking extension over B. So by the uniqueness axiom, any complete type over a model is stationary. A complete type over an arbitrary set need not be stationary. However, any strong type (over a set) is stationary; i.e., if $a_0 \equiv^s_A a_1$ and $a_i \mathop{\smile\!\!\!\!\!\!\,|}_A B$ $(i = 0, 1)$, then $a_0 \equiv_{AB} a_1$ (indeed even $a_0 \equiv^s_{AB} a_1$). Equivalently, any type over an algebraically closed set (in $\mathcal{M}^{\mathrm{eq}}$) is stationary. It follows that $p \in S(A)$ is stationary iff for $a, b \models p$, $a \equiv_{\mathrm{acl}(A)} b$ iff it is a strong type, i.e., for $a, b \models p$, $a \equiv^s_A b$. Hence, as we know $\mathrm{stp}(a/A)$ is identified as $\mathrm{tp}(a/\operatorname{acl}(A))$.

In the rest of this section T is simple This section confirms that for simple theories, the notion of Lascar types is a proper correspondent to that of strong types for stable theories, by showing type amalgamation for Lascar types, and type amalgamation over a model (Theorem 3.2.8, 3.2.9).

Lemma 3.2.1 (Extension for Ltp) *For any $A \subseteq B$ and a, there is b such that $a \equiv^L_A b$ and $b \downarrow_A B$.*

Proof Let $M \supseteq A$ be chosen so that $a \downarrow_A M$. Let $p = \mathrm{tp}(a/M)$. Let $q \in S(M \cup B)$ be a non-forking extension of p, and let b realize q. Then, $\mathrm{tp}(b/M) = \mathrm{tp}(a/M)$ so $\mathrm{Ltp}(b/A) = \mathrm{Ltp}(a/A)$, and also $\mathrm{tp}(b/B)$ does not fork over A. ■

Lemma 3.2.2 *Let $p(x, a)$ be a partial type over Aa which does not fork over A. If $I = \langle a_i | i < \omega \rangle$ is a Morley sequence of $\mathrm{tp}(a/A)$ with $a_0 = a$, then there is some b realizing $p(x, a)$ such that I is indiscernible over Ab. Furthermore, for any such b, $\mathrm{tp}(b/AI)$ does not fork over A.*

Proof Recall that dividing is the same as forking. We will work with dividing in this proof. Actually the existence of such b is already stated in Lemma 2.1.6. Let $q = \mathrm{tp}(b/AI)$. In order to show that q does not divide over A it is enough to prove that for given $m < \omega$, $\psi(x, c, a_0, \ldots, a_{m-1})$ $(c \in A)$ in q does not divide over A. Now, let

$$q' = \{\psi(x, c, a_{m\cdot r}, \ldots, a_{m\cdot r+m-1}) | r \in \omega\}.$$

Then since I is Ab-indiscernible, q' is a subset of q. Thus, q' is consistent. Furthermore, since I is a Morley sequence of $\mathrm{tp}(a/A)$, by symmetry and transitivity it follows that $\langle d_r | r < \omega \rangle$, where $d_r = c a_{m\cdot r} a_{m\cdot r+1} \ldots a_{m\cdot r+m-1}$, is a Morley sequence of $\mathrm{tp}(d_0/A)$. Hence, by Proposition 2.2.6, $\psi(x, c, a_0, \ldots, a_{m-1})$ does not divide over A. ■

Lemma 3.2.3 *Let $I = \langle a_i | i < \omega + \omega \rangle$ be an indiscernible sequence over A. Then $I_\omega = \langle a_{\omega+i} | i < \omega \rangle$ is a Morley sequence of $\mathrm{tp}(a_\omega/AI_0)$, where $I_0 = \langle a_i | i < \omega \rangle$.*

Proof Due to symmetry, it suffices to show that $p = \mathrm{tp}(a_\omega \ldots a_{\omega+i}/AI_0 a_{\omega+i+1})$ does not divide over AI_0 for given $i < \omega$. The rest of the proof will be similar to that of 2.4.2. Namely, as $p\lceil Aa_0 \ldots a_n a_{\omega+i+1}$ realized by $a_{n+1} \ldots a_{n+i+1} \in I_0$ cannot divide over AI_0, neither does p. ■

Proposition 3.2.4 *Let $p(x, a)$ be a partial (or complete) type over Aa that does not fork over A and let $I = \langle a_i | i < \omega \rangle$ be an A-indiscernible sequence with $a_0 = a$. Then there is a tuple $b \models p(x, a)$ such that I is Ab-indiscernible and $\mathrm{tp}(b/AI)$ does not fork over A.*

Proof Suppose that $I' = \langle a_{\omega+n} | n < \omega \rangle$ is a sequence such that $I^\frown I'$ is indiscernible over A. Then

$$I' \text{ is a Morley sequence of } \mathrm{tp}(a_\omega/AI) \quad (*)$$

by the preceding lemma. Now, there is a complete type $p'(x, a_\omega)$ over AIa_ω containing $p(x, a_\omega)$ such that $p'(x, a_\omega)$ does not fork over A (**), by the extension axiom. Then $p'(x, a_\omega)$ does not fork over AI, either. Hence, by Lemma 3.2.2 with (*), there exists some tuple $b' \models p'(x, a_\omega)$ such that $b' \downfree_{AI} I'$ and I' is $b'AI$-indiscernible (***). Then from (**) with transitivity, we have $b' \downfree_A II'$ and $b' \downfree_A I'$. Finally, we find the desired b such that bI is an A-automorphic image of $b'I'$. Thus, $p(b, a)$, $b \downfree_A I$ and from (***), I is Ab-indiscernible. ■

Corollary 3.2.5 *If a partial type $p(x, a)$ does not fork over A and $\langle a_i | i < \omega \rangle$ is an A-indiscernible sequence of* $\mathrm{tp}(a/A)$, *then $p(x, a_0) \cup \ldots \cup p(x, a_n)$ does not fork over A.*

Corollary 3.2.6 *There do not exist formulas $\varphi(x, y)$, $\psi(x, y')$ over A with an A-indiscernible sequence $\langle d_n | \ n \in \mathbb{Z} \rangle$ and c such that for each $i > 0$,* $\mathrm{tp}(cd_1/A) = \mathrm{tp}(cd_i/A)$ *and $\varphi(x, c) \wedge \psi(x, d_1)$ forks over A, while for $j \leq 0$,* $\mathrm{tp}(cd_0/A) = \mathrm{tp}(cd_j/A)$ *and $\varphi(x, c) \wedge \psi(x, d_j)$ does not fork over A.*

Proof Suppose that there *do* exist such formulas with a sequence $\langle d_n | \ n \in \mathbb{Z} \rangle$ and c. Let h_i be an A-automorphism sending d_n to d_{n+i} $(i, n \in \mathbb{Z})$. Let $c_i = h_i(c)$ $(c_0 = c)$. Then for each $i \geq j \in \mathbb{Z}$, $\mathrm{tp}(c_i d_j/A) = \mathrm{tp}(c_0 d_0/A)$ and for $i < j$, $\mathrm{tp}(c_i d_j/A) = \mathrm{tp}(c_0 d_1/A)$. Then as these are type-describable, by Ramsey and compactness, we can assume that $\langle c_i d_i | i \in \mathbb{Z} \rangle$ is A-indiscernible. Now

$$\varphi(x, c_i) \wedge \psi(x, d_j) \text{ does not fork over } A \text{ iff } i \geq j. \quad (*)$$

In particular $\varphi(x, c_0) \wedge \psi(x, d_0)$ does not fork over A. Then from 3.2.5, $\varphi(x, c_0) \wedge \psi(x, d_0) \wedge \varphi(x, c_1) \wedge \psi(x, d_1)$ must not fork over A. But this is a contradiction, as (*) implies $\varphi(x, c_0) \wedge \psi(x, d_1)$ forks over A. The proof is completed. ■

Proposition 3.2.7 *Suppose that $a \downfree_A b$, $a \downfree_A b'$ and $b \equiv^L_A b'$. If $\varphi(x, a) \wedge \psi(x, b)$ does not fork over A, then neither does $\varphi(x, a) \wedge \psi(x, b')$.*

Proof By 3.2.1, there is $b_0 \models \mathrm{Ltp}(b/Aa)$ such that $b_0 \downfree_{Aa} b'$ and, hence, $b_0 \downfree_A ab'$, $a \downfree_A b_0 b'$. Moreover, b_0, b' have the same Lascar type over A, and $\varphi(x, a) \wedge \psi(x, b_0)$ does not fork over A. Now, from 3.1.5, we can find tuples $b_1, \ldots, b_n = b'$, all having the same Lascar strong type over A such that b_i, b_{i+1} $(i = 0, \ldots, n-1)$ are the initial two components of some A-indiscernible. Now, by extension and transitivity, there is $a' \models \mathrm{tp}(a/Ab_0 b_n)$ such that $a' \downfree_A b_0 \ldots b_n$. It suffices to show that if $\varphi(x, a') \wedge \psi(x, b_i)$ does not fork over A, then neither does $\varphi(x, a') \wedge \psi(x, b_{i+1})$. To lead a contradiction assume that $\varphi(x, a') \wedge \psi(x, b_{i+1})$ forks over A. Now, note that there is an A-indiscernible sequence $\langle \ldots, c_{-1}, c_0, c_1, c_2, \ldots \rangle$, where $c_0 = b_i, c_1 = b_{i+1}$. Moreover, since $a' \downfree_A c_0 c_1$, from 2.1.6, we can further assume that $\langle \ldots, c_{-2} c_{-1}, c_0 c_1, c_2 c_3, \ldots \rangle$ is Aa'-indiscernible. Hence, for $n \geq 0$, $\mathrm{tp}(a' c_0/A) = \mathrm{tp}(a' c_{-2n}/A)$ and $\varphi(x, a') \wedge \psi(x, c_{-2n})$ does not fork over A, while $\mathrm{tp}(a' c_1/A) = \mathrm{tp}(a' c_{2n+1}/A)$ and $\varphi(x, a') \wedge \psi(x, c_{2n+1})$ forks over A. Then since

$$\langle \ldots, c_{-4}, c_{-2}, c_0, c_1, c_3, c_5, \ldots \rangle$$

is again A-indiscernible, it contradicts 3.2.6. We have proved the proposition. ∎

The following is type amalgamation for Lascar types, which is also called the independence theorem.

Theorem 3.2.8 (The independence theorem for Lascar strong types) *Suppose* A, b_i, d_i $(i = 1, 2)$ *satisfy*

(1) $b_1 \mathop{\smile\!\!\!\!\!\mid}_A b_2$, $d_i \mathop{\smile\!\!\!\!\!\mid}_A b_i$, *and*

(2) $\mathrm{Ltp}(d_1/A) = \mathrm{Ltp}(d_2/A)$.

Then there is $d \models \mathrm{tp}(d_1/Ab_1) \cup \mathrm{tp}(d_2/Ab_2)$ *such that* $d \mathop{\smile\!\!\!\!\!\mid}_A b_1 b_2$ *and* $\mathrm{Ltp}(d/Ab_i) = \mathrm{Ltp}(d_i/Ab_i)$.

Proof First we claim that under the conditions (1),(2), $p_1(x, b_1) \cup p_2(x, b_2)$ does not fork over A where $p_i(x, b_i) = \mathrm{tp}(d_i/Ab_i)$. Since $d_1 \equiv^L_A d_2$, there is $f \in \mathrm{Autf}_A(\mathcal{M})$ such that $f(d_2) = d_1$. Let $b_2' = f(b_2)$. Hence, $d_2 b_2 \equiv^L_A d_1 b_2'$ and thus $d_1 \mathop{\smile\!\!\!\!\!\mid}_A b_2'$. Moreover, according to extension for Lascar types, there is b_2'' such that $b_2'' \equiv^L_{Ad_1} b_2'$ and $b_2'' \mathop{\smile\!\!\!\!\!\mid}_{Ad_1} b_1$. Therefore, $\{d_1, b_1, b_2''\}$ is A-independent and, hence, $p_1(x, b_1) \cup p_2(x, b_2'')$, realized by d_1, not forking over A. Then since $b_2'' \equiv^L_A b_2$, from 3.2.7, $p_1(x, b_1) \cup p_2(x, b_2)$ does not fork over A either. The claim is proved.

Now, again by Ltp extension, we have $e_1 \equiv^L_{Ab_1} d_1$ such that $e_1 \mathop{\smile\!\!\!\!\!\mid}_{Ab_1} b_2 d_1$. Similarly, there is $e_2 \equiv^L_{Ab_2} d_2$ such that $e_2 \mathop{\smile\!\!\!\!\!\mid}_{Ab_2} e_1 b_1 d_2$. Then it easily follows that $e_1 b_1 \mathop{\smile\!\!\!\!\!\mid}_A e_2 b_2$, $d_i \mathop{\smile\!\!\!\!\!\mid}_A e_i b_i$ (*). Hence, by applying the claim above to (*), we find that $d \models \mathrm{tp}(d_i/Ab_i e_i)$ such that $d \mathop{\smile\!\!\!\!\!\mid}_A b_1 b_2 e_1 e_2$. Moreover, since $d \equiv_{Ab_i e_i} d_i$ and $e_i \equiv^L_{Ab_i} d_i$, according to 3.1.6 we have that $d \equiv^L_{Ab_i} d_i$. We have proved the theorem. ∎

Since over a model, equality of type is that of Lascar type, we have the following.

Theorem 3.2.9 (Type amalgamation over a model) *Suppose that a model* M *and* b_i, d_i $(i = 1, 2)$ *satisfy:*

(1) $b_1 \mathop{\smile\!\!\!\!\!\mid}_M b_2$, $d_i \mathop{\smile\!\!\!\!\!\mid}_M b_i$, *and*

(2) $\mathrm{tp}(d_1/M) = \mathrm{tp}(d_2/M)$.

Then there is $d \models \mathrm{tp}(d_1/Mb_1) \cup \mathrm{tp}(d_2/Mb_2)$ *such that* $d \mathop{\smile\!\!\!\!\!\mid}_M b_1 b_2$ *and* $\mathrm{Ltp}(d/Mb_i) = \mathrm{Ltp}(d_i/Mb_i)$.

Conditions (1),(2) in the independence theorem are essential. Let $(U; E, R)$ be a structure such that E is an equivalence relation on U giving two infinite classes, and R defines the bipartite random graph relation between the two classes. Then, any $d \in U$ realizes the same type over $\emptyset$ but if $\neg d_1 E d_2$, then $d_1 \not\equiv^L d_2$. Choose $b_1, b_2 \in U$ distinct from d_1, d_2 such that

$\neg b_1 E b_2$. Then, $b_1 \downarrow b_2$ and $d_i \downarrow b_i$ $(i = 1, 2)$. But certainly $\text{tp}(d_1/b_1) \cup \text{tp}(d_2/b_2)$ is inconsistent so not amalgamable. If $b_1 = b_2$ (so they are dependent), and $d_1 R b_1, \neg d_2 R b_2, b_i \downarrow d_i$ then even if $d_1 E d_2$ so that $d_1 \equiv^L d_2$, we again have inconsistent $\text{tp}(d_1/b_1) \cup \text{tp}(d_2/b_2)$.

If T is stable, type amalgamation holds for strong types (without the condition $b_1 \downarrow_A b_2$ in (1)).

Corollary 3.2.10 *Let $\{a_i | i \in \kappa\}$ be M-independent and $p \in S(M)$. Suppose that each $p_i \in S(Ma_i)$ is some non-forking extension of p, then $\bigcup\{p_i | i \in \kappa\}$ does not fork over M.*

Proof Use induction on $\alpha < \kappa$. Assume that $\bigcup\{p_i | i < \alpha\}$ does not fork over M. Then by extension, there is $q \in S(M \cup \{a_i | i < \alpha\})$ containing $\bigcup\{p_i | i < \alpha\}$, which does not fork over M. Now, by applying the independence theorem to q and p_α, we see that $\bigcup\{p_i | i < \alpha + 1\}$ does not fork over M either. ■

Type amalgamation is a core technique deepening the development of simplicity theory. We shall see its applications throughout the book.

Now an issue is whether Ltp is stp in simple theories. Similarly, does type amalgamation hold for strong types in simple theories? It can be observed that two questions are equivalent (3.2.12), but the answer is beyond the reach of current knowledge. This issue is related to the subjects dealt in Chapter 5.

Corollary 3.2.11 (Recall T is simple). *Assume (possibly infinite) tuples a_0, a_1 are given.*

(1) Suppose that $a_0 \downarrow_A a_1$ and $a_0 \equiv^L_A a_1$. Then there is a Morley sequence $\langle a_i | i < \omega \rangle$ of $\text{tp}(a_0/A)$.

(2) $a_0 \equiv^L_A a_1$ if and only if $d_A(a_0, a_1) \leq 2$.

(3) Equality $x \equiv^L_A y$ of Ltp *over A is the finest type-definable equivalence relation over A of arity $|x|$. Moreover, $x \equiv^L_A y$ if and only if $x' \equiv^L_a y'$ for each finite $a \in A$ and corresponding finite x', y' from x, y, respectively.*

Proof (1) Using induction on n by applying 3.2.10, one easily gets $I = \langle a_i | i < \omega \rangle$ such that for each $n < \omega$:

(i) $a_0 a_1 \equiv_A a_i a_j$ for each $i < j \leq n$, and

(ii) $\{a_i | i \leq n\}$ is A-independent.

Then by Ramsey argument we can additionally assume I is A-indiscernible.

(2) ($\Leftarrow$) Due to 2.1.5.

($\Rightarrow$) By 2.2.1 there is $b \equiv^L_A a_0$ such that $b \downarrow_A a_0 a_1$. Apply (1) to a_0, b and b, a_1.

(3) comes from (2) with 3.1.5, and compactness. ■

Hence, in simple T equality of Ltp over A is the finest type-definable bounded equivalence relation over A.

Now, we clearly have the following according to (the proof of) Corollary 3.2.11, and type amalgamation for Ltp.

Proposition 3.2.12 *The following are equivalent:*

(1) *For any* a, b *possibly infinite,* $\text{Ltp}(a/A) = \text{Ltp}(b/A)$ *if and only if* $\text{stp}(a/A) = \text{stp}(b/A)$.

(2) *For any finite tuples* a, b, $\text{Ltp}(a/A) = \text{Ltp}(b/A)$ *if and only if* $\text{stp}(a/A) = \text{stp}(b/A)$.

(3) *The independence theorem holds over* $\text{acl}(A)$.

(4) *Suppose that* $a_0 \equiv^s_A a_1$ *and* $a_0 \mathop{\smile\!\!\!\!\!\!|}_A a_1$. *Then there is an A-indiscernible* $\langle a_i | i < \omega \rangle$.

Remark 3.2.13 It is not hard to see, according to the independence theorem or Corollary 3.2.10, that if a complete type $p \in S(M)$ over a model satisfies the boundedness axiom (i.e., there is some cardinal $\lambda < \bar{\kappa}$ such that for any $B \supseteq M$ there are at most λ-many non-forking extensions of p over B), then it is stationary.

3.3 Characterizing Simple Theories

This section shows that type amalgamation over a model for simple theories is the appropriate substitute for uniqueness over a model for stable theories (Fact 2.1.4). Namely, if a theory has a ternary independence relation satisfying symmetry, transitivity, extension, finite and local characters, and type amalgamation over a model, then the theory must be simple and the independence is the non-forking independence.

Theorem 3.3.1 *In a theory* T, *let* $\mathop{\smile\!\!\!\!\!\!|}^*$ *be a ternary "invariant" relation among (small) sets, i.e.,* $A \mathop{\smile\!\!\!\!\!\!|}^*_C B$ *holds iff* $f(A) \mathop{\smile\!\!\!\!\!\!|}^*_{f(C)} f(B)$ *holds for any automorphism* f. *Suppose that* $\mathop{\smile\!\!\!\!\!\!|}^*$ *satisfies the following axioms (1) to (6). Then* T *is simple, and* $\mathop{\smile\!\!\!\!\!\!|}^*$ *characterizes non-forking, i.e.,* $A \mathop{\smile\!\!\!\!\!\!|}^*_C B$ *if and only if* $A \mathop{\smile\!\!\!\!\!\!|}_C B$.

(1) (Symmetry) $A \mathop{\smile\!\!\!\!\!\!|}^*_C B$ *iff* $B \mathop{\smile\!\!\!\!\!\!|}^*_C A$.

(2) (Transitivity) *Let* $A \subseteq B \subseteq C$. *Then* $D \mathop{\smile\!\!\!\!\!\!|}^*_A C$ *if and only if* $D \mathop{\smile\!\!\!\!\!\!|}^*_A B$ *and* $D \mathop{\smile\!\!\!\!\!\!|}^*_B C$.

(3) (Extension) *Let* $B \subseteq C$. *For any* A, *there is* $A' \equiv_B A$ *such that* $A' \mathop{\smile\!\!\!\!\!\!|}^*_B C$.

(4) (Local character) *For a set* B *and any finite* c, *there is a subset* A *of* B *such that* $|A| \leq |T|$ *and* $c \mathop{\smile\!\!\!\!\!\!|}^*_A B$.

(5) (Finite character) $A \mathop{\smile\!\!\!\!\!\!|}_C B$ *iff for any* $a \in A, b \in B$, *both* a, b *finite tuples, we have* $a \mathop{\smile\!\!\!\!\!\!|}^*_C b$.

(6) (The Independence Theorem over a model) *Assume that* M *is a model and* $B_1 \mathop{\smile\!\!\!\!\!\!|}^*_M B_2$, $D_i \mathop{\smile\!\!\!\!\!\!|}^*_M B_i$ $(i = 1, 2)$, *and* $D_1 \equiv_M D_2$. *Then there is* $D \equiv_{MB_i} D_i$ *such that* $D \mathop{\smile\!\!\!\!\!\!|}^*_M B_1 B_2$.

Proof Recall that T is simple iff dividing satisfies local character (2.3.7). Hence, by finite character, it is enough to show that for any set A and finite b, c, we have

$c \mathop{\smile\!\!\!\!\!|}^*_A b$ if and only if $\mathrm{tp}(c/Ab)$ does not divide over A: Note that the proof of 2.2.4 together with above (3) enables us to construct an A-indiscernible sequence $I = \langle b_i | i < \omega \rangle$, with $b_0 = b$, which is also $\mathop{\smile\!\!\!\!\!|}^*$-independent over A. From indiscernibility and finite character, we may further assume that the length of the sequence I is any large κ.

Now, first assume that $c \not\mathop{\smile\!\!\!\!\!|}^*_A b$. Suppose by the way of contradiction that $\bigcup\{p(x, b_i) : i \in \kappa\}$ is consistent. Let c' realize it. Then $c' \not\mathop{\smile\!\!\!\!\!|}^*_A b_i$ for all $i < \kappa$. Since $I = \langle b_i | i < \kappa \rangle$ is $\mathop{\smile\!\!\!\!\!|}^*$-independent over A, by symmetry and transitivity, we easily obtain that for each $i \in \kappa$, $c' \not\mathop{\smile\!\!\!\!\!|}^*_{A \cup \{b_j : j < i\}} b_i$. This contradicts local character. Therefore, $\mathrm{tp}(c/Ab)$ must divide over A.

Conversely, let us assume that $c \mathop{\smile\!\!\!\!\!|}^*_A b$. We want to show $p(x, b) := \mathrm{tp}(c/Ab)$ does not divide over A. Hence, given an arbitrary A-indiscernible $J = \langle b_i | i < \omega \rangle$ with $b = b_0$, we claim that $\bigcup\{p(x, b_i) | i < \omega\}$ is consistent: First, extend J to an A-indiscernible $\langle b_i | i \leq \kappa \rangle$ for some sufficiently large κ. Now by compactness and Ramsey arguments in Chapter 1, we can find (small) models M_i $(i < \kappa)$ such that for each $i < \kappa$,

(i) $A \cup \bigcup_{j<i} M_j b_j \subseteq M_i$,

(ii) $\langle b_j | i \leq j \leq \kappa \rangle$ is M_i-indiscernible.

By local character and transitivity, $a_\kappa \mathop{\smile\!\!\!\!\!|}^*_{M_i} \bigcup_{j<\kappa} M_j$ for some $i < \kappa$. Again by transitivity, $a_\kappa \mathop{\smile\!\!\!\!\!|}_{M_i} \bigcup\{a_j | i \leq j < \kappa\}$. Then it follows from (ii) and the automorphism invariance of $\mathop{\smile\!\!\!\!\!|}^*$, that $\langle a_j | i \leq j \leq \kappa \rangle$ is $\mathop{\smile\!\!\!\!\!|}^*$-independent over M_i. Hence, we find a model $M (\supseteq A)$, an A-automorphic image of M_i, such that $\langle a_i | i < \omega \rangle$ is indiscernible, $\mathop{\smile\!\!\!\!\!|}^*$-independent over M. Hence, applying (6) repeatedly (cf. Corollary 3.2.10), we can find some tuple c' realizing $\bigcup\{p(x, a_i) | i < \omega\}$ such that $c' \mathop{\smile\!\!\!\!\!|}^*_A I$. ■

The above characterization is one of the main methods for checking simplicity of a structure. Namely, given a structure, one finds an independent notion and shows it satisfying all the axioms listed in Theorem 3.3.1 to confirm that the structure is simple. See Section 2.6. We finish this section by stating a couple of remarks on 3.3.1.

Remark 3.3.2

(1) The axiom (4) can be replaced by the following weaker form. There is a cardinal $\kappa < \bar{\kappa}$ such that for any set B and finite c, it holds $c \mathop{\smile\!\!\!\!\!|}^*_A B$ for some $A \subseteq B$ with $|A| \leq \kappa$.

(2) In [95], [36], a class of theories called *rosy* defined in terms of *thorn forking* is newly introduced. The class of rosy theories is characterized by having an invariant relation $\mathop{\smile\!\!\!\!\!|}^*$ for $\mathcal{M}^{\mathrm{eq}}$ satisfying all the basic axioms (1) to (5) in 3.3.1 and *anti-reflexivity*: If $a \notin \mathrm{acl}(A)$, then $a \not\mathop{\smile\!\!\!\!\!|}^*_A a$. Hence, all simple theories and o-minimal theories are rosy.

3.4 Bibliographical Remarks

The Lascar strong automorphism group Autf($\mathcal{M}$) and the notion of Lascar strong types were introduced by Lascar [83], [81], [84]. Proposition 3.1.5 is observed in [67] and [84], [74]. Example 3.1.7 is a variation of Poizat's [83].

The independence theorem (type amalgamation) over a model and for Lascar types in simple theories are due to Kim and Pillay [74]. But the proofs of 3.2.7, 3.2.8 here are due to Shami [109]. Hrushovski and others show type amalgamation for specific structures having simple theories before the general results come out [30], [55], [58].

Section 3.3 is from [74]. H. Adler studied the relationship between several abstract axioms of an independence relation [1].

4

Hyperimaginaries and Canonical Bases

In this chapter we introduce and deal with hyperimaginaries. A hyperimaginary is an equivalence class of $\emptyset$-type-definable equivalence relation. In $\mathcal{M}^{\mathrm{eq}}$, if $E(x,y)$ is an $\emptyset$-definable equivalence relation, then any of its class a/E is already in there as a single object. More generally, if F is an $\emptyset$-type-definable equivalence relation we can still consider the hyperimaginary c/F as a sequence of imaginary element $\langle c_i/F_i | i \in I\rangle$, i.e., a possibly infinite tuple from $\mathcal{M}^{\mathrm{eq}}$, as far as F is equivalent to some conjunction of $\emptyset$-definable equivalence relations F_i on $\mathrm{tp}(c)$. This is indeed the case for stable theories (Proposition 5.0.3). It is *not* known yet whether the same holds for all simple theories. Even if this is the case for some subclass of simple theories (such as supersimple theories, Theorem 5.4.9), to show this, we need to deal with hyperimaginaries a priori.

In Chapter 3 we saw the need to deal with hyperimaginaries for simple theories. However, there are more impending reasons. Namely, *canonical bases* for Lascar types exist as hyperimaginaries. Given $p(x)$, a complete type over a model or a Lascar strong type, its *canonical base* is an invariant set A such that any automorphism fixes A setwise iff $p(x)$ and $f(p(x))$ are in the same parallelism class (Lemma 4.3.4). We shall see that such a canonical base exists in the form of a hyperimaginary. Arguments for hyperimaginaries and canonical bases interwoven together contribute deep theoretic developments for simple theories. Canonical bases are used almost everywhere throughout other chapters.

4.1 Hyperimaginaries

Definition 4.1.1 A *hyperimaginary* is an equivalence class $a_E = a/E$ of a type-definable equivalence relation $E(x,y)$ over $\emptyset$ of an arbitrary arity $|x| = |y| (< \bar{\kappa})$; i.e., a_E is the set on $\mathcal{M}^{|x|}$ type defined by $E(x,a)$. We may call a_E an *E hyperimaginary*. Given a hyperimaginary c_E, we say it is *finitary* (*countable*, resp.) if its arity($=|c|$) is finite (countable, resp.). In general we let $|c_E| := |c|$.

Two hyperimaginaries a_E, b_F are said to be *interdefinable* (or *equivalent*) if the set a_E is fixed by an automorphism f iff it fixes b_F. When there is little chance of confusion we may identify two interdefinable hyperimaginaries.

Remark 4.1.2

(1) Here, either understanding of $\mathcal{M}$ as just the home-sort or the total $\mathcal{M}^{\mathrm{eq}}$ is acceptable. A tuple c of any arity (including the empty tuple) from $\mathcal{M}$ is a hyperimaginary as c is $c/(x = y)$. The universe $\mathcal{M}^{\alpha}$ itself as a set is a hyperimaginary since it is an equivalence class of $x = x \wedge y = y$, which is interdefinable with $\emptyset$. A finite subset $\{a_0, \ldots, a_{n-1}\}$ of $\mathcal{M}$ is an imaginary as it is an equivalence class of $\bigvee_{\sigma \in \mathrm{Sym}(n)}(\bigwedge_{i<n} x_i = y_{\sigma(i)})$. Any sequence $\langle c_i/E_i | i \in I\rangle$ of imaginary elements is a hyperimaginary $\langle c_i | i \in I\rangle / \bigwedge E_i(x_i, y_i)$. This is even true when each c_i/E_i is a hyperimaginary; i.e., a sequence of hyperimaginaries is a hyperimaginary. But a hyperimaginary need not be equivalent to a sequence of imaginaries (Example 3.1.7).

(2) Given a hyperimaginary c/E, let $E'(x,y) \equiv E(x,y) \wedge \mathrm{tp}(x) = \mathrm{tp}(y)$. Then clearly c/E and c/E' are interdefinable. Moreover, if $E(x,y)$ type-defines an equivalence relation only on a partial type $\pi(x)$ over $\emptyset$, we can assume E is an equivalence relation on the whole universe by substituting

$$(E(x,y) \wedge \pi(x) \wedge \pi(y)) \vee x = y$$

for E.

(3) Hereafter, for convenience, given a type-definable equivalence relation $E(x,y)$, *we always assume that* E is closed under finite conjunctions, and each formula in E is symmetric (of course, it is reflexive), by possibly substituting $\{\varphi(x',y') \wedge \varphi(y',x') | \varphi(x',y') \in E\}$ for E. If necessary we may even assume E is closed under the projections of formulas; i.e., if $\varphi(x',y') \in E$, then $\exists x''y''\varphi(x',y') \in E$ $(x''y'' \subseteq x'y')$, too.

Lemma 4.1.3 *Every hyperimaginary is equivalent to a sequence of (at most) countable hyperimaginaries.*

Proof Let $E(x,y) = \{\varphi_i(x_i,y_i) | i \in I\}$ be a $\emptyset$-type-definable equivalence relation where x_i is a finite subtuple of possibly infinite x. For each $i \in I$, there must be $j \in I$ such that $\varphi_j(x_j,y_j) \wedge \varphi_j(y_j,z_j) \models \varphi_i(x_i,z_i)$ with $x_i \subseteq x_j$ and $y_i \subseteq y_j$. Let E_i be a countable subset of E closed under such choices having a countable arity. Clearly, each E_i forms an equivalence relation and $\bigwedge_{i \in I} E_i$ type-defines E too. ■

Hence as far as we are concerned with hyperimaginaries up to interdefinability, we may only deal with countable hyperimaginaries.

We now explain more terminologies (some might be a bit abusive) involved with the use of hyperimaginaries. Recall our convention that a *real* tuple or set means a tuple or a (small) set from $\mathcal{M}$, and an *imaginary* tuple means that from $\mathcal{M}^{\mathrm{eq}}$. Now, $\mathcal{M}^{\mathrm{heq}}$ denotes the set of all hyperimaginaries. Hence, as sets, $\mathcal{M} \subseteq \mathcal{M}^{\mathrm{eq}} \subseteq \mathcal{M}^{\mathrm{heq}}$. Similar to the "real" case, given a hyperimaginary c_E, if an automorphism f fixes the class c_E setwise then we call f a c_E *automorphism* or an *automorphism over* c_E. The set of all c_E automorphisms is denoted by $\mathrm{Aut}_{c_E}(\mathcal{M})$, which is a subgroup of $\mathrm{Aut}(\mathcal{M})$. As usual a relation is said to be "over c_E" or "c_E-invariant" if it is $\mathrm{Aut}_{c_E}(\mathcal{M})$-invariant.

Mostly, in this book, dealing with hyperimaginaries of $\emptyset$-type-definable equivalence relations would suffice. In particular, if $\bar{E}(x, y; A) \equiv \bigwedge_{i\in I} E_i(x_i, y_i; a_i)$ is a conjunction of real A-definable equivalence relations, then $c/\bar{E}$ is *interdefinable* with a sequence of imaginaries $cA/\bar{F}$, where

$$\bar{F}(xz, yw) \equiv \bigwedge_{i\in I} \forall u_i\,(E_i(u_i, x_i; z_i) \leftrightarrow E_i(u_i, y_i; w_i))$$

is a conjunction of definable equivalence relations over $\emptyset$. In general, if $\bar{E}$ is merely a type-definable equivalence relation that happens to be a hyperimaginary A/F-invariant, then $c/\bar{E}$ and another hyperimaginary $cA/\bar{F}'$ are interdefinable *over* A/F, where

$$F'(xz, yw) \equiv \left(\bar{E}(x, y; z) \wedge F(z, w) \wedge p(z) \wedge p(w)\right) \vee xz = yw$$

with $p(z) = \mathrm{tp}(A)$.

Remark & Definition 4.1.4 Let a_E, b_F be hyperimaginaries.

(1) By $\mathrm{tp}(a_E/b_F)$, a *type of* a_E *over* b_F, we mean a partial type over b

$$\exists yy'(E(x, y) \wedge F(b, y') \wedge q(yy')),$$

where $q(yy') = \mathrm{tp}(ab)$. Note, in particular, that $\mathrm{tp}(a_E) = \exists y(E(xy) \wedge p(y))$ for $p = \mathrm{tp}(a)$, and $\mathrm{tp}(a_E, b_F) = \mathrm{tp}((ab)_{E\wedge F}) = \exists yy'(E(x, y) \wedge F(x', y') \wedge q(yy'))$. The solution set of $\mathrm{tp}(a_E/b_F)$ is exactly the orbit of the set a_E under all b_F automorphisms, i.e., the union of all E-class conjugates of a_E over b_F. But different from real types, a formula in $\mathrm{tp}(a_E/b_F)$ *need not* be b_F-invariant.

A partial type $\pi(x)$ is said to be a *(partial)* E*-type* over b_F if it is b_F-invariant, and for any $c \models \pi$, it holds that $c' \models \pi$ as far as $E(c, c')$ holds (i.e., the solution set is some union of E classes). In this sense, we may write $\pi(x)$ as $\pi(x_E)$. Thus $\mathrm{tp}(a_E/b_F)$ is a *complete* E type over b_F. These definitions cover the usual case of complete or partial types. Note that if $\pi(x)$ is an E type over b_F, then due to 1.1.1, there is a type $\pi'(x)$ over b such that $\pi(x) \equiv \pi'(x)$. A set of (E) hyperimaginaries is said to be *hyperdefinable over* b_F if it is a solution set of an E type over b_F. More generally, an n-ary relation on $\pi_1(x_{E_1}) \times \cdots \times \pi_n(x_{E_n})$ is said to be *hyperdefinable over* b_F if some $R(x_{E_1}, \ldots, x_{E_n})$, an $E_1 \wedge \ldots \wedge E_n$ type over b_F, expresses the relation.

As in the real case, $S(b_F)$ denotes the *collection of all the types of hyperimaginaries over* b_F, $S_E(b_F)$ denotes the *set of all complete E types over* b_F, and $a_E \equiv_{b_F} c_E$ means $\mathrm{tp}(a_E/b_F) = \mathrm{tp}(c_E/b_F)$ (this need not imply that $\mathrm{tp}(a) = \mathrm{tp}(c)$).

Thus when we write $p(x) \in S(b_F)$ or $x \equiv_{b_F} y$, an underlying $\emptyset$-type-definable equivalence relation $E'(x, y)$ is assumed, so the variables x, y are actually $x_{E'}, y_{E'}$. For example, even in hyperimaginary context, $x \equiv_z y$ (more precisely $x_{E'} \equiv_{z_F} y_{E'}$) is type-definable by

$$\exists z_1 z_2 w_1 w_2 (F(z, z_1) \wedge F(z, z_2) \wedge \mathrm{tp}(z_1 w_1) = \mathrm{tp}(z_2 w_2) \wedge E'(w_1, x) \wedge E'(w_2, y)).$$

Hence, as in the real case, given imaginary tuples c, d we have $c \equiv_{a_E} d$ iff $c_i \equiv_{a_E} d_i$ for all corresponding finite subtuples c_i, d_i of c, d, respectively.

(2) We say b_F is *bounded over* a_E (write $b_F \in \mathrm{bdd}(a_E)$) if b_F has only boundedly many a_E-automorphism conjugates; i.e., the set $\{f(b_F) | f \in \mathrm{Aut}_{a_E}(\mathcal{M})\}$ is bounded. Similarly, we say b_F is *definable over* a_E (write $b_F \in \mathrm{dcl}(a_E)$) if b_F has no other a_E-automorphism conjugate; i.e., the set $\{f(b_F) | f \in \mathrm{Aut}_{a_E}(\mathcal{M})\}$ is a singleton. That a_E, b_F are *interbounded* means $a_E \in \mathrm{bdd}(b_F)$ and $b_F \in \mathrm{bdd}(a_E)$.

But by $\mathrm{bdd}(a_E)$, we mean the set of all *(at most) countable hyperimaginaries that are bounded over* a_E, so that it is a set of bounded cardinality. Then since $\mathrm{bdd}(a_E)$ can be considered as a *single* hyperimaginary, $\mathrm{tp}(b_F/\mathrm{bdd}(a_E))$ makes sense as in (1). By the same manner, $\mathrm{dcl}(a_E)$ denotes the set of all *countable hyperimaginaries definable over* a_E. Note also that $\mathrm{bdd}(\mathrm{bdd}(a_E)) = \mathrm{bdd}(a_E)$ and $\mathrm{dcl}(\mathrm{dcl}(a_E)) = \mathrm{dcl}(a_E)$. We similarly define $\mathrm{acl}(a_E)$ as the set of all *countable hyperimaginaries, each of which has finitely many conjugates over* a_E. That b_F is *algebraic over* a_E (written $b_F \in \mathrm{acl}(a_E)$) means b_F is definable over $\mathrm{acl}(a_E)$. On the other hand, we put

$$\mathrm{acl}^{\mathrm{eq}}(a_E) := \{x, a_E \mid x \text{ imaginary and } x \in \mathrm{acl}(a_E)\}.$$

In Corollary 5.1.15 we observe that $\mathrm{acl}(a_E)$ and $\mathrm{acl}^{\mathrm{eq}}(a_E)$ are interdefinable.

It turns out the notation "$b_F \in \mathrm{bdd}(a_E)$" is non-standard, or, as we understand it, that b_F's equivalent tuple (=sequence) of at most countable hyperimaginaries is from $\mathrm{bdd}(a_E)$. We understand "$b_F \in \mathrm{dcl}(a_E)$ or $\in \mathrm{acl}(a_E)$" similarly. (It is possible $b_F \in \mathrm{acl}(a_E)$ but b_F has infinitely many a_E conjugates.) A type $\mathrm{tp}(b_F/a_E)$ is said to be *bounded* (*algebraic*, resp.) if $b_F \in \mathrm{bdd}(a_E)$ ($b_F \in \mathrm{acl}(a_E)$, resp.).

Note that by compactness, if an imaginary tuple $c \in \mathrm{bdd}(a_E)$ then $c \in \mathrm{acl}(a_E)$. This need not hold if c is even a finitary hyperimaginary. The point is that $x \neq y$ is a formula for finite real tuples, but $x_E \neq y_E$ (i.e., $\neg E(x, y)$) is *not* even type-definable in general.

If the real tuple a is from a model M, then $\mathrm{bdd}(a_E) \subseteq \mathrm{dcl}(M)$. Let $b_F \in \mathrm{bdd}(M)$ and let $b' \equiv_M b$. Then since $d_M(b, b') = 1$ (by taking a common coheir sequence as in 1.1.6), $F(b, b')$ must hold. Hence $b_F \in \mathrm{dcl}(M)$.

(3) We abuse the notation a bit further so that we may simply write "$b_F \subseteq a_E$" to mean $b_F \in \mathrm{dcl}(a_E)$. Now, if $p(x) = p(x_{E'}) \in S(a_E)$ and $b_F \subseteq a_E$, then $p(x) \lceil b_F$ denotes $\mathrm{tp}(c_{E'}/b_F)$ for some (any) $c_{E'} \models p(x)$.

From now on, *a, b, c, …; A, B, …* used for tuples and sets are representing hyperimaginaries But when we write a formula $\varphi(x, c)$, the parameter c is *always* real or imaginary (unless said otherwise, c is finite, but in a certain case c can be infinite to mean that the parameter of φ is a finite subtuple of c). Whether a given tuple is an imaginary or a hyperimaginary will be cleared up to context. Occasionally to emphasize, we may write hyperimaginaries in boldface letters.

Note that given a hyperimaginary $e = a/E$, it makes sense to write e/F as long as the equivalence relation F is coarser than E. In this case e/F simply means a/F, and this does not depend on the choice of a.

Lemma 4.1.5 *Let a be an imaginary tuple, and let $\mathbf{c}, \mathbf{e}$ be hyperimaginaries.*

(1) *Suppose that $\mathbf{c} \in \operatorname{dcl}(a)$ and $a \in \operatorname{bdd}(\mathbf{c})$. Then $\mathbf{c}$ is equivalent to some imaginary tuple.*

(2) *Suppose that $\mathbf{c} \in \operatorname{bdd}(\mathbf{e})$. Then the set C of all $\mathbf{e}$-conjugate of $\mathbf{c}$ is equivalent to a hyperimaginary $\mathbf{e}/F$ for some type-definable F.*

(3) *If $\mathbf{c} \in \operatorname{dcl}(\mathbf{e})$, then for some type-definable F, $\mathbf{c}$ is equivalent to $\mathbf{e}/F$.*

Proof (1) Let $\langle a_i |\ i \in I\rangle$ be a sequence of all finite subtuples of a. For each $i \in I$, let a_i' be the finite set of all $\mathbf{c}$ conjugates of a_i, identified as a single imaginary element. We claim $a' = \langle a_i' | i \in I\rangle$ is an imaginary tuple equivalent to $\mathbf{c}$: Clearly, $a' \in \operatorname{dcl}(\mathbf{c})$. Now if an automorphism f fixes a', then it fixes each set a_i'. Hence, $\operatorname{tp}(a_i/\mathbf{c}) = \operatorname{tp}(f(a_i)/\mathbf{c})$ and from 4.1.4(1), $\operatorname{tp}(a/\mathbf{c}) = \operatorname{tp}(f(a)/\mathbf{c})$. Then, since $\mathbf{c} \in \operatorname{dcl}(a)$, it follows $\mathbf{c} = f(\mathbf{c})$. Thus, $\mathbf{c} \in \operatorname{dcl}(a')$.

(2) Suppose that $\mathbf{c} = c/E$, where c is real; $E = \{\varphi_i(z^i, w^i) | i \in I\}$; and that c^i from c corresponds to finite z^i. Let $p(z, x) = \operatorname{tp}(c, \mathbf{e})$ and let $p_i(z^i, x) := \operatorname{tp}(c^i, \mathbf{e})$. Note that $p(z, x) = \bigcup p_i(z^i, x)$. Let $c_0^i, \ldots, c_{n_i}^i$ be a maximal set of realizations of $p_i(z^i, \mathbf{e})$ such that $\neg\varphi_i(c_j^i, c_k^i)$ for $j \neq k \leq n_i$. Since $\mathbf{c} \in \operatorname{bdd}(\mathbf{e})$, such a finite number n_i exists. Now there is a partial type

$$r\left(z_j^i, x \mid i \in I,\ j \leq n_i\right)$$

saying that $p_i(z_j^i, x)$ and $\neg\varphi(z_j^i, z_k^i)$ hold, for each $i \in I$ and $j \neq k \leq n_i$. In particular, $(\tilde{c}, \mathbf{e})$ realizes r, where $\tilde{c} = (c_0^i, \ldots, c_{n_i}^i \mid i \in I)$. Let

$$F(x, y) = \exists \tilde{z}(r(\tilde{z}, x) \wedge r(\tilde{z}, y))$$

be a type-definable relation on $\operatorname{tp}(\mathbf{e}) \times \operatorname{tp}(\mathbf{e})$. We claim that for any automorphism f, it setwise fixes C iff $F(\mathbf{e}, f(\mathbf{e}))$ holds: If f setwise fixes C, then since $f(c_j^i) \models p_i(z^i, \mathbf{e})$, $(f(\tilde{c}), \mathbf{e})$ realizes r, too. Then as $(f(\tilde{c}), f(\mathbf{e})) \models r$, $F(\mathbf{e}, f(\mathbf{e}))$ holds. Conversely, suppose that $(\tilde{d}, \mathbf{e}), (\tilde{d}, f(\mathbf{e})) \models r$ with $\tilde{d} = (d_0^i, \ldots, d_{n_i}^i \mid i \in I)$. Choose an arbitrary $b/E \in C$ with $b \equiv_{\mathbf{e}} c$, and let $b^i \subseteq b$ correspond to z^i. Then $d_0^i, \ldots, d_{n_i}^i, f(b^i)$ all realize $p_i(z^i, f(\mathbf{e}))$ since $b^i \models p_i(z^i, \mathbf{e})$. Hence, by the maximality of n_i, $\varphi(d_{j_0}^i, f(b^i))$ holds for some $j_0 \leq n_i$. Since $d_{j_0}^i \models p_i(z^i, \mathbf{e})$ as well, by compactness there is $b' \models E(z, f(b)) \cup p(z, \mathbf{e})$. Thus, $f(b/E) = f(b)/E = b'/E$ is in C again. Now if $F(\mathbf{e}, f(\mathbf{e}))$, then $F(f(\mathbf{e}), \mathbf{e})$ and $F(\mathbf{e}, f^{-1}(\mathbf{e}))$ hold. Hence, $f^{-1}(C) \subseteq C$ follows too. We have proved the claim.

Now by the claim, F must be an equivalence relation (on tp(e) or by 4.1.2(2) on the whole universe), and e/F is equivalent to the set C.

(3) is an obvious consequence of (2). ■

A consequence of Lemma 4.1.5(3) is that any hyperimaginary a_E with imaginary tuple a is interdefinable with a hyperimaginary b_F, where b is a real tuple. Hence, up to interdefinability, any hyperimaginary is an $\emptyset$-type-definable equivalence relation class of a real tuple.

4.2 Non-forking Independence of Hyperimaginaries

Once we adopt hyperimaginaries as objects within our scope of considerations we need to talk about diving between hyperimaginaries. This section develops arguments for that. As in the *real* case, non-forking independence notion among hyperimaginaries makes perfect sense in simple theories. We begin with indiscernible sequences of hyperimaginaries. We work with hyperimaginaries, and recall that tuples $a, b, c, \ldots$ are hyperimaginaries.

Definition 4.2.1 Let a_0, b be hyperimaginaries. As in the real case, we say a sequence $\langle a_i \mid i < \omega \rangle$ is *indiscernible over* b if $a_0, \ldots, a_n \equiv_b a_{i_0}, \ldots, a_{i_n}$ for any $i_0 < \cdots < i_n < \omega$.

Definition 4.2.2 Let $\pi(x, a_0)$ be a partial type over a_0. We say $\pi(x, a_0)$ *divides over* c if there is a c-indiscernible $\langle a_i | i < \omega \rangle$ such that $\bigcup_{i<\omega} \pi(x, a_i)$ is inconsistent. Then as in real case there is some k such that $\pi(x, a_0)$ k*-divides over* c; i.e., $\bigcup_{i<\omega} \pi(x, a_i)$ is k-inconsistent. $\pi(x, a_0)$ *forks over* c if it implies a finite disjunction of formulas, each of which divides over c.

We too write $a \perp_b c$ if tp(a/bc) *does not divide over* b.

Remark 4.2.3

(1) Let $I = \langle a_i | \ i < \omega \rangle$ be b-indiscernible, and let $a_0 = c_0/E$ and $b = d/F$, where c_0 and d are real. We claim that there is some d'-indiscernible $J = \langle c_i | i < \omega \rangle$ such that $b = d'/F$ with $d \equiv d'$ and $a_i = c_i/E$. Of course, then I is d'-indiscernible too. Based on Ramsey and compactness, we can find a d-indiscernible $J' = \langle c_i' | i < \omega \rangle$ such that $c_0 I \equiv_b c'_0 \langle c'_i/F | \ i < \omega \rangle$. Then $d' = f(d)$ and $J = f(J')$, where f is some b automorphism sending $c_0' \langle c_i'/F | \ i < \omega \rangle$ to $c_0 I$.

Indeed I is bdd(b)-indiscernible: There is a model M containing d' such that J is M-indiscernible. Thus, by 4.1.4(2), both J and I are bdd(b)-indiscernible.

Hence, $a \notin$ bdd(b) iff $a \not\perp_b a$. In particular, any type over bdd(b) does not divide over b.

(2) Let $\pi(x), \pi'(x)$ be partial types over b, and $\pi(x) \models \pi'(x)$. Then if $\pi'(x)$ divides (forks, resp.) over b, then $\pi(x)$ divides (forks, resp.) over any $a \subseteq b$. In particular if $\pi(x) \equiv \pi'(x)$ then $\pi(x)$ divides (forks, resp.) over b iff $\pi'(x)$ divides (forks, resp.) over b. Hence, dividing and forking depend only on the solution set of the partial type.

(3) (*Extension*) Let $q(x)$ be a partial E type over c that does not fork over b. Then there is a completion of q over c as an E type that does not fork over b. Let $x'(\subseteq x)$ be finite, and let $c = v/F$ with real v. Given a formula $\varphi(x', d)$ let $\varphi_E(x, d)$ denote a *partial type* $\exists y(E(x, y) \wedge \varphi(y', d))$. Now, it follows that

$$q(x) \cup \left\{\neg\psi_E(x', v') \mid \psi(x', y') \text{ forks over } b\ (v' \subseteq v)\right\}$$

is consistent, which is clearly an E type. Choose $q'(x)$ some completion of it over v as an E type. Then $q'(x)$ so $q'(x)\lceil c$ too should not fork over b.

(4) If $\pi(x)$ is an E type over b, when we write $a \models \pi(x)$, unless said otherwise, a is an E hyperimaginary (occasionally, we may need to look $\pi(x)$ as an E'-type, where E' is finer or coarser than E). However, in rank evaluations for $D_{\varphi,k}(\pi(x))$ or $D(\pi(x))$, $\pi(x)$ is measured as a partial *real* type (and x is finite if the D-rank is computed). When we talk about the SU-rank of hyperimaginaries, the situation will be a bit more subtle and we will get back to this in 4.2.13(2).

Our goal is to show in simple theories, hyperimaginaries also satisfy non-forking extension, symmetry, and transitivity. So we redo Section 2.2 for hyperimaginaries. As the proofs will only need some variations of the real case, we would rather sketch those.

Lemma 4.2.4 *Assume T is simple.*

(1) tp(a/b) *does not fork over b.*

(2) (Extension) *Given a, b, c there is* $a' \equiv_b a$ *such that* tp(a'/bc) *does not fork over b.*

Proof (1) Let $p(x) = \text{tp}(a/b)$, which is a partial real type. Suppose not, so $p(x) \models \bigvee_{i\leq n} \varphi_i(x', c_i)$ and each $\varphi_i(x', c_i)$ k_i-divides over b. Let $\psi(x', z, y)(y = y_0 \ldots y_n)$ be a code formula for $\varphi_0(x', y_0), \ldots, \varphi_n(x', y_n)$. By 2.3.4(3), there is $r \leq n$ such that $D(p(x), \psi, k) = D(p(x) \cup \{\varphi_r(x', c_r)\}, \psi, k) = m < \omega$, where $k := \max\{k_0, \ldots, k_n\}$. But then by 2.3.3(1), $D(p(x), \psi, k) \geq m + 1$, a contradiction.

(2) follows from (1) and 4.2.3(4). ■

The notion of *Morley sequence* for hyperimaginaries is defined similarly. We repeat here for convenience.

Definition 4.2.5 Let $a \subseteq b$ and $p \in S(b)$. By a *Morley sequence of p over a*, we mean a b-indiscernible sequence $I = \langle a_i : i \in \omega\rangle$ in p such that for every $i \in \omega$, tp$(a_i/ba_0 \ldots a_{i-1})$ does not fork over a. When $a = b$, we omit mentioning "over a".

Based on non-forking extension with 2.1.5 and possibly 4.2.3(1), we have the following:

Proposition 4.2.6 *If* $p(x) \in S(b)$ *does not fork over* $a(\subseteq b)$, *then there exists a Morley sequence of p over a. Thus, for simple T, a Morley sequence of a complete type exists.*

Now, results on dividing for hyperimaginaries analogous to 2.1.6, 2.1.7, and 2.2.5 follow. The proofs can be carried out in the same fashion so we omit.

Lemma 4.2.7 *$a \downarrow_b c$ iff for any b-indiscernible I containing c, there is $a' \equiv_{bc} a$ such that I is $a'b$-indiscernible.*

Lemma 4.2.8 *If $a \downarrow_b c$ and $d \downarrow_{ab} c$, then $da \downarrow_b c$ holds.*

Lemma 4.2.9 *$I = \langle a_i : i \in \omega \rangle$ be a Morley sequence of $\text{tp}(a_0/b)$. Suppose $\langle a^j | j < \omega \rangle$ is some b-indiscernible sequence of $\text{tp}(a_0/b)$. Then there is a sequence $J \equiv_b I$ such that $\langle a^j \rangle ^\frown J$ is b-indiscernible for each $j < \omega$.*

Proposition 4.2.10 *Assume T is simple. Let $\pi(x, b)$ be a partial type over b. The following are equivalent:*

(1) *$\pi(x, b)$ divides over a.*

(2) *For any Morley sequence I of $\text{tp}(b/a)$, $\bigcup\{\pi(x, b') \mid b' \in I\}$ is inconsistent.*

(3) *For some Morley sequence I of $\text{tp}(b/a)$, $\bigcup\{\pi(x, b') \mid b' \in I\}$ is inconsistent.*

(4) *$\pi(x, b)$ forks over a.*

Proof (2)⇒(3)⇒(1) Clear by the definition and 4.2.6.

(1)⇒(2) Assume (1) is witnessed by an a-indiscernible $\langle b^j | j < \omega \rangle$ of $\text{tp}(b/a)$ ($b = b^0$), so by a formula $\varphi(x', u)$ and a number k. Suppose (2) does not hold. Thus, there is Morley $I = \langle b_i | i < \omega \rangle$ of $\text{tp}(b/a)$ such that $\bigcup\{\pi(x, b_i) \mid i \in \omega\}$ is realized by, say, c. Let $p(x, aI) = \text{tp}(c/aI)$. Now by 4.2.9, we can assume that $b_0 = b^0$ and J is $I'(= \langle b_i | 0 < i \rangle)$-indiscernible. Hence, both $\pi(x, b_0)$ and $p(x, aI)$ divide over aI' witnessed by the same φ, k. Since $I \equiv_a I'$, we have $D(p(x, aI), \varphi, k) = D(p(x, aI'), \varphi, k) = m < \omega$. But by 2.3.3(1), $D(p(x, aI'), \varphi, k) \geq m + 1$, a contradiction.

(1)⇒(4) Clear.

(4)⇒(1) By the same arguments in the proof of 2.2.7. ■

Theorem 4.2.11 *T is simple.*

(Symmetry) *$a \downarrow_b c$ iff $c \downarrow_b a$.*

(Transitivity) *$a \downarrow_b cd$ iff $a \downarrow_b c$ and $a \downarrow_{bc} d$.*

(Countable character) *$a_0 \downarrow_b a_1$ iff for any $a'_i \subseteq a_i$, $a'_0 \downarrow_b a'_1$ iff for any countable $a'_i \subseteq a_i$, $a'_0 \downarrow_b a'_1$.*

(Finite character) *$(a_i | i \in I_0) \downarrow_c (b_j | j \in I_1)$ iff for each finite $I'_i \subseteq I_i$, $(a_i | i \in I'_0) \downarrow_c (b_j | j \in I'_1)$.*

(Local character) *For any $a = c/E, b$ with real c, there is $b' = d/F \subseteq b$ with real d of size $\leq |T| + |c|$ such that $a \downarrow_{b'} b$.*

Proof Symmetry and transitivity come from 4.2.6, 4.2.10, 4.2.3(2), and 4.2.8 with the same proof for the real case.

Countable character and finite character are indeed true for any theory due to the definition with 4.2.3(1)(2), and compactness applied to 4.1.3.

For local character, assume it is false. Put $\kappa = |T| + |c|$. By 4.1.3 there is a sequence of countable hyperimaginaries $\langle b_i | i < |d| \rangle$ equivalent to b. By re-indexing we can assume $a \not\downarrow_{b'_i} b_i$ for $i < \kappa^+$, where $b'_i = \langle b_j | j < i \rangle$. For each i, there is $(\varphi_i(x_i, y_i), k_i)$

witnessing this. As there are κ-many such choices (x_i is a finite subvariable for c), there is a fixed $(\varphi(x,y),k)$ appearing infinitely many times. Then by 2.3.3(1) and 2.3.7, the values $D_{\varphi,k}(\mathrm{tp}(a/b_i')) < \omega$ must be infinitely decreasing, a contradiction. ■

Proposition 4.2.12 *T is simple. Then $a \downarrow_b c$ iff for any ψ, k, $D(\mathrm{tp}(a/bc), \psi, k) = D(\mathrm{tp}(a/b), \psi, k)$.*

Proof By 1.3.3(1), if $a \not\downarrow_b c$ witnessed by ψ, k, then $D(\mathrm{tp}(a/bc), \psi, k) < D(\mathrm{tp}(a/b), \psi, k)$.

Conversely, assume $a \downarrow_b c$. Let $a = u_E$, $b = v_{E'}$, $c = w_F$ with real u, v, w, and let $D_{\psi,k}(p(x)) = n$, where $p(x) = \mathrm{tp}(a/b)$. As $p(x)$ is a real partial type over v, by 2.3.4(4) there is a real complete type $p'(x) \supseteq p(x)$ over v such that $D_{\psi,k}(p') = n$, too. Then for $u' \models p'$ and $a' = u'/E$, it follows $a' \equiv_b a$. By moving a' to a over b, we can clearly assume that $p'(x) = \mathrm{tp}(u/v)$. By extension and transitivity, we can further assume $uva \downarrow_b c$ and $uva \downarrow_{bc} w$. Hence, $uva \downarrow_b cw$ and $u \downarrow_v w$ follow. Hence, by 2.3.9 we have $D_{\psi,k}(u/vw) = n$. But $\mathrm{tp}(u/vw) \models \mathrm{tp}(u_E/vw) \models \mathrm{tp}(a/bc) \models \mathrm{tp}(a/b)$ (as real partial types). Hence, $D_{\psi,k}(a/bc) = n$ too. ■

Remark 4.2.13

(1) The above proposition enables us to observe the type definability of $\downarrow$ for the following cases. Note first that $D(\mathrm{tp}(x'_E/cz_F), \varphi(x,y), k) \geq n$ is a hyperdefinable relation over c. In Lemma 2.3.2(2) simply replace $p(x'_{\bar{u}})$ by $(x'_{\bar{u}})_E \equiv_{cz_F} x'_E$, which is hyperdefinable as in 4.1.4(1). Now let T be simple. Then given a complete type $p(x'_E, c)$ over c,

$$\pi(x'_E, z_F) \equiv p(x'_e, c) \wedge x'_E \downarrow_c z_F$$

is hyperdefinable over c by

$$p(x'_E, c) \cup \left\{D(x'_E/cz_F, \varphi, k) \geq n_{p,\varphi,k} : \varphi(x,y) \in \mathcal{L}; x \subseteq x'; k \geq 2\right\},$$

where $n_{p,\varphi,k} := D(p(x'_E, c), \varphi, k)$.

In particular, for complete types $p_i(z_{E_i})$ $(0 < i)$ over c, that

$$\left\{z_{E_i} \middle| 0 \leq i\right\} \text{ is } c\text{-independent and } p_i(z_{E_i}) \text{ hold for } i > 0,$$

is hyperdefinable over c. Moreover, that

$$\left\langle (z_i)_{E_1} \middle| i < \omega \right\rangle \text{ is a Morley sequence of } p_1 \text{ over } c'$$

is also c-hyperdefinable if p_1 does not fork over $c' \subseteq c$.

(2) **Supersimplicity and hyperimaginaries.** It follows by the similar reason for local character (4.2.11), T is supersimple iff for any finitary c and $A = (a_i | i \in I)$, there is a finite subset I_0 of I such that $c \downarrow_{(a_i | i \in I_0)} A$. Note that for hyperimaginaries, the SU-rank makes sense. In particular T is supersimple iff $SU(p) < \infty$ for any complete

type p of a finitary hyperimaginary. Moreover, 2.5.16(2) and the Lascar inequalities 2.5.19 hold, too. In particular, in simple T, $a \in \mathrm{bdd}(b)$ iff $SU(a/b) = 0$, and if $SU(a/b) < \infty$ then for $e \in \mathrm{bdd}(ab)$, both $SU(ae/b)$ and $SU(e/b) < \infty$ since

$$SU(a/eb) + SU(e/b) \leq SU(ae/b) \leq SU(e/ab) \oplus SU(a/b).$$

When we talk about the SU-rank, an extra attention is required. If $p(x) = \mathrm{tp}(a/b)$ and $q(x) = \mathrm{tp}(c/b)$, where $a = u/E$, $b = v/F$ with real u, v, then even if $p(x) \equiv q(x)$ as partial real types, $SU(a/b)$ and $SU(c/b)$ *need not* be equal. Even although $p'(x) = \mathrm{tp}(u/b) \models p(x)$, in general $SU(u/b) \geq SU(a/b)$. Note that $SU(u/b)$ is measured in X, the solution set of $\mathrm{tp}(u/b)$, while $SU(a/b)$ is measured in X/E.

Hereafter when we write $SU(a/b)$, that a is an E hyperimaginary for a fixed E will always be clear in context. Also for a *partial E-type* $\pi(x)$ over A, we let

$$SU(\pi(x)) := \sup \{SU(p(x)) | \, p(x)(\in S_E(A)) \models \pi(x)\}.$$

Again the SU-rank of $\pi(x)$ is measured as an E type will be clear from context. For E types $\pi(x)$ and $\pi'(x)$, if $\pi(x) \equiv \pi'(x)$ then $SU(\pi(x) = SU(\pi'(x))$.

We are now talk about type amalgamation of hyperimaginary types.

Definition 4.2.14

(1) For a hyperimaginary c, by $\mathrm{Autf}_c(\mathcal{M})$ we mean a subgroup of $\mathrm{Autf}(\mathcal{M}) \cap \mathrm{Aut}_c(\mathcal{M})$ generated by

$$\{f \in \mathrm{Aut}_c(\mathcal{M}) \mid f \in \mathrm{Aut}_M(\mathcal{M}) \text{ for some model } M \text{ with } \mathrm{dcl}(M) \ni c\}.$$

(2) We say two E hyperimaginaries a, b have the *same Lascar type over c* (written $a \equiv^L_c b$) if $b = f(a)$ for some $f \in \mathrm{Autf}_c(\mathcal{M})$.

We can similarly define $d_c(a, b)$ for hyperimaginaries as in 2.1.4, and the proof of Proposition 3.1.5 can go through the hyperimaginary context, so we have the following:

Lemma 4.2.15 *The following are equivalent (a, b are E hyperimaginaries):*

(1) $a \equiv^L_c b$.

(2) $d_c(a, b) < \omega$.

(3) $\models F(a, b)$ *for any c-invariant bounded equivalence relation F coarser than E.*

Conclusively, $x_E \equiv^L_c y_E$ *is an invariant-bounded equivalence relation over c coarser than E, and is the finest among those.*

In the rest of this section assume T **is simple.**

Remark 4.2.16 Now the proofs in Section 3.2 (3.2.1 to 3.2.8) go through the hyperimaginary case so we obtain Lascar type amalgamation for hyperimaginaries. In particular the following analogous facts hold for hyperimaginaries:

(1) Let $\pi(x, a)$ be an E type over a that does not fork over b and let $I = \langle a_i | i < \omega \rangle$ be some b-indiscernible sequence with $a_0 = a$. Then there is a hyperimaginary $c \models \pi(x, a)$ such that I is bc-indiscernible and $c \mathop{\smile\!\!\!\!\!\!|}_b I$.

(2) There do not exist E types $\pi_0(x, c), \pi_1(x, d_0)$ and an a-indiscernible sequence $\langle d_n | n \in \mathbb{Z} \rangle$ such that for each $i > 0$, $cd_1 \equiv_a cd_i$ and $\pi_0(x, c) \wedge \pi_1(x, d_1)$ forks over a, while for $j \leq 0$, $cd_0 \equiv_a cd_j$ and $\pi_0(x, c) \wedge \pi_1(x, d_j)$ does not fork over a.

(3) Suppose that $a \mathop{\smile\!\!\!\!\!\!|}_c b$, $a \mathop{\smile\!\!\!\!\!\!|}_c b'$ and $b \equiv^L_c b'$. If $\pi_0(x, a) \wedge \pi_1(x, b)$ does not fork over c, then neither does $\pi_0(x, a) \wedge \pi_1(x, b')$.

Theorem 4.2.17 *(Type amalgamation for Lascar types of hyperimaginaries) Suppose a, b_i, d_i $(i = 1, 2)$ satisfy*

(1) $b_1 \mathop{\smile\!\!\!\!\!\!|}_a b_2$, $d_i \mathop{\smile\!\!\!\!\!\!|}_a b_i$, and

(2) $\text{Ltp}(d_1/a) = \text{Ltp}(d_2/a)$.

Then there is d such that $d \mathop{\smile\!\!\!\!\!\!|}_a b_1 b_2$ and $\text{Ltp}(d/ab_i) = \text{Ltp}(d_i/ab_i)$.

Theorem 4.2.18 *(Type amalgamation over a model) Suppose that a model M and hyperimaginaries b_i, d_i $(i = 1, 2)$ satisfy*

(1) $b_1 \mathop{\smile\!\!\!\!\!\!|}_M b_2$, $d_i \mathop{\smile\!\!\!\!\!\!|}_M b_i$, and

(2) $\text{tp}(d_1/M) = \text{tp}(d_2/M)$.

Then there is $d \mathop{\smile\!\!\!\!\!\!|}_M b_1 b_2$ such that $d \equiv^L_{Mb_i} d_i$.

The following is analogous to 3.2.11.

Corollary 4.2.19

(1) Let $a_0 \equiv^L_b a_1$ and $a_0 \mathop{\smile\!\!\!\!\!\!|}_b a_1$. Then there is Morley $\langle a_i | i < \omega \rangle$ of $\text{tp}(a_0/b)$.

(2) $d_c(x, y) \leq 2$ iff $x \equiv^L_c y$ (hence is hyperdefinable over c).

(3) $a \equiv^L_c b$ iff $a \equiv_{\text{bdd}(c)} b$. Hence we identify $\text{Ltp}(a/c)$ as $\text{tp}(a/\text{bdd}(c))$.

Proof We only need to prove (3). Assume $a \equiv^L_c b$. As pointed out in 4.1.4(2), for any model M with $c \in \text{dcl}(M)$, it follows $\text{bdd}(c) \subseteq \text{dcl}(M)$. Hence from the definition of Lascar types, we have $a \equiv_{\text{bdd}(c)} b$.

Conversely if $a \equiv_{\text{bdd}(c)} b$, then due to (2), $a/\equiv^L_c = b/\equiv^L_c \in \text{bdd}(c)$, whence $a \equiv^L_c b$. ■

As explained prior to 4.1.4, if $F(x, y)$ hyperdefines $x_E \equiv^L_c y_E$, then each F class is equivalent over c to a hyperimaginary.

4.3 Canonical Bases

In this section T *is simple*, and $a, b, \ldots, A, B, \ldots$, are hyperimaginaries which we may simply call tuples. We aim to prove the existence of the canonical base of an amalgamation base.

Definition 4.3.1 We say a complete hyperimaginary type $p(x)$ over A is an *amalgamation base* if type amalgamation holds for p; i.e., given any $B_0, B_1 \supseteq A$ with $B_0 \mathop{\smile\!\!\!\!\!\!\;|}_A B_1$ and any non-forking extensions $q_i(x) \in S(B_i)$ of p, $q_0 \cup q_1$ does not fork over A.

Remark 4.3.2 Any type over a model or a boundedly closed set, i.e., any Lascar type is an amalgamation base. Conversely if $\mathrm{tp}(a/A)$ is an amalgamation base then since $\mathrm{bdd}(A) \mathop{\smile\!\!\!\!\!\!\;|}_A \mathrm{bdd}(A)$, we have $\mathrm{tp}(a/A) \equiv \mathrm{tp}(a/\mathrm{bdd}(A))$.

If T is stable then a complete type is an amalgamation base iff it is stationary.

Definition 4.3.3 Let p be an amalgamation base over A.

(1) By a *canonical base* of p, denoted by $\mathrm{Cb}(p)$, we mean a hyperimaginary such that, for

$$S = \left\{B \subseteq A \mid p \text{ does not fork over } B, \text{ and } p\lceil B \text{ is an amalgamation base}\right\},$$

$\mathrm{Cb}(p) \in S$, and $\mathrm{Cb}(p) \subseteq B$ for all $B \in S$.

(2) We say p a *canonical type* if $p \equiv p\lceil\mathrm{Cb}(p)$.

(3) By a *parallelism class* of p, denoted by $\mathcal{P}_p$ (or simply $\mathcal{P}$ when p is clear), we mean a collection of amalgamation bases generated from p via taking non-forking extensions or unforked restrictions.

Assume temporarily any amalgamation base has a canonical base (which indeed is the case, Proposition 4.3.8). The following are direct consequences from the definition.

Lemma 4.3.4

(1) Clearly,

$$\mathcal{P} = \left\{q \mid q \text{ is an amalgamation base, and there are amalgamation bases } r_0, \ldots, r_n \text{ such that } p = r_0, \ldots, r_n = q \text{ and for } i < n, r_i, r_{i+1} \text{ are parallel}\right\}.$$

Thus for $q \in \mathcal{P}$, $\mathcal{P} = \mathcal{P}_q$.

(2) The canonical base of p is uniquely determined (up to interdefinability) by $\mathcal{P}$. Namely, for every $q \in \mathcal{P}$, $\mathrm{Cb}(p)$ and $\mathrm{Cb}(q)$ are interdefinable, and the canonical types $p\lceil\mathrm{Cb}(p)$, $q\lceil\mathrm{Cb}(q)$ are equivalent. Hence,

$$\mathcal{P} = \{q \mid q \text{ is an amalgamation base, and a non-forking extension of } p\lceil\mathrm{Cb}(p)\}.$$

(3) For each $q \in \mathcal{P}$, $\mathrm{Cb}(p) \subseteq \mathrm{dom}(q)$. Moreover, if p does not fork over $B(\subseteq A)$, then $\mathrm{Cb}(p) \subseteq \mathrm{bdd}(B)$.

(4) For an automorphism f, f fixes $\mathrm{Cb}(p)$ pointwise iff it fixes $\mathcal{P}$ setwise iff it fixes (the set of realizations of) $p\lceil\mathrm{Cb}(p)$ setwise.

Proof We sketch the proof. Let $p = \text{tp}(a/A)$. For (2), first let q in $\mathcal{P}$ be a non-forking extension or restriction of p. Then the equivalence of $\text{Cb}(p), \text{Cb}(q)$ easily comes from forking transitivity and the smallestness of the canonical base in S (4.3.3(1)). It follows that $p\lceil\text{Cb}(p) \equiv q\lceil\text{Cb}(q)$, as well. By iterating the argument, indeed we get theses properties for any $q \in \mathcal{P}$. For (3), if p does not fork over $B(\subseteq A)$, then $a \mathop{\smile\!\!\!\!\!\!\mid}_{\text{bdd}(B)} \text{bdd}(A)$ and so the amalgamation base $\text{tp}(a/\text{bdd}(B)) \in \mathcal{P}$. Hence, (3) follows. Lastly, (4) comes from (2). ■

The following observation will be useful.

Remark 4.3.5 Let $p(x, a) \in S(a)$. For each $\varphi_i(x_i, y_i) \in \mathcal{L}, x_i \subseteq x$ and $k_i \in \omega$, let $n_i = D_{\varphi_i,k_i}(p(x, a))$. Then for $b \models \text{tp}(a)$, $p(x, a)||p(x, b)$; i.e., they are parallel iff for each φ_i, k_i, $D_{\varphi_i,k_i}(p(x, a) \cup p(x, b)) \geq n_i$. Hence, $p(x, a)||p(x, b)$ is a hyperdefinable relation on $\text{tp}(a)$:

($\Rightarrow$) Clear from 4.2.12.

($\Leftarrow$) Note first that due to 2.3.4(5), $p(x, a) \cup p(x, b)$ does not fork over a. Hence, by extension, there is a completion $q'(x, ab)$ of $p(x, a) \cup p(x, b)$ that does not fork over a. Then q' has the same D_{φ_i,k_i}-ranks with $p(x, a)$ and so with $p(x, b)$. Therefore, q' does not fork over b either; i.e., q' is a common non-forking extension of $p(x, a)$ and $p(x, b)$.

As similarly in 4.2.13(1), $D_{\varphi_i,k_i}(p(x, a) \cup p(x, b)) \geq n_i$ is expressed by a type.

We point out an important criterion verifying the type definability of the transitive closure of a symmetric relation.

Proposition 4.3.6 *Suppose that R is a hyperdefinable relation on a partial E-type $\pi(x)$ (say over $\emptyset$) such that*

(a) R is symmetric;

(b) For a hyperimaginary $u \models \pi(x)$, there is $u' \models \pi$ such that $R(u, u')$ holds;

(c) For hyperimaginaries $u, v, w \models \pi(x)$, if $R(u, v), R(v, w)$ and $u \mathop{\smile\!\!\!\!\!\!\mid}_v w$, then $R(u, w)$ holds.

Now, let $\tilde{R}$ be the transitive closure of R. Then for $a, b \models \pi$, the following are equivalent:

(1) $\tilde{R}(a, b)$.

(2) For some $c \models \pi$, $R(c, a)$ and $R(c, b)$.

(3) There is $c \models \pi$ such that $c \mathop{\smile\!\!\!\!\!\!\mid}_b a$, $c \mathop{\smile\!\!\!\!\!\!\mid}_a b$, and $R(c, a), R(c, b)$.

Hence, $\tilde{R}$ is a hyperdefinable equivalence relation on $\pi(x)$ (coarser than E, so we can see $\pi(x)$ as an $\tilde{R}$-type, too).

Proof It suffices to show that (1) implies (3). Assume that $\tilde{R}(a, b)$ holds. Now, let $\{(\varphi_i(x_i, y_i), k_i)|\ i < \alpha\}$ be a list of all pairs, where $\varphi_i \in \mathcal{L}$, finite $x_i \subseteq x$ and $k_i \in \omega$. Define $n_i = n_i^a$ inductively as $n_i = \max\{D(c/a, \varphi_i(x_i, y_i), k_i)$:

$$c \models \pi, R(c, a) \text{ and } D(c/a, \varphi_j(x_j, y_j), k_j) \geq n_j \text{ for } j < i\}.$$

(See the convention of terminology for ranks in Chapter 2.) Since $D(z/a, \varphi_j(x_j, y_j), k_j) \geq n_j$ is type-definable (4.2.13(1)), each $n_i (\geq 0$ by (b)) is well-defined. Now, on $\pi(x)$, put

$$R_1^a(x, y) = R_1(x, y) \equiv R(x, y) \wedge \bigwedge_i D(x/y, \varphi_i, k_i) \geq n_i^a.$$

Then R_1 is a hyperdefinable relation. Hence, by compactness and (b), there is c' such that $R_1(c', a)$. (†).

Claim 1 For each i, $n_i^a = n_i^b$, and hence $R_1^a \equiv R_1^b$: It suffices to show the case when $\models R(a, b)$, for the general case follows by iteration. By the extension axiom, there is $c_a \models \pi$ such that $c_a \mathop{\smile\!\!\!\!\!\mid}_a b$ and $R_1(c_a, a)$. Then, by (c), $R(c_a, b)$ holds. Hence,

$$D(c_a/b, \varphi_i, k_i) \geq D(c_a/ab, \varphi_i, k_i) = D(c_a/a, \varphi_i, k_i) \geq n_i^a.$$

By maximality, $n_i^b \geq D(c_a/b, \varphi_i, k_i)$. Similarly we can show that $n_i^a \geq n_i^b$ by choosing c_b such that $c_b \mathop{\smile\!\!\!\!\!\mid}_b a$ and $R_1^b(c_b, a)$. Claim 1 is proved.

Claim 2 There is $d \models \pi$ such that $d \mathop{\smile\!\!\!\!\!\mid}_b a$, $d \mathop{\smile\!\!\!\!\!\mid}_a b$, and $R_1(d, a), R_1(d, b)$; so this shows (3): We show this by induction on the length of R-chain between a and b. The above (†) gives the induction step for length 0 (although $R(a, a)$ may not hold). Assume now that there is b_n such that a, b_n are related by an R-chain of length n, and that $R(b_n, b)$ holds. By the induction hypothesis for n, there is d_n such that $d_n \mathop{\smile\!\!\!\!\!\mid}_{b_n} a$, $d_n \mathop{\smile\!\!\!\!\!\mid}_a b_n$ and $R_1(d_n, a), R_1(d_n, b_n)$ (*). By extension, we can assume $d_n \mathop{\smile\!\!\!\!\!\mid}_{ab_n} b$ (**). Then, from (*), $d_n \mathop{\smile\!\!\!\!\!\mid}_{b_n} ab$ (***). Hence by (c), $R(d_n, b)$ holds. Thus according to Claim 1 and (***),(*),

$$n_i \geq D(d_n/b, \varphi_i, k_i) \geq D(d_n/b_n b, \varphi_i, k_i) = D(d_n/b_n, \varphi_i, k_i) \geq n_i.$$

Hence, both $R_1(d_n, b)$ and $d_n \mathop{\smile\!\!\!\!\!\mid}_b b_n$ hold. Then it follows from (***) that $d_n \mathop{\smile\!\!\!\!\!\mid}_b a$. Also from (*),(***), we have $d_n \mathop{\smile\!\!\!\!\!\mid}_a b$. The proof is completed. (Note that as in 4.1.2(2), we can assume $\tilde{R}$ is an equivalence relation on the whole universe.) ■

We are ready to prove our *goal*, the existence of the canonical base. For the rest of this section, fix an amalgamation base $p(x, u)$ over u. Let R be the reflexive and symmetric *parallelism relation* on $\text{tp}(u)$; i.e., for $b, c \models \text{tp}(u)$, $R(b, c)$ holds iff $p(x, b) || p(x, c)$. By 4.3.5, R is hyperdefinable.

Remark 4.3.7 If $R(a, c), R(c, b)$ and $a \mathop{\smile\!\!\!\!\!\mid}_c b$, then $R(a, b)$ holds: Now, there are $q_1(x, ac)$, $q_2(x, bc)$ common non-forking extensions of $p(x, a), p(x, c)$ and $p(x, b), p(x, c)$, respectively. Then since $p(x, u)$ is an amalgamation base and $a \mathop{\smile\!\!\!\!\!\mid}_c b$, there is $d \models q_1(x, ac) \cup q_2(x, ab) \supseteq p(x, a) \cup p(x, b) \cup p(x, c)$ such that $d \mathop{\smile\!\!\!\!\!\mid}_c ab$, and clearly $d \mathop{\smile\!\!\!\!\!\mid}_a c$, $d \mathop{\smile\!\!\!\!\!\mid}_b c$. Hence, $d \mathop{\smile\!\!\!\!\!\mid}_{ca} b$, $d \mathop{\smile\!\!\!\!\!\mid}_a b$ and similarly we have $d \mathop{\smile\!\!\!\!\!\mid}_b a$; i.e., $R(a, b)$ holds.

Therefore from 4.3.6, the transitive closure $\tilde{R}$ of R is a type-definable equivalence relation. Let $e = u/\tilde{R}$, The following proposition shows that $e = \text{Cb}(p(x, u))$.

Proposition 4.3.8

(1) *For* $a, b \models \mathrm{tp}(u/e)$, *if* $a \downarrow_e b$ *then* $R(a, b)$.

(2) $p(x, u)$ *does not fork over* e.

(3) $p(x, u)\lceil e$ *is an amalgamation base.*

(4) *If* $p(x, u)$ *does not fork over* c *with* $c \subseteq u$, *and* $p(x, u)\lceil c$ *is an amalgamation base, then* $e \subseteq c$.

Proof (1) Since $\tilde{R}(a, b)$, due to 4.3.6 there is c such that $R(c, a)$, $R(c, b)$, and $c \downarrow_a b$. Then since $a \downarrow_e b$ and $e = c/\tilde{R} \in \mathrm{dcl}(c)$, $a \downarrow_c b$ holds. Then from 4.3.7, $R(a, b)$.

(2) By extension, there is $v \models \mathrm{tp}(u/e)$ such that $v \downarrow_e u$. Hence, by (1), $p(x, v)$ and $p(x, u)$ have a common non-forking extension realized by, say d. So $d \downarrow_v u$, and $d \downarrow_e u$. Hence, $p(x, u)$ does not fork over e.

(3) First, we claim the following:

Claim Let $q(x, b)$ $(e \subseteq b)$ be a non-forking extension of $p(x, u)\lceil e$. For $v \models \mathrm{tp}(u/e)$, if $v \downarrow_e b$, then $p(x, v)$ and $q(x, b)$ are parallel: Let $d \models q(x, b)$, $d' \models p(x, u)$. Then since $d \equiv_e d'$, there is u' such that $du' \equiv_e d'u$. Thus, $d \downarrow_e u'$. Moreover, by extension, we can assume $u' \downarrow_{de} bv$. Hence, $u' \downarrow_e dbv$. Now since $d \downarrow_e b$, we have $d \downarrow_e bu'$. Hence, $q(x, b), p(x, u')$ have a common non-forking extension realized by d. Moreover, due to (1), $p(x, v)$ and $p(x, u')$ are parallel as well. Then from $b \downarrow_e u'v$, $b \downarrow_{u'} v$ and that $p(x, u')$ is an amalgamation base, there is $d'' \models p(x, u') \cup p(x, v) \cup q(x, b)$ such that $d'' \downarrow_{u'} bv$ and $d'' \downarrow_e bv$. Therefore, the claim follows.

Now, suppose that each $q_i(x, b_i)$ $(e \subseteq b_i,\ i = 1, 2)$ is a non-forking extension of $p(x, u)\lceil e$, and $b_1 \downarrow_e b_2$. There is $v \models \mathrm{tp}(u/e)$ such that $v \downarrow_e b_1 b_2$. By the claim, $p(x, v)$ and $q_i(x, b_i)$ have a common non-forking extension, say $q_i'(x, vb_i)$. Then since $b_1 \downarrow_v b_2$ and $p(x, v)$ is an amalgamation base, q_1' and q_2' (and so q_1 and q_2) have a common non-forking extension, as well. We have proved (3).

(4) Let f be an automorphism fixing c. Let $u' = f(u)$. By extension, there is $v \models \mathrm{tp}(u/c)$ such that $v \downarrow_c uu'$. Then since $p(x, u)\lceil c$ is an amalgamation base, both $R(v, u)$ and $R(v, u')$ hold. Hence, $\tilde{R}(u, u')$ and e is fixed by f; i.e., $e \subseteq c$. ■

We finish this section by stating several remarks related to canonical bases. The notion of canonical bases will be used almost everywhere throughout the rest of this book. If the reader wants to see more practical uses of canonical bases, s/he may jump ahead to Chapter 7.

Remark 4.3.9

(1) Let $p_e(x) = p(x, u)\lceil e$, and let f be an automorphism. Then $f(e) = e$ iff for some $v \equiv u$, $p(x, u) \| p(x, v)$ and $p(x, f(u)) \| p(x, v)$ iff $p_e(x) \wedge f(p_e(x))$ does not fork over e. If $p_e(x) \wedge f(p_e(x))$ does not fork over e, then by the $D_{\psi,k}$-rank argument, it does not fork over $f(e)$ either. Hence, $p_e(x) \| f(p_e(x))$ and $f(p_e(x)) \in \mathcal{P}$,

so by 4.3.4(4), $f(e) = e$. The other directions are obvious or a consequence of Proposition 4.3.6.

(2) It is *not* known whether any amalgamation base $\mathrm{tp}(c/A)$ with *imaginary* tuples c, A has an imaginary canonical base. As mentioned previously, this is the main reason we work with hyperimaginaries in simple theories. But as is well known, if T is stable, any stationary (hyper)imaginary type $p(x)$ has its imaginary canonical base e; moreover, for any automorphism f, $f(e) = e$ iff $p(x) \| f(p(x))$. In general, this does not hold in simple T (compare with (1)). Consider the random graph with the edge relation E. Given elements $a \neq b$, there is an automorphism f sending a to b. Choose c distinct from a, b such that $E(c, a) \wedge \neg E(c, b)$. Certainly $\emptyset$ is the canonical base of $q(x) := \mathrm{tp}(c/ab)$, but $q \cup f(q)$ is inconsistent.

(3) Given $\mathrm{tp}(a/A)$, $\mathrm{Cb}(a/A)$ denotes $\mathrm{Cb}(\mathrm{Ltp}(a/A)) = \mathrm{Cb}(\mathrm{tp}(a/\operatorname{bdd}(A)))$.

(4) $\mathrm{Cb}(a/A) \in \operatorname{dcl}(aA)$: If an automorphism f fixes aA, then it fixes the solution set of $\mathrm{Ltp}(a/A)$; hence, by (1) it fixes $\mathrm{Cb}(a/A)$.

(5) Let $\langle a_i |\ i \leq \omega \rangle$ be c-indiscernible. Then $\mathrm{tp}(a_\omega/cI)$, where $I = \langle a_i |\ i < \omega \rangle$ is an amalgamation base. It suffices to show that for any $a'_\omega \equiv_{cI} a_\omega$, we have $d_{cI}(a'_\omega, a_\omega) = 1$. But already $d_{ca_0 \ldots a_n}(a'_\omega, a_\omega) = 1$, so by compactness it follows.

Hence, given Morley $\langle a_i | i < \omega \rangle$ in $\mathrm{tp}(a/A)$, by 2.4.2 (which is true for hyperimaginaries), $\mathrm{Cb}(a/A) \in \operatorname{dcl}(a_i | i < \omega)$; if, in addition, T is supersimple, then according to (4), $\mathrm{Cb}(a/A) \in \operatorname{dcl}(a_i | i < k)$ for some $k < \omega$.

(6) Let a and a sequence of hyperimaginaries $b = \langle b_i | i \in I \rangle$ be equivalent. For any C, it follows that $e = \mathrm{Cb}(a/C)$ and $e' = \langle e_J | J (\subseteq I) \text{ finite} \rangle$, where $e_J = \mathrm{Cb}(b_J/C)$ and $b_J = \langle b_i | i \in J \rangle$ are interdefinable. Since $a \mathop{\smile\!\!\!\!\!\!\;|}_e C$; $b_J \mathop{\smile\!\!\!\!\!\!\;|}_e C$ and $\mathrm{tp}(a/e)$ so $\mathrm{tp}(b_J/e)$ are Lascar types, $e' \subseteq e$. Conversely, since $b_J \mathop{\smile\!\!\!\!\!\!\;|}_{e'} C$, by finite character $a \mathop{\smile\!\!\!\!\!\!\;|}_{e'} C$. By applying compactness to 4.2.19(2) and that $e_{J_1} \ldots e_{J_n} \subseteq e_{J_1 \cup \cdots \cup J_n}$, it follows that $\mathrm{tp}(a/e') = \mathrm{Ltp}(a/e')$. Hence, $e \subseteq e'$.

(7) Let both c and c_E be hyperimaginaries (see the remark before 4.1.5). Now let $F(x, y)$ be $E(x, y) \wedge x \equiv^L y$ (x, y correspond to c). Then c_F is hyperdefined by $x \equiv^L_{c_E} c$. We claim that $c_F = \mathrm{Cb}(c/c_E)$. We use 4.2.19 freely. Let $e = \mathrm{Cb}(c/c_E)$. Since F is finer than E, clearly $c_E \in \operatorname{dcl}(c_F)$. Hence, $c \mathop{\smile\!\!\!\!\!\!\;|}_{c_F} c_E$. Also $c_F \in \operatorname{bdd}(c_E)$. Thus, by 4.2.19(3) $\mathrm{tp}(c/c_F)$ is a Lascar type, so $e \in \operatorname{dcl}(c_F)$. Conversely, note now $c \mathop{\smile\!\!\!\!\!\!\;|}_e c_E$. Since $c_E \in \operatorname{dcl}(c)$, we have $c_E \in \operatorname{bdd}(e)$, so $c_F \in \operatorname{bdd}(e)$, too. Then $\mathrm{tp}(c/e) \equiv \mathrm{tp}(c/\operatorname{bdd}(e)) \models \mathrm{tp}(c/c_F) \models F(x, c)$. Therefore, $c_F \in \operatorname{dcl}(e)$.

4.4 Forking Relationships between Types

This section introduces some technical but important notions and lemmas necessary for the theoretical developments of simplicity theory. We work with hyperimaginary types in **simple** T.

4.4.1 *Regular Types and Weight*

Definition 4.4.1

(1) We say $p \in S(A)$ and $q \in S(B)$ are *almost orthogonal* (write $p \perp^a q$) if whenever a realizes a non-forking extension of p to AB and b realizes a non-forking extension of q to AB then $a \downarrow_{AB} b$.

(2) p and q are *orthogonal* (write $p \perp q$) if whenever p' is a non-forking extension of p and q' is a non-forking extension of q then $p' \perp^a q'$.

(3) p is a *regular type* if p is not bounded and orthogonal to all of its forking extensions.

(4) T is said to be *unidimensional* if any two complete unbounded types are non-orthogonal.

Remark 4.4.2

(1) If $SU(p) = \omega^\alpha$, then $p \perp q$ for any type q with $SU(q) < \omega^\alpha$: This easily follows by the Lascar inequality 2.5.19(4).
Hence any complete type of SU-rank ω^α is regular.

(2) Let $a \downarrow_A B$ with $A \subseteq B$. Then $\mathrm{tp}(a/A)$ is regular iff $\mathrm{tp}(a/B)$ is regular: $(\Rightarrow)$ By the definition. $(\Leftarrow)$ Suppose not, say $a_1 \downarrow_A C_1$; $a_1' \not\downarrow_A C_0$; $a_1' \downarrow_{C_0} C_1$; but $a_1' \not\downarrow_{C_1} a_1$, where $A \subseteq C_0 \subseteq C_1$ and both a_1 and $a_1' \models \mathrm{tp}(a/A)$. There is no harm to assume $A = \mathrm{bdd}(A)$ and $B \downarrow_A C_1$. By type-amalgamation, there is $a_0 \models \mathrm{tp}(a/B) \cup \mathrm{tp}(a_1/C_1)$ such that $a_0 \downarrow_A BC_1$; and there is a_0' such that $a_0 a_0' \equiv_{C_1} a_1 a_1'$ and by extension $a_0 a_0' \downarrow_{C_1} B$. This altogether violates $\mathrm{tp}(a/B)$ being regular.

(3) Assume $a \downarrow_A B$ with $A \subseteq B$. Clearly, $\mathrm{tp}(a/A) \perp q$ implies that $\mathrm{tp}(a/B) \perp q$. But differently from stable theories, the converse does not hold in the random graph (V, R). Choose distinct points $a, b, c, d \in V$ such that aRb; dRc; $\neg aRc$ hold. Now $a \downarrow_b c$, but $\mathrm{tp}(a/b) \not\perp \mathrm{tp}(d/c)$, whereas $\mathrm{tp}(a/bc) \perp \mathrm{tp}(d/c)$.
In general, for $q = \mathrm{tp}(b/C)$, if $B \downarrow_A C$ and $\mathrm{tp}(a/B) \perp q$, then using type-amalgamation as before one can show that $\mathrm{tp}(a/A) \perp q$ holds too.

(4) By basic forking axioms, we have the following: If $\mathrm{tp}(e_0/e) \perp p$ then $\mathrm{tp}(e'/e) \perp p$ holds for any $e' \in \mathrm{bdd}(e_0 e)$. If $\mathrm{tp}(e_0/e) \perp p$, $\mathrm{tp}(e_1/e) \perp p$, and $e_0 \downarrow_e e_1$, then $\mathrm{tp}(e_0 e_1/e) \perp p$.

The notion of regular type is related to that of (pre)geometry.

Definition 4.4.3 Let S be a set. If an operation $\mathrm{cl} : \mathcal{P}(S) \to \mathcal{P}(S)$ satisfies the following properties, then we say that (S, cl) is a *pregeometry*.

(1) For $A \subseteq S$, $A \subseteq \mathrm{cl}(A) = \mathrm{cl}(\mathrm{cl}(A))$.

(2) For $A \subseteq B \subseteq S$, $\mathrm{cl}(A) \subseteq \mathrm{cl}(B)$.

(3) For $A \subseteq S$ and $a, b \in S$, if $a \in \mathrm{cl}(Ab) \setminus \mathrm{cl}(A)$, then $b \in \mathrm{cl}(Aa)$.

(4) If $a \in \mathrm{cl}(A)$, then $a \in \mathrm{cl}(A_0)$, for some finite $A_0 \subseteq A$.

Let $(S, \mathrm{cl}), (S_1, \mathrm{cl}_1)$ be pregeometries. Obviously, a bijection $f : S \to S_1$ is said to be an *isomorphism* between them if $\mathrm{cl}_1(f(A)) = f(\mathrm{cl}(A))$ for $A \subseteq S$. An isomorphism on (S, cl) itself is called an *automorphism*. We simply call (S, cl) a *geometry* if $\mathrm{cl}(\emptyset) = \emptyset$ and $\mathrm{cl}(a) = \{a\}$ for all $a \in S$. A canonical geometry $(\hat{S}, \hat{\mathrm{cl}})$ is associated with (S, cl), where $\hat{S} = \{\mathrm{cl}(a) | a \in S \setminus \mathrm{cl}(\emptyset)\}$, a partition of $S \setminus \mathrm{cl}(\emptyset)$, and for $A \subseteq S$ we put $\hat{A} = \{\mathrm{cl}(a) | a \in A\}$, and $\hat{\mathrm{cl}}(\hat{A}) = \{\mathrm{cl}(b) | b \in \mathrm{cl}(A)\}$. We say $(S, \mathrm{cl}), (S_1, \mathrm{cl}_1)$ are *equivalent* if $(\hat{S}, \hat{\mathrm{cl}})$ and $(\hat{S_1}, \hat{\mathrm{cl}_1})$ are isomorphic.

Some more terminologies for (S, cl): We say that $A(\subseteq S)$ is *independent* if $a \notin \mathrm{cl}(A \setminus \{a\})$ for all $a \in A$. Given $B \subseteq S$, a subset B_0 is said to be a *basis* for B if $B \subseteq \mathrm{cl}(B_0)$ and B_0 is independent. It follows, in particular from (4) and (3), that there exists a basis, and any two bases for B have the same cardinality, denoted by $\dim(B)$. Any $A \subseteq S$ gives a *localized* pregeometry on S defined by $\mathrm{cl}_A(B) = \mathrm{cl}(A \cup B)$, and a notion of dimension *over* A ($\dim(-/A)$). For $A, B, C \subseteq S$, if $\dim(A'/C) = \dim(A'/B \cup C)$ for any finite $A' \subseteq A$, then we say that A is *independent from* B *over* C (in the sense of the pregeometry). Clearly, independence and dimension notions are carried to its canonical geometry. In particular A is independent from B over C iff $\hat{A}$ is independent from $\hat{B}$ over $\hat{C}$. In Chapter 8, we shall further study related issues on pregeometries.

Remark 4.4.4 Let $p \in S(A)$ and let $\mathcal{D}$ be the solution set of p. Define a closure for $\mathcal{B}$, a subset of $\mathcal{D}$ as $a(\in \mathcal{D}) \in \mathrm{cl}(\mathcal{B})$ iff $a \not\perp_A B_0$ for some small subset B_0 of $\mathcal{B}$. Then p is regular iff $(\mathcal{D}, \mathrm{cl})$ forms a pregeometry: ($\Rightarrow$) 4.4.3(2)(3)(4) easily follow from finite character and symmetry. For (1), assume $a_i \not\perp_A B$ and $b \not\perp_A Ba_1 \ldots a_n$ for a_i $(i = 1, \ldots, n)$, $b \in D$, and small $B \subseteq D$. We want to see $b \not\perp_A B$. If not, then by transitivity there must exist smallest $1 \leq k \leq n$ such that $b \not\perp_{ABa_1 \ldots a_{k-1}} a_k$. Hence, $\mathrm{tp}(b/ABa_1 \ldots a_{k-1})$ is a non-forking extension of p, whereas $\mathrm{tp}(a_k/ABa_1 \ldots a_{k-1})$ is a forking extension of p. Therefore, p cannot be regular, a contradiction.

($\Leftarrow$) If p is not regular, then there are B (which need not be a subset of $\mathcal{D}$) and $a, b \models p$ such that $a \perp_A B$, $b \not\perp_A B$, and $a \not\perp_{AB} b$. Let $I = \langle a_i b_i | \, i < \omega \rangle$ be a sequence in $\mathcal{D}$ such that Iab forms a Morley sequence of $\mathrm{tp}(ab/AB)$. Then due to 2.4.2, we have $ab \perp_I AB$. By basic forking calculation, it easily follows that $a \perp_A I$, $b \not\perp_A I$ and $a \not\perp_A Ib$. Hence, $\mathrm{cl}(\mathrm{cl}(I)) \neq \mathrm{cl}(I)$.

Proposition 4.4.5 *Let p be an unbounded type. If T is supersimple or $SU(p) < \infty$, then p is not orthogonal to a regular type of SU-rank $< \infty$.*

Proof Let $q \in S(B)$ be an unbounded type having minimal SU-rank such that $q \not\perp p \in S(A)$. Since $p \not\perp p$, such q exists, and either case we can assume that $SU(q) < \infty$ (if T is supersimple, choose q among types of finite real tuples). We claim q is regular: Suppose not, say $q \not\perp q'$ for some forking extension $q' \in S(C_1)$ of q. Now there is no harm to assume $B = \mathrm{bdd}(B)$. Suppose $q \not\perp p$ is witnessed over $C_0 \supseteq A \cup B$ with $a_0 \not\perp_{C_0} b_0$, where $\mathrm{tp}(a_0/C_0)$ and $\mathrm{tp}(b_0/C_0)$ are, respectively, non-forking extensions of p and q. Also $q \not\perp q'$ is witnessed over C_2 with $b_1 \not\perp_{C_2} b'_1$, where $\mathrm{tp}(b_1/C_2)$ and

$\mathrm{tp}(b'_1/C_2)$ are, respectively, non-forking extensions of q and q'. We can assume that $C_0 \mathop{\smile\!\!\!\!\!\mid}_B C_2$. Then by type-amalgamation, there is $b \models \mathrm{tp}(b_0/C_0) \cup \mathrm{tp}(b_1/C_2)$ such that $b \mathop{\smile\!\!\!\!\!\mid}_B C_0C_2$. Now, there are a, b' such that $ab \equiv_{C_0} a_0b_0$ and $bb' \equiv_{C_2} b_1b'_1$, and by extension we can even let $ab \mathop{\smile\!\!\!\!\!\mid}_{C_0} C_2$ and $bb' \mathop{\smile\!\!\!\!\!\mid}_{C_2} C_0$. In summary, a, b, b' are realizations of non-forking extensions of p, q, q', respectively, to $C = C_0C_2$ such that $a \not\mathop{\smile\!\!\!\!\!\mid}_C b$ and $b \not\mathop{\smile\!\!\!\!\!\mid}_C b'$ (*). Now, $SU(b'/C) = SU(q') < SU(q)$ as in 4.2.13(2). Thus, by minimality of q, it follows that $a \mathop{\smile\!\!\!\!\!\mid}_C b'$, and $a \not\mathop{\smile\!\!\!\!\!\mid}_{Cb'} b$. Then since $\mathrm{tp}(a/Cb')$ is a non-forking extension of p, again by minimality of q, we have $SU(q) = SU(b/C) \leq SU(b/Cb')$. This contradicts (*). ■

Definition 4.4.6

(1) The *preweight* of $p = \mathrm{tp}(a/A)$ denoted by $\mathrm{prwt}(p) = \mathrm{prwt}(a/A)$ is the supremum of $\{\kappa \text{ cardinal} \mid \text{there is } A\text{-independent } \{b_i | i < \kappa\} \text{ such that } a \not\mathop{\smile\!\!\!\!\!\mid}_A b_i \text{ for all } i < \kappa\}$.
The *weight* of p, written $\mathrm{wt}(p)$, is $\sup \mathcal{W}$ where $\mathcal{W} := \{\mathrm{prwt}(q) | \ q \text{ is a non-forking extension of } p\}$. We say the weight of p is *strictly* κ if $\mathrm{wt}(p) = \kappa = \max \mathcal{W}$.

(2) We say *a dominates b* over A, written $a \trianglerighteq_A b$, if for any c, $c \not\mathop{\smile\!\!\!\!\!\mid}_A b$ implies that $c \not\mathop{\smile\!\!\!\!\!\mid}_A a$; a, b are *equidominant* over A if $a \trianglerighteq_A b$ and $b \trianglerighteq_A a$; two complete types $p_i(x_i) \in S(A_i)$ $(i = 0, 1)$ are *equidominant* if for some non-forking extensions $q_i \in S(B)$ of p_i and $a_i \models p_i$, a_0, a_1 are equidominant over A.

Remark 4.4.7

(1) If $a = d/E$ with real d, by the similar proof of local character (4.2.11) applied to the remark before 2.2.9, one can see that $\mathrm{wt}(a/c) \leq |T| + |d|$. By the similar reason in 4.2.13(2) if T is supersimple then for a type p of a finitary hyperimaginary, $\mathrm{wt}(p) \leq \omega$ but it cannot be strictly ω. We shall see that indeed $\mathrm{wt}(p)$ should be finite (4.4.8, 4.4.12).
The rest below will follow without difficulty by the basic independence axioms.

(2) If $a \mathop{\smile\!\!\!\!\!\mid}_A B$ and $A \subseteq B$, then $\mathrm{prwt}(a/A) \leq \mathrm{prwt}(a/B)$, and $\mathrm{wt}(a/A) = \mathrm{wt}(a/B)$. We only point out the reason for the latter. By the definition $\mathrm{wt}(a/A) \geq \mathrm{wt}(a/B)$. Conversely if $a \mathop{\smile\!\!\!\!\!\mid}_A C$ with $A \subseteq C$ and $a \not\mathop{\smile\!\!\!\!\!\mid}_C c$ for each c in I which is C-independent, then we can assume $IC \mathop{\smile\!\!\!\!\!\mid}_{aA} B$ and it follows that $a \mathop{\smile\!\!\!\!\!\mid}_B C$, I is CB-independent, and $a \not\mathop{\smile\!\!\!\!\!\mid}_{BC} c$.

(3) Let $\{a_1, \ldots, a_n\}$ be A-independent. Then $\mathrm{wt}(a_1 \ldots a_n/A) = \mathrm{wt}(a_1/A) + \cdots + \mathrm{wt}(a_n/A)$: It suffices to show when $n = 2$. If $a_i \mathop{\smile\!\!\!\!\!\mid}_A B_i$ and $a_i \not\mathop{\smile\!\!\!\!\!\mid}_{B_i} c_i (\in I_i)$ $(i = 1, 2)$, then we can assume that $a_1I_1B_1 \mathop{\smile\!\!\!\!\!\mid}_A a_2I_2B_2$, and this shows that $\mathrm{wt}(a_1/A) + \mathrm{wt}(a_2/A) \leq \mathrm{wt}(a_1a_2/A)$. Conversely, assume $\kappa_i = \mathrm{wt}(a_i/A)$, $a_1a_2 \mathop{\smile\!\!\!\!\!\mid}_A B$, and AB-independent sequence I of length $\geq \kappa_1 + \kappa_2$ is given. Then I is decomposed as I_1I_2J such that $|I_i| \leq \kappa_i$ and $a_1 \mathop{\smile\!\!\!\!\!\mid}_B I_2J$; $a_2 \mathop{\smile\!\!\!\!\!\mid}_{Ba_1} J$. Hence, $a_1a_2 \mathop{\smile\!\!\!\!\!\mid}_B J$, and it follows that $\mathrm{wt}(a_1/A) + \mathrm{wt}(a_2/A) \geq \mathrm{wt}(a_1a_2/A)$.

(4) If $a \trianglerighteq_A b$ and $a \mathop{\smile\!\!\!\!\!\mid}_A B$, then $a \trianglerighteq_{AB} b$. If $a \trianglerighteq_A b$ and $c \mathop{\smile\!\!\!\!\!\mid}_A ab$ then $ca \trianglerighteq_A cb$. Hence, it follows if a_i, b_i $(i = 0, 1)$ are equidominant over A and $a_0b_0 \mathop{\smile\!\!\!\!\!\mid}_A a_1b_1$, then a_0a_1 and

b_0b_1 are equidominant over A. Moreover, if types p, q are equidominant, then they have the same weight. Two non-orthogonal types of weight 1 are equidominant.

(5) If $a_0 \rhd_A a_1$, then there is $C \supseteq A$ such that $C \mathop{\smile\!\!\!\!\!\!|}_A a_i$ $(i = 0, 1)$, and $a_0 \rhd_C a_0a_1$. Due to simplicity (e.g., 2.3.7), there must be some $C \supseteq A$ such that $a_i \mathop{\smile\!\!\!\!\!\!|}_A C$ and $\mathrm{tp}(a_0/Ca_1)$ has *no* forking extension, so for any d, if $a_i \mathop{\smile\!\!\!\!\!\!|}_A Cd$ then $a_0 \mathop{\smile\!\!\!\!\!\!|}_{Ca_1} d$. It follows that $a_0 \rhd_C a_0a_1$.

Theorem 4.4.8 *Let T be supersimple, and let $SU(p) < \infty$ (e.g., a type of finitary one). Then $\mathrm{wt}(p)$ is finite. More precisely, if $SU(p) = \sum_{j<\ell} \omega^{\alpha_j}.n_j$ in normal form, then $\mathrm{wt}(p) \leq \sum_{j<\ell} n_j$.*

Proof It suffices to show $\mathrm{prwt}(p) \leq \sum_{j<\ell} n_j$. Let $p = \mathrm{tp}(b)$ (for notational convenience we suppress the parameter). Clearly, $\mathrm{prwt}(b)$ cannot be strictly ω. Let independent $I = \langle a_i | i < m\rangle$ such that $b \not\mathop{\smile\!\!\!\!\!\!|} a_i$ be given. Due to supersimplicity we can suppose that $SU(a_i) < \infty$. Put $\alpha_{-1} = \alpha_0 + 1$, and $A_i = a_0 \ldots a_{i-1}$. Of course, $SU(b/A_i) > SU(b/A_{i+1})$. Now there is the maximal $k < \ell$ such that for all $i < m$, $SU(b/A_i) < SU(b/A_{i+1}) + \omega^{\alpha_{k-1}}$. Hence it follows that for all $i < m$, $SU(b/A_i) \geq \sum_{j<k} \omega^{\alpha_j}.n_j$ (*). Now put

$$I_k := \{a_i \in I |\ SU(a_i/A_i) \geq SU(a_i/A_ib) + \omega^{\alpha_k}\}.$$

Then by 2.5.19(4), we have $SU(b/A_i) \geq SU(b/A_{i+1}) + \omega^{\alpha_k}$ for all $a_i \in I_k$. It follows then from (*) that $|I_k| \leq n_k$.

Let $I' = I - I_k$ and we re-index I' as $\langle a'_i | i < m'\rangle$ for some $m' \leq m$, and also put $A'_i = a'_0 \ldots a'_{i-1}$ $(i < m')$. Then we easily get $SU(a'_i/A'_i) < SU(a'_i/A'_ib) + \omega^{\alpha_k}$. Then according to the contraposition of 2.5.19(4), for each $i < m'$, $SU(b/A'_i) < SU(b/A'_{i+1}) + \omega^{\alpha_k}$. Therefore, by choosing the new maximal $(k <)k' < \ell$ such that for all $i < m'$ $SU(b/A'_i) < SU(b/A'_{i+1}) + \omega^{\alpha_{k'-1}}$, we can repeat the above process, which must terminate in finitely many steps. Hence, $|I| = m \leq \sum_{j<\ell} n_j$ and the proof is completed. ■

Proposition 4.4.9 *Every regular type has weight* 1.

Proof According to 4.4.2(2) it suffices to show the preweight of a regular type is 1. Suppose not; say for a regular $\mathrm{tp}(a)$ (we suppress the parameter), there is independent $\{B, C\}$ such that $a \not\mathop{\smile\!\!\!\!\!\!|} B$ and $a \not\mathop{\smile\!\!\!\!\!\!|} C$. Let Ia be a Morley sequence of $\mathrm{tp}(a/B)$ where $I = \langle a_i | i < \omega\rangle$. By extension we can assume $I \mathop{\smile\!\!\!\!\!\!|}_{Ba} C$ so $I \mathop{\smile\!\!\!\!\!\!|}_B aC$ and $I \mathop{\smile\!\!\!\!\!\!|} C$ (*). Moreover, by 2.4.2, $a \mathop{\smile\!\!\!\!\!\!|}_I B$ holds; hence, $a \not\mathop{\smile\!\!\!\!\!\!|} I$. Due to finite character, there is the minimal n such that $a \not\mathop{\smile\!\!\!\!\!\!|} a_0 \ldots a_n$. Let $J = a_0 \ldots a_{n-1}$. Due to minimality $a \mathop{\smile\!\!\!\!\!\!|} J$, so $a_n \mathop{\smile\!\!\!\!\!\!|} J$ and $a_n \not\mathop{\smile\!\!\!\!\!\!|}_J a$ (**). Now $\mathrm{tp}(a/CJ) \perp \mathrm{tp}(a) = \mathrm{tp}(a_n)$ and from (*),(**), $a_n \mathop{\smile\!\!\!\!\!\!|} CJ$. Therefore, we must have $a \mathop{\smile\!\!\!\!\!\!|}_{CJ} a_n$. Then again by (*), $a \mathop{\smile\!\!\!\!\!\!|}_J a_n$, which contradicts (**). ■

Recall that a *product of types* $p_i \in S(A_i)$ $(i \in I)$ is some type of the form $\mathrm{tp}((a_i|i \in I)/A)$, where $(a_i|i \in I)$ is A-independent and each $\mathrm{tp}(a_i/A)$ is a non-forking extension of p_i.

Theorem 4.4.10 *Assume that* $\mathrm{wt}(p) \leq \omega$ *but not strictly* ω. *Then p is equidominant with some finite product of types of weight* 1. *Hence, p has finite weight. Moreover, if in T any type of weight* 1 *is non-orthogonal to a regular type, then p is equidominant with a finite product of regular types.*

Proof By suppressing the parameter we let $p = \mathrm{tp}(a)$. We claim the following first.

Claim 4.4.11 Given independent $\{c_0, \ldots, c_n\}$ $(0 \leq n)$ with $a \not\downarrow c_i$ for all $i \leq n$, there are $A \downarrow a$ and e with $\mathrm{wt}(e/A) = 1$ such that $Ae \downarrow c_0 \ldots c_{n-1}$, and $a \not\downarrow_A e$.

Proof Assume not. We inductively find sets $B_n \subseteq B_{n+1} \subseteq \ldots$ and additional tuples c_i, c'_i $(i > n)$ satisfying the following:

(i) $a \downarrow B_i$

(ii) $a \not\downarrow_{B_i} c_j$ $(j = 0, \ldots, i)$ and $a \not\downarrow_{B_i} c'_i$

(iii) $c_0 \ldots c_{i-1} \downarrow B_i c_i c'_i$ and $c_i \downarrow_{B_i} c'_i$.

We put B_n is the suppressed parameter of p so the empty set. Then, the above vacuously hold when $i = n$. Assume we have found $B_n \subseteq \ldots \subseteq B_i$ and tuples c_i, c'_i satisfying the above. We shall find $B_{i+1} \supseteq B_i$ and c_{i+1}, c'_{i+1} $(n < i+1)$. Due to simplicity (as in 4.4.7(5)), there is $B \supseteq B_i$ such that $B \downarrow a$; $Bc'_i \downarrow c_0 \ldots c_i$; and $\mathrm{tp}(c'_i/B)$ has no forking extension (†). Now by the induction hypothesis, $\{c_0, \ldots, c_i; c'_i\}$ is B-independent; $a \not\downarrow_B c_j$ $(j = 0, \ldots, i)$; and $a \not\downarrow_B c'_i$. Thus, if we put $A = B$ and $e = c'_i$, then Claim 4.4.11 holds as far as $\mathrm{wt}(c'_i/B) = 1$. Since we have denied the claim, $\mathrm{wt}(c'_i/B) > 1$, so there are $B_{i+1} (\supseteq B) \downarrow_B c'_i$ and c_{i+1}, c'_{i+1} with $c_{i+1} \downarrow_{B_{i+1}} c'_{i+1}$; $c'_i \not\downarrow_{B_{i+1}} c_{i+1}$ and $c'_i \not\downarrow_{B_{i+1}} c'_{i+1}$ (‡). By extension we can further assume that $B_{i+1}c_{i+1}c'_{i+1} \downarrow_{Bc'_i} ac_0 \ldots c_i$; thus, $B_{i+1} \downarrow_B c'_i a$ and $B_{i+1} \downarrow a$, we achieve (i) for $i+1$. Now by the second of (†), we have $c_0 \ldots c_i \downarrow B_{i+1} c'_i c_{i+1} c'_{i+1}$, too. Hence, (iii) for $i+1$ is achieved. Moreover, from (‡), we have $c'_i \not\downarrow_B B_{i+1} c_{i+1}$; $c'_i \not\downarrow_B B_{i+1} c'_{i+1}$; hence, by the choice of B as in (†), we must have $a \not\downarrow B_{i+1} c_{i+1}$ and $a \not\downarrow B_{i+1} c'_{i+1}$; (ii) for $i+1$ is obtained.

Now we take $B := \bigcup_{i>n} B_i$. Then $a \downarrow B$; $\{c_i|i < \omega\}$ is B-independent; and $a \not\downarrow_B c_i$ for all $i < \omega$. This contradicts that $\mathrm{wt}(a)$ is not strictly ω. The proof of Claim 4.4.11 is completed. ∎

Since we can assume p is unbounded, there are A, e as in the above Claim 4.4.11, so it follows that $a \rhd_A e$. Then since $\mathrm{wt}(a)$ is not strictly ω, there exist a set of tuples $\{e = e_0, \ldots, e_m\}$ and a set $A_0 \supseteq A$ satisfying that

$$a \downarrow_A A_0; \{e_0, \ldots, e_m\} \text{ is } A_0\text{-independent; for all } i \leq m,$$
$$a \not\downarrow_{A_0} e_i,\ \mathrm{wt}(e_i/A_0) = 1; \text{ and } a \rhd_{A_0} \bar{e}(= e_0 \ldots e_m);$$

and which are *maximal* (no $A_1 \supseteq A_0, \{e_0, \ldots, e_{m+1}\}$ can replace the above). Now as in 4.4.7(5), by extending A_0 if necessary, we can, in addition, assume that $a \rhd_{A_0} a\bar{e}$ (*).

We now claim that $\bar{e} \rhd_{A_0} a$ holds too. Suppose not; say there is some $e' \mathop{\smile\!\!\!\!\!\!|}_{A_0} \bar{e}$ with $e' \not\mathop{\smile\!\!\!\!\!\!|}_{A_0} a$. By applying Claim 4.4.11 to $\mathrm{tp}(a/A_0)$, there are $A_1(\supseteq A_0) \mathop{\smile\!\!\!\!\!\!|}_{A_0} a$ and e_{m+1} such that $\{e_0, \ldots, e_{m+1}\}$ is A_1-independent, $\mathrm{wt}(e_i/A_1) = 1$, and $a \not\mathop{\smile\!\!\!\!\!\!|}_{A_1} e_i$ $(i \leq m+1)$. In order not to fail maximality, we must have $a \not\rhd_{A_1} \bar{e}e_{m+1}$. But if $d \mathop{\smile\!\!\!\!\!\!|}_{A_1} a$ then $dA_1 \mathop{\smile\!\!\!\!\!\!|}_{A_0} a$ and, by (*) $d \mathop{\smile\!\!\!\!\!\!|}_{A_1} a\bar{e}$ and $d \mathop{\smile\!\!\!\!\!\!|}_{A_1\bar{e}} a$ (**); moreover, $e_{m+1} \mathop{\smile\!\!\!\!\!\!|}_{A_1} \bar{e}$ so $e_{m+1} \not\mathop{\smile\!\!\!\!\!\!|}_{A_1\bar{e}} a$ and since $\mathrm{wt}(e_{m+1}/A_1\bar{e}) = 1$, it follows from (**), $e_{m+1} \mathop{\smile\!\!\!\!\!\!|}_{A_1\bar{e}} d$ so $d \mathop{\smile\!\!\!\!\!\!|}_{A_1} \bar{e}e_{m+1}$; a contradiction.

Hence, by 4.4.7(2)(3)(4), p has finite weight. For the moreover clause, it is routine to check, based on non-forking extension and 4.4.7(2)(4), 4.4.9, that we can replace each weight 1 type by a regular type, and the product of the regular types is equidominant to p. We have proved Theorem 4.4.10. ■

The following comes from Proposition 4.4.5 and Theorem 4.4.10. We reconfirm that in supersimple T any finitary type has finite weight.

Corollary 4.4.12 *If T is supersimple then any type of finitary hyperimaginary is equidominant with some product of regular types.*

4.4.2 *Internality and Analysability*

Definition 4.4.13 Let $p(x) \in S(A)$, and let Σ be a collection of types setwise invariant under A-automorphisms.

(1) We say $p(x)$ is Σ-*internal*, or *internal* in Σ, if for $a \models p$, there are $B \supseteq A$ and a sequence $\langle b_i | \ i < \alpha \rangle$ of realizations of types over B in some subset of Σ such that $a \mathop{\smile\!\!\!\!\!\!|}_A B$ and $a \in \mathrm{dcl}(B, b_i : i < \alpha)$. If $a \in \mathrm{bdd}(B, b_i : i < \alpha)$ we say p is *almost* Σ-*internal.*

(2) We say p is *analyzable* in Σ, or Σ-*analyzable* if for $a \models p$, there are $a_i \in \mathrm{dcl}(aA)$ $(i < \beta)$ such that $\mathrm{tp}(a_i/A\{a_j | \, j < i\})$ is Σ-internal for all $i < \beta$ and $a \in \mathrm{bdd}(Aa_i | i < \beta)$. The sequence $\langle a_i | i < \beta \rangle$ is called a Σ-*analysis* of a over A of *length* β.

The type p being internal or analyzable in Σ is independent from the choice of $a \models p$. Note that $q \in \Sigma$ can be a complete type over B_0, which is also a *partial* type over $B \supseteq B_0$. Due to A-invariance, usually Σ need not be *small*, i.e., can be a family of size $\bar{\kappa}$. Obviously, if $\mathrm{tp}(a/A)$ is (almost) Σ-internal and $a \mathop{\smile\!\!\!\!\!\!|}_{A'} A$ with $A' \subseteq A$, $\mathrm{tp}(a/A')$ is (almost) Σ-internal. When we extend the parameter while keeping internality we have more freedom.

Remark 4.4.14

(1) Given A-invariant set of types Σ, if $\mathrm{tp}(c/A)$ is Σ-internal then for any $c' \in \mathrm{dcl}(cA)$ and $B \supseteq A$, we have $\mathrm{tp}(c'/B)$ is Σ-internal ("almost" version follows similarly). It is an easy consequence of extension. Since $\mathrm{tp}(c/A)$ is Σ-internal, there are $B'(\supseteq A) \mathop{\smile\!\!\!\!\!\!|}_A c$ and b_i $(i \in I)$ realizing types in Σ such that $c \in \mathrm{dcl}(B'b_i | \, i \in I)$, so

$c' \in \mathrm{dcl}(B'b_i|i \in I)$, too. We can assume that $B \mathop{\smile\!\!\!\!\!\!\mid}_{Ac}(B'c_i|\ i \in I)$. Then $c' \mathop{\smile\!\!\!\!\!\!\mid}_B B'$, and the remark follows.

(2) Assume that $\mathrm{tp}(a_0/A)$ is (almost, resp.) Σ-internal. If $d := \mathrm{Cb}(a_0/B)$ for $B \supseteq A$, then $\mathrm{tp}(d/A)$ is (almost, resp.) Σ-internal too. We prove the being internal case ("almost" case follows by the same proof). Since $\mathrm{tp}(a_0/A)$ is Σ-internal, there are $C_0(\supseteq A)$ with $a_0 \mathop{\smile\!\!\!\!\!\!\mid}_A C_0$ and realizations c_i $(i \in I)$ of types in Σ over C such that $a_0 \in \mathrm{dcl}(C_0\bar{c}_0)$, where $\bar{c}_0 = \langle c_i|i \in I\rangle$. Let $J = \langle a_j|\ j < \omega\rangle$ be a Morley sequence of $\mathrm{tp}(a_0/B)$. Then by 4.3.9(5), $d \in \mathrm{dcl}(J)$. By extension we can assume that $C_0\bar{c}_0 \mathop{\smile\!\!\!\!\!\!\mid}_{a_0A} B$. Then again by extension we can choose $C_j\bar{c}_j$ $(j < \omega)$ such that $C_j\bar{c}_ja_j \equiv_A C_0\bar{c}_0a_0$ and $C_j\bar{c}_j \mathop{\smile\!\!\!\!\!\!\mid}_{Aa_j} B \cup \{C_k\bar{c}_ka_k|k < j\}$. Then $B \mathop{\smile\!\!\!\!\!\!\mid}_A\{C_j|\ j < \omega\}$, so $d \mathop{\smile\!\!\!\!\!\!\mid}_A\{C_j|\ j < \omega\}$ and $d \subseteq J \subseteq (C_j\bar{c}_j|\ j \in \omega)$; thus, we are done.

The following lemma says a connection between non-orthogonality and internality.

Lemma 4.4.15 *Suppose that* $\mathrm{tp}(a/A)$ *is non-orthogonal to a type* q. *Then there is* $a_0 \in \mathrm{dcl}(Aa) \setminus \mathrm{bdd}(A)$ *such that* $\mathrm{tp}(a_0/A)$ *is internal in the family of all A-conjugates of* q. *On the other hand, if* $\mathrm{tp}(a/A)$ *is almost* Σ*-internal and* $a \notin \mathrm{bdd}(A)$, *then* $\mathrm{tp}(a/A)$ *is non-orthogonal to an extension of a type in* Σ.

Proof According to non-orthogonality, there are $b \models q$, and B containing $A \cup \mathrm{dom}(q)$ such that $a \mathop{\smile\!\!\!\!\!\!\mid}_A B$ and $a \not\mathop{\smile\!\!\!\!\!\!\mid}_B b$. Let $\langle b_iB_i|\ i<\omega\rangle$ be a Morley sequence of $\mathrm{tp}(bB/aA)$, and let $e = \mathrm{Cb}(bB/aA)$. Then by 4.3.9(5), $e \in \mathrm{dcl}(b_iB_i : i < \omega)$. From $a \mathop{\smile\!\!\!\!\!\!\mid}_A(B_i|\ i < \omega)$ and $e \in \mathrm{bdd}(aA)$, we have $e \mathop{\smile\!\!\!\!\!\!\mid}_A(B_i|i < \omega)$ and hence $\mathrm{tp}(e/A)$ is internal in the family of A-conjugates of q. Now due to 4.1.5(b), there is a (desired) hyperimaginary a_0 equivalent to the set of all boundedly many aA-conjugates of e. Clearly, $a_0 \in \mathrm{dcl}(Aa)$, and $\mathrm{tp}(a_0/A)$ is internal in the family of A-conjugates of q as well. Moreover, $a_0 \notin \mathrm{bdd}(A)$ since, otherwise, $e \in \mathrm{bdd}(A)$ (as a_0, e are interbounded) and $bB \mathop{\smile\!\!\!\!\!\!\mid}_A a$, a contradiction.

The last sentence comes easily by forking transitivity. ∎

In the rest of this subsection, T is **supersimple**. Recall that the D-rank is measured only on a type of a finite real variable (4.2.3(4)). The results below will be used in Section 5.4.

Lemma 4.4.16 *Let* $SU(a/a_0) < \infty$. *Then there are* $n < \omega$, *hyperimaginaries* $a_0 \subseteq \ldots \subseteq a_n \subseteq aa_0$, *and ordinals* α_i *with* $i < n$ *such that:*

(1) $a \in \mathrm{bdd}(a_n)$;

(2) For each $i < n$, $a_{i+1} \notin \mathrm{bdd}(a_i)$ *and* $\mathrm{tp}(a_{i+1}/a_i)$ *is internal in the family of formulas of D-rank* α_i; *and*

(3) $\mathrm{tp}(a/a_i)$ *is orthogonal to all types of D-rank* $< \alpha_i$ *(hence, by 4.4.2(4) so is* $\mathrm{tp}(a_{i+1}/a_i)$*).*

Proof If $SU(a/a_0) = 0$, then $a \in \mathrm{bdd}(a_0)$ and we are done with $n = 0$. If $SU(a/a_0) > 0$, then there is an imaginary b (with $D(b/a_0) < \infty$) such that $a \not\mathop{\smile\!\!\!\!\!\!\mid}_{a_0} b$. Hence, there is the least ordinal α_0 such that $\mathrm{tp}(a/a_0)$ is non-orthogonal to some *type* of D-rank α_0. We can clearly assume the type is a real type over a real parameter, say A, so that each formula in the type is A-invariant. Then from 4.4.15, there is $a_1 \in \mathrm{dcl}(aa_0) \setminus \mathrm{bdd}(a_0)$ (*) such that $\mathrm{tp}(a_1/a_0)$ is internal in the family of all *formulas* of D-rank α_0. As we can put a_0 into a_1, there is no harm to assume that $a_0 \subseteq a_1$. Moreover, from (*), $a_1 \not\perp_{a_0} a$. Hence, $SU(a/a_1) < SU(a/a_0)$, and the process will stop after producing $a_1, \ldots, a_n$ with $0 = SU(a/a_n) < \ldots < SU(a/a_0)$. ■

Definition 4.4.17 For hyperimaginaries a, a_0 with $SU(a/a_0) < \infty$, define $R(a/a_0)$ as the least ordinal α such that for some $n < \omega$, there are $a_0, \ldots, a_n$ and ordinals $\alpha_0, \ldots, \alpha_{n-1}$ as in 4.4.16, with $\alpha_i \leq \alpha$ for all $i < n$.

Remark 4.4.18

(1) Lemma 4.4.16 says if $SU(a/a_0) < \infty$ then $\mathrm{tp}(a/a_0)$ is analyzable in finite length, in the family of formulas of D-rank $\leq R(a/a_0)$. Moreover, an analysis $a_0 \subseteq \ldots a_i$ with α_i for a over a_0 in 4.4.16 is that for aa_0 over a_0, too, and vice versa. Hence, $R(a/a_0) = R(aa_0/a_0)$.

(2) If $\mathrm{tp}(a/A)$ of SU-rank $< \infty$ is unbounded, then $R(a/A) > 0$. Hence, $R(a/A) = 0$ iff $a \in \mathrm{bdd}(A)$.

(3) If $R(a/a_0) < \alpha$ and $\mathrm{tp}(b/b_0)$ is orthogonal to all types of D-rank $< \alpha$, then $\mathrm{tp}(a/a_0) \perp \mathrm{tp}(b/b_0)$. Since $R(a/a_0) < \alpha$, there are $a_0 \subseteq \ldots \subseteq a_n$ and $\alpha_0, \ldots, \alpha_{n-1} < \alpha$ witnessing $R(a/a_0) < \alpha$. Now, by forking transitivity, it suffices to see that $\mathrm{tp}(b/b_0)$ and $\mathrm{tp}(a_{i+1}/a_i)$ are orthogonal ($i < n$). Now, since $\mathrm{tp}(a_{i+1}/a_i)$ is internal in the types of D-rank $< \alpha$, there are $B(\perp_{a_i} a_{i+1})$ containing a_i and real $e = \langle e_j | \ j < \beta \rangle$ such that $D(e_j/B \cup \{e_l | \ l < j\}) < \alpha$, and $a_{i+1} \in \mathrm{dcl}(Be)$. By extension we can clearly assume $B \perp_{a_{i+1}} b_0$, so $B \perp_{a_i} b_0$ holds (*). Now $\mathrm{tp}(b/b_0)$ is orthogonal to each $\mathrm{tp}(e_j/B \cup \{e_l | \ l < j\})$, and again by transitivity, so is to each of $\mathrm{tp}(e/B)$ and $\mathrm{tp}(a_{i+1}/B)$. Then due to (*) and 4.4.2(3), we have $\mathrm{tp}(b/b_0) \perp \mathrm{tp}(a_{i+1}/a_i)$.

Remark 4.4.19 This remark talks about how to find a bound of the R-rank of given tuples from the R-rank data of other tuples. There are several cases as below but the idea is all the same. Basically, given an analysis $a_0 \subseteq \ldots a_n \subseteq aa_0$ with α_i as in 4.4.16, we often want to find a new analysis $b_0 \subseteq \ldots b_m \subseteq bb_0$ with $R(b/b_0) \leq R(a/a_0)$. Assume $b_0 \subseteq \ldots b_j$ with $\beta_0, \ldots, \beta_{j-1} \leq R(a/a_0)$ have been found. To find b_{j+1} and β_j, consider the first i such that $a_i \notin \mathrm{bdd}(b_j)$. Now, by 4.4.14(1), $\mathrm{tp}(a_i/\mathrm{bdd}(b_j))$ and so $\mathrm{tp}(a_i/b_j)$ are also internal in the families of formulas of D-rank α_{i-1}. Then if $a_i \in \mathrm{bdd}(bb_0)$ (which is usually given), we have that $\mathrm{tp}(b/b_j)$ is non-orthogonal to a type of D-rank $\leq \alpha_{i-1}$ according to 4.4.15 and 4.4.2(4). Then, as in the proof of 4.4.16, we can find $(b_j \subseteq) b_{j+1} \subseteq bb_0$, forming the next component of the analysis for bb_0, while keeping $\beta_j \leq \alpha_{i-1}$.

(1) If $SU(e/a_0) < \infty$, where $e = \langle e_i | i \in I \rangle$, and $R(e_i/a_0) < \alpha$ for all $i \in I$, then $R(e/a_0) < \alpha$, too. We can clearly assume $\mathrm{tp}(e/a_0)$ is unbounded, and write $e = \langle e_0, e_1, \ldots \rangle$ with $e_i \notin \mathrm{bdd}(a_0)$. Now there is an analysis $a_0 \subseteq a_1^0 \ldots$ for $e_0 a$. As stated earlier, $\mathrm{tp}(e/a_0)$ is non-orthogonal to a type of D-rank $< \alpha$, and we

find $a_1 \subseteq ea_0$ and $\alpha_0 < \alpha$ such that $a_0 \subseteq a_1$ form an initial analysis for ea_0. Now if $e \in \mathrm{bdd}(a_1)$, we are done. Otherwise, there is $e_i \notin \mathrm{bdd}(a_1)$, and let $a_0 \subseteq a_1^i \dots$ be an analysis for $e_i a_0$. There now is the first j such that $a_j^i \notin \mathrm{bdd}(a_1)$. Then, again as said, $\mathrm{tp}(a_j^i/a_1)$ (so $\mathrm{tp}(e/a_1)$, too) is non-orthogonal to a type of D-rank $< \alpha$; and we can continue to find $(a_1 \subseteq) a_2 \subseteq ea_0$ and $\alpha_1 < \alpha$. Since $SU(e/a_2) < SU(e/a_1) < SU(e/a_0) < \infty$, this stops in finitely many steps and $R(e/a_0) < \alpha$.

(2) If $e \in \mathrm{bdd}(aa_0)$, then $R(e/a_0) \leq R(a/a_0)$. The proof will be similar, but it needs an extra argument. Let $a_0 \subseteq \dots \subseteq a_n \subseteq aa_0$ be an analysis with $\alpha_0, \dots, \alpha_{n-1} \leq \alpha := R(a/a_0)$. Now there is the maximal j such that $e \mathop{\smile\!\!\!\!\!\!\mid}_{a_0} a_j$. If $j = n$ then $e \in \mathrm{bdd}(a_j) = \mathrm{bdd}(aa_0)$, whence $e \in \mathrm{bdd}(a_0)$ and we are done. So we can assume that $j < n$. We claim that there is $d \in \mathrm{bdd}(ea_j) \setminus \mathrm{bdd}(a_j)$ such that $\mathrm{tp}(d/a_j)$ is internal in the families of formulas of D-rank $\leq \alpha$. Let $d := \mathrm{Cb}(a_{j+1}/a_j e)$. Then by 4.4.14(2), it suffices to show $d \notin \mathrm{bdd}(a_j)$. If, otherwise, then $e \mathop{\smile\!\!\!\!\!\!\mid}_{a_j} a_{j+1}$, contradicting the choice of j.

Now by the claim and 4.4.15, 4.4.2(3), $\mathrm{tp}(e/a_j)$ and $\mathrm{tp}(e/a_0)$ are non-orthogonal to a type of D-rank $\leq \alpha$, and again as in the proof of 4.4.16, we find that $e_1 \subseteq ea_0$ and such that $a_0 \subseteq e_1$ with $\beta_0 (\leq \alpha)$ form an initial analysis for ea_0. To find a next component, again choose the maximal $a_{j'}$ such that $e \mathop{\smile\!\!\!\!\!\!\mid}_{e_1} a_{j'}$. As above we can assume that $j' < n$, and for $d' := \mathrm{Cb}(a_{j'+1}/a_{j'} e e_1) \in \mathrm{bdd}(e; e_1 a_{j'}) \setminus \mathrm{bdd}(e_1 a_{j'})$ we have $\mathrm{tp}(d'/e_1 a_{j'})$ is internal in the families of formulas of D-rank $\leq \alpha$. Hence, again each $\mathrm{tp}(e/e_1 a_{j'})$, $\mathrm{tp}(e/e_1)$ is non-orthogonal to a type of D-rank $\leq \alpha$, and we can proceed. Note that if $SU(a/a_0) < \infty$ so is $SU(e/a_0)$ (4.2.13(2)).

(3) Assume that $\mathrm{tp}(a/a_0)$ is internal in the family of formulas of D-rank $< \alpha$, and $SU(a/a_0) < \infty$. Then $R(a/a_0) < \alpha$. If $\mathrm{tp}(a/a_0)$ is unbounded, then by the assumption it is non-orthogonal to a type of D-rank $< \alpha$. Then the rest of the finite process of getting an analysis $a_0 \subseteq a_1 \subseteq \dots \subseteq a_{i+1}$ with $\alpha_i < \alpha$ is the same, while applying 4.4.14(1) to $\mathrm{tp}(a/a_i)$.

(4) If $R(b_0/a_0) = \alpha$ and $R(b/b_0) = \beta$, then $R(b/a_0) \leq R(bb_0/a_0) \leq \max\{\alpha, \beta\}$. The proof will be similar. Note that due to 2.5.19(1), $SU(bb_0/a_0) < \infty$; and by (2) $R(b/a_0) \leq R(bb_0/a_0)$. There are analyses $a_0 \subseteq a_1 \dots \subseteq a_n$ with $\alpha_0, \dots, \alpha_{n-1}$ for b_0 over a_0; and $b_0 \subseteq b_1 \dots \subseteq b_m$ with $\beta_0, \dots, \beta_{m-1}$ for b over b_0. Put for $i > 0$, $b_i = a_{n+i}$; and for $i \geq 0$, $\beta_i = \alpha_{n+i}$. Since $\mathrm{tp}(a_1/a_0)$ is non-orthogonal to a type of D-rank α_0, so is $\mathrm{tp}(bb_0/a_0)$. Then, as before, we have that $a_1' \subseteq bb_0 a_0$ and $\alpha_0' \leq \alpha_0$ such that $a_0 \subseteq a_1'$ form an initial analysis for $bb_0; a_0$. Now there is the first $a_i \notin \mathrm{bdd}(a_1')$. Then again, $\mathrm{tp}(a_i/a_1')$ (and so $\mathrm{tp}(bb_0/a_1')$) is orthogonal to a type of D-rank $\leq \alpha_{i-1}$. Then we continue while keeping $R(bb_0/a_0) \leq \alpha$ or β.

(5) In (4), if the condition $R(b_0/a_0) = \alpha$ is replaced by that $\mathrm{tp}(b_0/a_0)$ is internal in the family of formulas of D-rank $\leq \beta$, then still $R(b/a_0) \leq R(bb_0/a_0) \leq \beta$ as far as $SU(bb_0/a_0) < \infty$.

Lemma 4.4.20 *Let* $b = \langle e \in \mathrm{dcl}(aa_0) | R(e/a_0) < \alpha \rangle$ *and* $SU(a/a_0) < \infty$. *Then:*

(1) $R(b/a_0) < \alpha$, *and*

(2) $\mathrm{tp}(a/b) \perp$ *any type of D-rank* $< \alpha$.

(3) $\mathrm{tp}(b/a_0) \perp \mathrm{tp}(a/b)$.

Proof (1) Note that as in 4.2.13(2), $SU(b/a_0) < \infty$. Hence, (1) is obtained from 4.4.19(1).

(2) Suppose not; then by 4.4.15, there is $c \in \mathrm{dcl}(ab) \setminus \mathrm{bdd}(b)$ such that $\mathrm{tp}(c/b)$ is internal in the family of formulas of D-rank $< \alpha$. Note that $cb \subseteq aba_0 \subseteq aa_0$. Hence, by 4.2.13(2), $SU(cb/a_0) < \infty$. Then due to (1) and 4.4.19(5), we have that $R(c/a_0) < \alpha$. Thus, $c \in b$; a contradiction.

(3) comes from 4.4.18(3). ■

4.5 Bibliographical Remarks

Hyperimaginaries are introduced in [45], [75]. The treatment of hyperimaginary forking and type amalgamation in simple theories is from [45]. Lemma 4.1.5 is from [13]. The existence of canonical bases is due to B. Hart, B. Kim, and A. Pillay [45]. Section 4.1 is a generalization of notions and results in stability theory [113], [80], [82], [60] to simplicity theory [45], [120]. Section 4.2 is from [13] and missing details are supplemented from [15].

5

Elimination of Hyperimaginaries

Chapter 4 introduces the notion of hyperimaginaries. This chapter now deals with elimination of hyperimaginaries.

Definition 5.0.1 We say *elimination of hyperimaginaries* (in short, *e.h.i.*) holds in T if any hyperimaginary is interdefinable with a sequence of imaginaries.

Remark 5.0.2 Elimination of hyperimaginaries holds in T iff for any $\emptyset$-type-definable equivalence relation $E(x,y)$ and a complete *real* type $p(x)$ over $\emptyset$, there is a set $E_p(x,y)$ of formulas over $\emptyset$, each of which defines an equivalence relation, such that $p(x) \wedge p(y) \models E(x,y) \leftrightarrow E_p(x,y)$: ($\Leftarrow$) This direction is obvious. ($\Rightarrow$) For $a \models p$, a/E is inter-definable with say $e = (d_i/E_i | i \in I)$ where each d_i/E_i is an imaginary element. We can identify e as $(d_i)/E'_p$, where $E'_p \equiv \bigwedge E_i$. Consider a type-definable equivalence relation $E_p(x,y) \equiv \exists zw(E'_p(z,w) \wedge xz \equiv yw)$, which by compactness can be considered as a conjunction of definable equivalence relations. It follows that $p(x) \wedge p(y) \models E(x,y) \leftrightarrow E_p(x,y)$.

We now come up with a natural question whether in simple theories, working with hyperimaginaries is really necessary. More precisely the question asks whether elimination of hyperimaginaries holds in simple theories. If so, we can indeed work within $\mathcal{M}^{\text{eq}}$ as traditionally by substituting an equivalent imaginary sequence for a hyperimaginary. The full answer is not known yet, but positive (or partial) answers are given to several subclasses of simple theories, in particular to that of all supersimple theories, and of all stable theories. The intriguing fact is to show elimination of hyperimaginaries for such subclasses; understanding hyperimaginary forking must come first, as in Chapter 4.

We observe another equivalent condition of elimination of hyperimaginaries for simple theories.

Proposition 5.0.3 *Assume T is simple. T has elimination of hyperimaginaries iff for any real finite tuple c, and a real submodel M, $\mathrm{Cb}(c/M)$ is interdefinable with a sequence of imaginaries.*

Proof (⇒) Clear.

(⇐) Let a hyperimaginary $a = u/E$ with real u be given. Let $e = \mathrm{Cb}(u/a)$. There clearly is a model M such that $u \mathop{\smile\!\!\!\!\!\!\,|}_e M$ and $a \in \mathrm{dcl}(M)$. Note that $e = \mathrm{Cb}(u/M)$. By the assumption and 4.3.9(6) we can suppose e is an imaginary tuple. Now by 4.3.9(7), $e \in \mathrm{bdd}(a)$, and $a \in \mathrm{dcl}(e)$. Hence, by 4.1.5(1), a is interdefinable with some sequence of imaginaries. ■

Therefore any stable theory eliminates hyperimaginaries.

Corollary 5.0.4 *Assume T is supersimple. For any real finite tuple c and a model M,* $\mathrm{Cb}(c/M)$ *is finitary. Hence, T has e.h.i. if it has elimination of finitary hyperimaginaries.*

Proof Since T is supersimple, there is finite $b \in M$ such that $c \mathop{\smile\!\!\!\!\!\!\,|}_b M$. Now by 4.3.9(4), $\mathrm{Cb}(c/M) \in \mathrm{dcl}(cb)$. Hence, by 4.1.5(3), $\mathrm{Cb}(c/M)$ is finitary. The conclusion follows from Proposition 5.0.3. ■

Supersimple theories indeed eliminate finitary hyperimaginaries, thus having e.h.i. (Theorem 5.4.9).

In the following section we introduce the Lascar group, which is an independent subject but also closely related to the issue of elimination of hyperimaginaries.

5.1 The Lascar Group

In this section we work with an *arbitrary* theory T, and pay our attention to the quotient group $\mathrm{Aut}_A(\mathcal{M})/\,\mathrm{Autf}_A(\mathcal{M})$ for a hyperimaginary A. Note that by Definition 4.2.14 (or Lemma 4.2.15), $\mathrm{Autf}_A(\mathcal{M})$ is a normal subgroup of $\mathrm{Aut}_A(\mathcal{M})$.

Definition 5.1.1 (Lascar group) $\mathrm{Gal_L}(\mathcal{M}, A) := \mathrm{Aut}_A(\mathcal{M})/\,\mathrm{Autf}_A(\mathcal{M})$.

In the rest of this section **we fix the hyperimaginary** A. But when there is no risk of confusion **we may omit** A for notational convenience, so $A \subseteq M$ for any model mentioned below, and $\mathrm{Aut}(\mathcal{M}), \mathrm{Autf}(\mathcal{M})$ below indeed mean $\mathrm{Aut}_A(\mathcal{M}), \mathrm{Autf}_A(\mathcal{M})$, respectively. For maps f, g we use $f.g$ to denote their composition (as declared in Chapter 1).

Remark 5.1.2 We argue that the Lascar group depends only on T and A, and we write $\mathrm{Gal_L}(T, A)$ for $\mathrm{Gal_L}(\mathcal{M}, A)$. We have $|\,\mathrm{Gal_L}(T, A)| \leq 2^{|T|+|A|}$.

(1) Let M be a (small) elementary submodel of $\mathcal{M}$. For $f, g \in \mathrm{Aut}(\mathcal{M})$, if $f(M) \equiv_M g(M)$ then $f.\mathrm{Autf}(\mathcal{M}) = g.\mathrm{Autf}(\mathcal{M})$ in $\mathrm{Gal_L}(\mathcal{M})$. There is $h \in \mathrm{Autf}(\mathcal{M})$ such that h fixes M pointwise and sends $f(\mathcal{M})$ to $g(\mathcal{M})$. Then since $s := f^{-1}.h^{-1}.g$ fixes M, too, $s \in \mathrm{Autf}(\mathcal{M})$. Thus, the claim follows.

(2) In (1), we can choose M having size $|T| + |A|$. Since there are at most $2^{|T|+|A|}$-many types over M, we have $|\,\mathrm{Gal_L}(\mathcal{M}, A)| \leq 2^{|T|+|A|}$.

(3) Let $\mathcal{M}' \succ \mathcal{M}$ be saturated and $|\mathcal{M}'| > |\mathcal{M}|$. There is a canonical isomorphism from $\mathrm{Gal_L}(\mathcal{M})$ to $\mathrm{Gal_L}(\mathcal{M}')$.

Let $f \in \mathrm{Aut}(\mathcal{M})$. Any two automorphisms of $\mathcal{M}'$ extending f are in the same coset in $\mathrm{Gal_L}(\mathcal{M}')$. This induces a homomorphism $\alpha : \mathrm{Aut}(\mathcal{M}) \to \mathrm{Gal_L}(\mathcal{M}')$. We claim that $\mathrm{Ker}(\alpha) = \mathrm{Autf}(\mathcal{M})$. If $f \in \mathrm{Autf}(\mathcal{M})$ then clearly any extension of f in $\mathrm{Aut}(\mathcal{M}')$ is in $\mathrm{Autf}(\mathcal{M}')$. Conversely, let $f' \in \mathrm{Autf}(\mathcal{M}')$ extend $f \in \mathrm{Aut}(\mathcal{M})$. Then for a small model $M \prec \mathcal{M}$, by 4.2.15, $M \equiv^L f'(M) = f(M)$ in $\mathcal{M}$. Hence, there is $g \in \mathrm{Autf}(\mathcal{M})$ such that $f(M) = g(M)$. Then $g^{-1}.f \in \mathrm{Autf}(\mathcal{M})$ since it fixes M, so $f \in \mathrm{Autf}(\mathcal{M})$ too.

Let us also write $\alpha : \mathrm{Gal_L}(\mathcal{M}) \to \mathrm{Gal_L}(\mathcal{M}')$ for the induced injection. We claim that α is surjective. Let $g \in \mathrm{Aut}(\mathcal{M}')$. For a small model $M \prec \mathcal{M}$, there is $M' \prec \mathcal{M}$ such that $g(M) \equiv_M M'$. Now there is $f \in \mathrm{Aut}(\mathcal{M})$ sending M to M'. Then by (1), $\alpha(f) = g.\mathrm{Autf}(\mathcal{M}')$.

In the rest of this section, for $f, g \in \mathrm{Aut}(\mathcal{M})$, we write $f \approx g$ if $f.\mathrm{Autf}(\mathcal{M}) = g.\mathrm{Autf}(\mathcal{M})$.

We shall endow $\mathrm{Gal_L}(T)$ with a quotient topology to make it a compact (*but not necessarily Hausdorff*) topological group. Let $\pi : \mathrm{Aut}(\mathcal{M}) \to \mathrm{Gal_L}(= \mathrm{Gal_L}(T, A))$ be the canonical projection. Fix a model $M \prec \mathcal{M}$ and let

$$S_M(M) := \{\mathrm{tp}(f(M)/M) | f \in \mathrm{Aut}(\mathcal{M})\},$$

equipped with its Stone topology. Then by 5.1.2(1), π factors through the surjection $\mu : \mathrm{Aut}(\mathcal{M}) \to S_M(M)$, sending $f \mapsto \mathrm{tp}(f(M)/M)$; i.e., there is a canonical surjection $\nu = \nu_M : S_M(M) \to \mathrm{Gal_L}$ such that $\pi = \nu.\mu$. We use these maps. We give $\mathrm{Gal_L}$ the quotient topology under the map ν. This topology is independent from the choice of M. It suffices to show for a model $N (\prec \mathcal{M})$, an elementary extension of M (since any two models have a common extension). Now the map $\nu_N : S_N(N) \to \mathrm{Gal_L}$ factors through the restriction map $S_N(N) \to S_M(M)$, sending $\mathrm{tp}(f(N)/N)$ to $\mathrm{tp}(f(M)/M)$. Since the restriction map is continuous and both $S_N(N)$ and $S_M(M)$ are compact Hausdorff, ν_M and ν_N induce the same quotient topology.

Lascar originally introduced the topology on $\mathrm{Gal_L}$ in terms of ultrafilters. It is known that his and the quotient topology coincide. Contrary to what the reader might expect, the proof that $\mathrm{Gal_L}$ is a topological group is quite subtle. The only known complete and correct proof can be found in [125], and the proof goes through when working over some hyperimaginary A.

Proposition 5.1.3 $\mathrm{Gal_L}$ *is a compact topological group.*

We now introduce two canonical subgroups of $\mathrm{Gal_L}(T, A)$. Note that $\overline{\{\mathrm{id}\}}$ is a closed normal subgroup of $\mathrm{Gal_L}(T, A)$. The connected component of $\mathrm{Gal_L}(T, A)$ containing id is denoted by $\mathrm{Gal_L^0}(T, A)$; it is also closed and normal.

Definition 5.1.4

(1) $\mathrm{Autf_{KP}}(\mathcal{M}, A) := \pi^{-1}(\overline{\{\mathrm{id}\}})$,

$$\mathrm{Autf_S}(T,A) := \pi^{-1}(\mathrm{Gal}^0_\mathrm{L}(T,A)).$$
$$\mathrm{Gal_{KP}}(T,A) := \mathrm{Gal_L}(T,A)/\overline{\{\mathrm{id}\}} = \mathrm{Aut}_A(\mathcal{M})/\,\mathrm{Autf_{KP}}(\mathcal{M},A),$$
$$\mathrm{Gal_S}(T,A) := \mathrm{Gal_L}(T,A)/\,\mathrm{Gal}^0_\mathrm{L}(T,A) = \mathrm{Aut}_A(\mathcal{M})/\,\mathrm{Autf_S}(\mathcal{M},A).$$

(2) We say T is *G-compact* over A if $\mathrm{Gal_L}(T,A)$ is Hausdorff.

KP stands for Kim–Pillay and S stands for Shelah or "strong." Again, in the following we may omit the subscript A. In the rest, *arities* refers to that of *real* tuples.

Remark 5.1.5

(1) Recall that for topological groups $H \lhd G$, the quotient group G/H is Hausdorff iff H is closed in G. Hence, both $\mathrm{Gal_{KP}}(T)$ and $\mathrm{Gal_S}(T)$ are compact Hausdorff. Moreover, $\mathrm{Gal_S}(T)$ is totally disconnected, so it is a profinite group.
Now T is G-compact iff $\{\mathrm{id}\}$ is closed iff $\mathrm{Autf}(\mathcal{M}) = \mathrm{Autf_{KP}}(\mathcal{M})$ (then $\mathrm{Gal_{KP}}(T)$ and $\mathrm{Gal_L}(T)$ are canonically isomorphic).

(2) For the rest of this section we endow $\mathrm{Aut}(\mathcal{M})$ with a topology having basic open sets of the form $\{f \in \mathrm{Aut}(\mathcal{M}) | f(a) = b\}$ with some real n-tuples $a, b \in \mathcal{M}$. (However, the topologies of $\mathrm{Gal_S}$ and $\mathrm{Gal_{KP}}$ are always the quotient topologies obtained from $\mathrm{Gal_L}$.)

(3) Let Γ be a subgroup of $\mathrm{Aut}_A(\mathcal{M})$. Fix F an $\emptyset$-type-definable equivalence relation on $\mathcal{M}^\alpha$. We write $\boldsymbol{E}^F_\Gamma$ to denote an equivalence relation such that for F-hyperimaginaries c, d, we have $\boldsymbol{E}^F_\Gamma(c,d)$ iff $d = f(c)$ for some $f \in \Gamma$. So $\boldsymbol{E}^F_\Gamma$ is an equivalence relation on $\mathcal{M}^\alpha/F$, or equivalently an equivalence relation on $\mathcal{M}^\alpha$ *coarser* than F. When we write $\boldsymbol{E}^=_\Gamma(x,y)$, $=$ means the finest *real* equivalence relation $x = y$. We omit F if F is clear from context. Note that if Γ is normal then $\boldsymbol{E}_\Gamma$ is A-invariant, but in general it *need not* be. When $\Gamma = \mathrm{Autf}(\mathcal{M},A)$, we know that $\boldsymbol{E}_\Gamma$ is $\equiv^L_A$, and $\boldsymbol{E}_{\mathrm{Autf_{KP}}(\mathcal{M},A)}$ is denoted by $\equiv^{KP}_A$.
Note that Γ is closed iff $\Gamma = \overline{\Gamma} = \{f \in \mathrm{Aut}_A(\mathcal{M})|$ for each finite real tuple $a \in \mathcal{M}, \boldsymbol{E}^=_\Gamma(a, f(a))$; i.e., f stabilizes all $\boldsymbol{E}^=_\Gamma$ classes of finite arities$\}$.

Lemma 5.1.6

(1) Let H' be a closed normal subgroup of $\mathrm{Gal_L}(\mathcal{M},A)$, *and let $H = \pi^{-1}(H')$. Then given F, $\boldsymbol{E}^F_H$ is a type-definable bounded equivalence relation over A.*

(2) The restriction map π is continuous, so both $\mathrm{Autf_S}(\mathcal{M},A)$ *and* $\mathrm{Autf_{KP}}(\mathcal{M},A)$ *are closed in* $\mathrm{Aut}(\mathcal{M},A)$.

Proof (1) Let $a = u_F \models p(x) \in S_F(A)$ where u is a real tuple of arity α. Fix a model $(u \subseteq)M = M_p \prec \mathcal{M}$. Since H' is closed, there is $\Phi(x', M)$ over M type-defining $\nu^{-1}(H')$; i.e., $f \in H$ iff $\Phi(f(M), M)$ holds. Note that H being a group implies that $\Phi(x', y')$ type-defines an equivalence relation on $\mathrm{tp}(M)$. Moreover, $x' \equiv^L_A M \to \Phi(x', M)$. Thus, $\Phi(x', y')$ is a bounded equivalence relation on $\mathrm{tp}(M/A)$. Note also that since H is normal it easily follows that $\Phi(x', y') \equiv \boldsymbol{E}^=_H(x', y')$ on $\mathrm{tp}(M/A)$. Then by taking existential quantifiers to $\Phi(x',y')$, clearly $E_H^{=}(x,y)$ is type-definable on $q(x) = \mathrm{tp}(u/A)$, too. Hence, $E_H^F(x,y)$ is type-defined by

$$\Psi_p(x,y) \equiv \exists zw\,(E_H^{=}(z,w) \wedge q(z) \wedge q(w) \wedge F(z,x) \wedge F(w,y))$$

on $\mathrm{tp}(a/A) = p(x)$. We put that

$$\Psi_p'(x,y) \equiv \big(\Psi_p(x,y) \wedge p(x) \wedge p(y)\big) \vee x \equiv_A y.$$

Therefore, clearly $E_H^F(x,y) \leftrightarrow \bigwedge\{\Psi_p' | p(x) \in S_F(A)\}$.

(2) Let $f \in \mathrm{Aut}(\mathcal{M})$, and let U be an open subset of $\mathrm{Gal_L}(T)$ containing $\pi(f) = f.\mathrm{Autf}(\mathcal{M})$. Since ν is continuous, $\nu^{-1}(U) \subseteq S_M(M)$ contains a basic open neighborhood

$$V_{\varphi(x)} = \{\mathrm{tp}(g(M)/M) \ni \varphi(x) : g \in \mathrm{Aut}(\mathcal{M})\}$$

of $\mu(f) = \mathrm{tp}(f(M)/M)$, where $\varphi(x)$ is some formula over M. Let $a \in M$ be a finite tuple corresponding to x. Then simply

$$\mu^{-1}(V_{\varphi(x)}) = \{g \in \mathrm{Aut}(\mathcal{M}) : g(a) \models \varphi(x)\}.$$

Since $f \in \mu^{-1}(V_{\varphi(x)})$, in particular, $\models \varphi(f(a))$ holds. Therefore, a basic open neighborhood $\{h \in \mathrm{Aut}(\mathcal{M}) | h(a) = f(a)\}$ of f is contained in $\mu^{-1}(V_{\varphi(x)})$. Hence, π is continuous. ■

Now fix a closed normal subgroup $H' \lhd \mathrm{Gal_L}$ and let $H = \pi^{-1}(H')$. We write $x \equiv^H y$ if $E_H^{=}(x,y)$ holds, which on any arity, according to 5.1.6(1), is a bounded type-definable equivalence relation (over A).

Proposition 5.1.7

(1) $H = \{f \in \mathrm{Aut}(\mathcal{M}) | f$ *stabilizes all the* $\equiv^H$*-classes of any arities*$\} = \{f \in \mathrm{Aut}(\mathcal{M}) | f$ *stabilizes all the* $\equiv^H$*-classes of finite real arities*$\}$.

(2) *The following are equivalent.*

  (a) $c_0 \equiv^H c_1$ *for real tuples.*

  (b) $c_0' \equiv^H c_1'$ *for each corresponding finite subtuple* c_i' *of* c_i $(i = 0, 1)$.

Proof (1) According to 5.1.6(2), H is closed in $\mathrm{Aut}(\mathcal{M})$. Hence, it comes from 5.1.5(3).

(2) Clearly, (a) implies (b). Assume (b) holds. In this proof all the tuples are real. Note that there is $h' \in H$, sending $h'(c_0') = c_1'$. Hence, for any finite d_0, there is $d_1 = h'(d_0)$ with $c_0'd_0 \equiv^H c_1'd_1$. Thus, by compactness there is a sufficiently saturated M_i containing c_i such that for each finite corresponding $b_i \in M_i$, $b_0 \equiv^H b_1$ (*). In particular, $\mathrm{tp}(M_0) = \mathrm{tp}(M_1)$ and there is $h \in \mathrm{Aut}(\mathcal{M})$, sending M_0 to M_1. Let $d/\equiv^H$ with arbitrary finite d be given. We claim h fixes $d/\equiv^H$, so by (1), $h \in H$, and (a)

follows: Due to the saturation of M_0 and the boundedness of $\equiv^H$, there is $d' \in M_0$ such that $\mathrm{tp}(d) = \mathrm{tp}(d')$ and $d \equiv^H d'$ holds. Then by (*), h clearly fixes $d'/\equiv^H = d/\equiv^H$. ■

In [85], using compact Lie group theory, Lascar and Pillay showed that any bounded hyperimaginary e is equivalent to a sequence of finitary bounded hyperimaginaries. As a corollary of Proposition 5.1.7, we obtain the result directly when $\mathrm{Aut}_e(\mathcal{M})$ is a normal subgroup of $\mathrm{Aut}(\mathcal{M})$.

Corollary 5.1.8 *Let $e = c_F$ with real c be a hyperimaginary bounded over A. Assume additionally $\mathrm{Aut}_{eA}(\mathcal{M}) \lhd \mathrm{Aut}_A(\mathcal{M})$. Then there are finitary hyperimaginaries e_i $(i \in I)$ such that e and $(e_i | i \in I)$ is interdefinable over A.*

Proof In the proof we again omit A. Let $p(x) = \mathrm{tp}(e)$. We may reset F as $(p(x) \wedge p(y) \wedge F(x, y)) \vee x \equiv y$, so that F, as a whole, is a type-definable bounded equivalence relation (over A). Choose a model M containing c. Note that F type-defines a bounded equivalence relation on $\mathcal{M}^{|M|}$, too. Moreover, $\mathrm{Aut}_e(\mathcal{M}) = \{f \in \mathrm{Aut}(\mathcal{M}) : \models F(f(c), c)\}$. Hence, $\pi(\mathrm{Aut}_e(\mathcal{M}))$ is a closed subgroup of $\mathrm{Gal}_{\mathrm{L}}$. By the assumption it is normal as well. Hence, our corollary follows from 5.1.7(2). ■

Corollary 5.1.9

(1) *For real x, y, $x \equiv_A^{KP} y$ is the finest bounded type-definable equivalence relation over A. Precisely, for real u_i $(i = 0, 1)$ the following are equivalent:*

(a) *$u_0 \equiv_A^{KP} u_1$ holds.*

(b) *$u_0 \equiv_A u_1$; and $E'(u_0, u_1)$ holds for any $\emptyset$-type-definable equivalence relation E' such that $u_0/E' \in \mathrm{bdd}(A)$.*

(c) *$u'_0 \equiv_A^{KP} u'_1$ for each corresponding finite subtuple u'_i of u_i.*

(2) $\mathrm{Autf}_{\mathrm{KP}}(\mathcal{M}, A)$

$= \{f \in \mathrm{Aut}(\mathcal{M}, A) | f$ *stabilizes all the bounded A-type-definable equivalence classes*$\}$

$= \{f \in \mathrm{Aut}(\mathcal{M}, A) | f$ *stabilizes all the bounded A-type-definable equivalence classes of finite arities*$\}$.

(3) *The following are equivalent:*

(a) *$a \equiv_A^{KP} b$, where $a = u_F$, $b = v_F$ with real u, v.*

(b) *$F_A^{KP}(u, v)$ holds, where*

$$F_A^{KP}(x, y) := \exists z z'(z \equiv_A^{KP} z' \wedge F(z, x) \wedge F(z', y)) \equiv \exists z(F(z, y) \wedge z \equiv_A^{KP} x)$$

($z \equiv_A^s x$ is, of course, equality of KP-types over A of real tuples). $F_A^{KP}(x, y)$ is a bounded type-definable equivalence relation over A coarser than F, and is the finest among those.

(c) *$a \equiv_{\mathrm{bdd}(A)} b$.*

Proof (1)(a)$\Leftrightarrow$(b) This comes from the argument in the proof of 5.1.6(1). Note that for a type $\Psi(x, M)$ type-defining $\nu^{-1}(\overline{\{\text{id}\}})$, $\Psi(x, y)$ must be the finest bounded type-definable equivalence relation on $\text{tp}(M/A)$.

The equivalence of (1)(c) to others, and (2) are due to 5.1.7.

(3)(a)$\Leftrightarrow$(b) Again this is from the argument in the proof of 5.1.6(1). Note that if $u' \equiv_A^{KP} v'$ and $F(u, u'), F(v, v')$ with real u', v' so that there is $f \in \text{Autf}_{\text{KP}}(\mathcal{M}, A)$ with $v' = f(u')$, then $f(u) \equiv_A^{KP} u$ and $F(f(u), v)$ holds. Hence, one can easily verify that Ψ_p there with $H = \text{Autf}_{\text{KP}}(\mathcal{M}, A)$ is equivalent to F_A^{KP} above on $p(x) = \text{tp}(a/A)$. Note also that F_A^{KP} is coarser than both $\equiv^s$ and F. It remains to show that F_A^{KP} is the finest one as stated. There clearly is the finest one; let it be F'. If $F_A^{KP}(u', v')$ holds, then there is u'' such that $F(u', u'') \wedge u'' \equiv_A^{KP} v'$. Hence, $F'(u', u'')$ and $F'(u'', v')$; so $F'(u', v')$ must hold.

(c)$\Rightarrow$(b) Obvious. Note only that due to (1)(b), $u/\equiv_A^{KP}$ is equivalent over A to a *hyperimaginary*.

(b)$\Rightarrow$(c) Assume (b). Let $a_0 \in \text{bdd}(A)$. By compactness it suffices to show that $c_0 \equiv_{Aa_0} c_1$. Suppose $\{a_i | i \in I\}$ is the set of all A-conjugates of a_0. Write $p(x, a_0) = p(x_F, a_0) = \text{tp}(c_0/Aa_0)$ and $p(x, a_i)$ is the conjugate of $p(x, a_0)$ over A. Now there is a maximal subset $J \subseteq I$ containing 0 such that $\{p(x, a_i) | i \in J\}$ is realized by c_0. Put $a_J = \langle a_i | i \in J\rangle$ and let $p'(xz) = \text{tp}(c_0 a_J/A)$; $q(z) = \text{tp}(a_J/A)$. Consider

$$\bar{E}(x, y) := \exists z(p'(x, z) \wedge p'(y, z) \wedge q(z)) \vee x_F \equiv_A y_F.$$

Due to the maximality of J, $\bar{E}$ is an A-type-definable equivalence relation, bounded, and coarser than F. Thus by (b), $\bar{E}(c_0, c_1)$ holds. Note that $c_0 \models \bigwedge_{i \in J} p(x, a_i)$, so $c_1 \models p(x, a_0)$, and $c_0 \equiv_{Aa_0} c_1$ as desired. ■

Proposition 5.1.10 *The following are equivalent:*

(1) *T is G-compact over A.*

(2) $\text{Autf}(\mathcal{M}, A)$ *is closed in* $\text{Aut}_A(\mathcal{M})$*, and for each finite arity, $x \equiv_A^L y$ is type-definable over A.*

(3) *For any arity, $x \equiv_A^L y$ iff $x \equiv_A^{KP} y$.*

(4) *For any arity, $x \equiv_A^L y$ is type-definable.*

(5) *For any F, $x_F \equiv_A^L y_F$ iff $x_F \equiv_A^{KP} y_F$.*

Proof (1)$\Rightarrow$(2),(3) By 5.1.6, 5.1.9.

(2)$\Rightarrow$(1) Let $f \in \text{Autf}_{\text{KP}}(\mathcal{M})$. Then by (2) and 5.1.9, f fixes all $\equiv^L$ classes of finite arities. Then since $\text{Autf}(\mathcal{M})$ is closed, from 5.1.5(3), $f \in \text{Autf}(\mathcal{M})$.

(3)$\Rightarrow$(1) Let $f \in \text{Autf}_{\text{KP}}(\mathcal{M})$. Due to (3), for a model M, we have $M \equiv^L f(M)$, and there is $g \in \text{Autf}(\mathcal{M})$ such that $f(M) = g(M)$. Thus, by 5.1.2(1), $f \in \text{Autf}(\mathcal{M})$.

(3)$\Leftrightarrow$(4), (1)$\Rightarrow$(5)$\Rightarrow$(3) Clear. ■

Recall that $x \equiv^L_A y$ iff $d_A(x, y)$ is finite (4.2.15). Newelski [94] shows that T is G-compact over real A iff there is $n < \omega$ such that for any arity, ($x \equiv^L_A y$ iff $d_A(x, y) \leq n$).

Due to 4.2.19, we have the following:

Proposition 5.1.11 *Simple T is G-compact over any hyperimaginary.*

Now, we talk about strong types in the hyperimaginary context that is somewhat more subtle, even the definition, than real case.

Definition 5.1.12 We say two hyperimaginaries a, b have the same strong or Shelah type over A, written $a \equiv^s_A b$ or $\mathrm{stp}(a/A) = \mathrm{stp}(b/A)$, if $\boldsymbol{E}_{\mathrm{Autf_S}(\mathcal{M},A)}(a, b)$ holds.

Proposition 5.1.13

(1) $\mathrm{Gal}^0_{\mathrm{L}}(T, A)$ *is the intersection of all closed (normal) subgroups of* $\mathrm{Gal}_{\mathrm{L}}(T, A)$ *having finite index.*

(2) $\mathrm{Autf_S}(\mathcal{M}, A) = \{f \in \mathrm{Aut}_A(\mathcal{M}) | f$ *stabilizes all strong types over A of finite arities*$\}$.

Proof (1) We recall 5.1.5(1). Clearly, $\mathrm{Gal}^0_{\mathrm{L}}$ is contained in any closed subgroup of $\mathrm{Gal}_{\mathrm{L}}$ of finite index. Moreover, $\mathrm{Gal_S}$ is a profinite group. In a profinite group, the identity is the intersection of all normal closed subgroups of finite index. Hence, (1) follows.

(2) follows from 5.1.5(3), 5.1.6(2). ■

According to Lemma 5.1.6, $x \equiv^s_A y$ is bounded type-definable over A. We have a more precise collection of formulas type-defining it.

Proposition 5.1.14 *Let u, v be real tuples. The following are equivalent:*

(1) *$u \equiv_A v$; and for any $\emptyset$-definable equivalence relation E, if $u/E \in \mathrm{acl}(A)$, then $E(u, v)$ holds.*

(2) $u \equiv^s_A v$.

(3) *For any A-type-definable equivalence relation E, if u/E has finitely many conjugates over A, then $E(u, v)$ holds.*

(4) $u \equiv_{\mathrm{acl}(A)} v$.

(5) $u \equiv_{\mathrm{acl^{eq}}(A)} v$.

(6) *$u'_0 \equiv^s_A u'_1$ for each corresponding finite subtuple u'_i of c_i ($i = 0, 1$).*

Proof (1)⇒(2) By 5.1.13(1), $\mathrm{Gal}^0_{\mathrm{L}}(T, A) = \bigcap_i H_i$, where H_i is a closed normal subgroup of $\mathrm{Gal}_{\mathrm{L}}(T, A)$ having finite index, so H_i is open as well. Due to the similar argument as in the proof of 5.1.6(1), there is a formula $F_i(x', y')$ over $\emptyset$ defining an equivalence relation on $\mathrm{tp}(u)$, hence on $\mathcal{M}^{|u|}$ by compactness, such that $u/F_i \in \mathrm{acl}(A)$. Since H_i is normal, it again follows that on $p(x) = \mathrm{tp}(u/A)$, $F_i(x', y')$ defines $\boldsymbol{E}_{H_i}(x, y)$, in particular, $p(x) \models F_i(x', u') \leftrightarrow \boldsymbol{E}_{H_i}(x, u)$ (the corresponding finite $u' \subseteq u$). Therefore, (1)⇒(2) follows.

(2)⇒(3) Let $E(x, y)$ type-define an equivalence relation over A such that $e = u/E \in \mathrm{acl}(A)$. Then $\mathrm{Aut}_{eA}(\mathcal{M})$ is a subgroup of $\mathrm{Aut}_A(M)$ containing $\mathrm{Autf}_A(\mathcal{M})$. Now,

$E(x,u)$ clearly defines $E_{\mathrm{Aut}_{eA}(\mathcal{M})}(x,u)$ on $\mathrm{tp}(u/A)$. Hence, $\pi(\mathrm{Aut}_{eA}(\mathcal{M}))$ is closed (may not be normal) and it has finite index in $\mathrm{Gal}^0_{\mathrm{L}}(T,A)$ as automorphisms in a coset send e to a different E class. Therefore, $\mathrm{Gal}^0_{\mathrm{L}}(T,A) \subseteq \pi(\mathrm{Aut}_{eA}(\mathcal{M}))$, and (2)⇒(3) follows.

(3)⇒(1); and (4)⇒(5)⇒(1) Obvious.

(3)⇒(4) The proof is the same as that of Corollary 5.1.9(3)(b)⇒(c). This time $J \subseteq I$ are finite, and q is algebraic.

(1)⇔(6) follows by compactness. ■

Corollary 5.1.15 $\mathrm{acl}(A)$ *and* $\mathrm{acl}^{\mathrm{eq}}(A)$ *are interdefinable.*

Proof Let $u/E \in \mathrm{acl}(A)(u$ real$)$. It suffices to show u/E is definable over $\mathrm{acl}^{\mathrm{eq}}(A)$. According to 5.1.14, $u_E \in \mathrm{dcl}(u/\equiv^s)$ and $u/\equiv^s$ is definable over $\mathrm{acl}^{\mathrm{eq}}(A)$. ■

Example 5.1.16 We give a couple of examples related to Proposition 5.1.14. In (1) that $u \equiv_A v$ is essential. Let $\mathcal{L} = \{=\}$, and let $u \neq v$ be singletons in the infinite $\mathcal{M}$. By quantifier elimination, the second clause of (1) holds with $A = v$, but $u \not\equiv_v v$ much less $u \equiv^s_v v$.

Also if A is real as well, then $u \equiv^s_A v$ iff for any A-definable finite equivalence relation E, $E(u,v)$ holds. *This no longer holds in hyperimaginary context.* Even the right-hand side need not imply that $u \equiv_A v$. The difference comes from that in real context any formula in $\mathrm{tp}(u/v)$ is v-invariant, but not in general if v is a hyperimaginary. Recall Example 3.1.7. Choose $b, d \in C$ such that $b_E \neq d_E$ there. Then for any d_E-invariant finite definable equivalence relation F (there almost no such except trivial one), $F(b,d)$ holds. But *not* even $b \equiv_{d_E} d$ holds.

In the same manner the properties of strong types of hyperimaginaries follow.

Proposition 5.1.17 *Let $a = u_F, b = v_F$ with real u, v be F-hyperimaginaries. The following are equivalent:*

(1) $a \equiv^s_A b$.

(2) $F^s_A(u,v)$ *holds where* $F^s_A(x,y)$ *is a bounded type-definable equivalence relation over A coarser than F such that*

$$F^s_A(x,y) := \exists z z'(z \equiv^s_A z' \wedge F(z,x) \wedge F(z',y)) \equiv \exists z(F(z,y) \wedge z \equiv^s_A x),$$

($z \equiv^s_A x$ is equality of strong types over A of real tuples).

(3) $a \equiv_{\mathrm{acl}(A)} b$.

Proof (1)⇔(2) By the similar reason in the proof of 5.1.9(3)(a)⇔(b). (1)⇔(3) comes from 5.1.14. ■

Proposition 5.1.18 *The following are equivalent:*

(1) $\mathrm{Autf}_{\mathrm{KP}}(\mathcal{M},A) = \mathrm{Autf}_{\mathrm{S}}(\mathcal{M},A)$.

(2) $\mathrm{Gal_{KP}}(T, A)$ *is compact Hausdorff and totally disconnected.*

(3) $\equiv^{KP}_A$ *is equivalent to* $\equiv^s_A$ *for any arity.*

(4) $\equiv^{KP}_A$ *is equivalent to* $\equiv^s_A$ *for finite arities.*

(5) $\equiv^{KP}_A$ *is equivalent to* $\equiv^s_A$ *for any hyperimaginary variables.*

(6) $\mathrm{acl}(A)$ *and* $\mathrm{bdd}(A)$ *are interdefinable.*

In Example 3.1.7, $\mathrm{Autf_{KP}} \neq \mathrm{Autf_S}$. Some non G-compact examples (so $\mathrm{Autf} \neq \mathrm{Autf_{KP}}$) are constructed in [17]. In an example, $\equiv^L$ is different from $\equiv^{KP}$ for a finite tuple; while they can be equal for all finite arities in another non-G-compact example. In [125], given any compact Hausdorff topological group G, a corresponding theory T_G is constructed so that $\mathrm{Gal_L}(T_G) = G$.

That "Ltp $\equiv$ stp in real (hyperimaginary, resp.) context" means for any tuple c and a set A both real (hyperimaginaries, resp.), it holds that $\mathrm{Ltp}(c/A) \equiv \mathrm{stp}(c/A)$.

Proposition 5.1.19

(1) *If T has elimination of hyperimaginaries, then the equivalent conditions in 4.1.18 hold. In particular if simple T has e.h.i. then* Ltp $\equiv$ stp.

(2) *Assume T is simple. The following are equivalent.*

  (a) Ltp $\equiv$ stp *in real context.*

  (b) Ltp $\equiv$ stp *in hyperimaginary context.*

  (c) *Type amalgamation holds over any algebraically closed set.*

Proof We only show (2)(a)$\Rightarrow$(b). Others are direct consequences of 5.1.18 with 5.2.19 or 5.2.12. Assume Ltp $\equiv$ stp in real context. Let u_E be a hyperimaginary and let v, v' be real. It suffices to show $v \equiv^s_{u_E} v'$ implies that $v \equiv^L_{u_E} v'$. Choose a model M containing u. Now E clearly type-defines an equivalence relation on M as well, and u/E and M/E are interdefinable. Hence, there is no harm to suppose that u is some enumeration of the model. Now let $F(x, y)$ be $E(x, y) \wedge x \equiv^L y$. Then as in 4.3.9(7), $u_F = \mathrm{Cb}(u/u_E)$; and u_F, u_E are interbounded.

We claim that $v \equiv^s_{u_F} v'$ (even $v \equiv_{u_F} v'$) implies $v \equiv^L_{u_F} v'$: Suppose that $v \equiv_{u_F} v'$. We can clearly assume $u \downarrow_{u_F} v$, Take v'' such that $v'' \equiv^L_{u_F} v'$ and $v'' \downarrow_{u_F} v$. Now there is u' such that $u'v'' \equiv_{u_F} uv$. Hence, by type amalgamation there is $u'' \models \mathrm{tp}(u/u_F v) \cup \mathrm{tp}(u'/u_F v'')$. Then $u''v'' \equiv u''v$, and as $u''(\supseteq u_F)$ is a model, $v \equiv^L_{u_F} v''$ so $v \equiv^L_{u_F} v'$, as desired.

By the claim and 5.1.18, $\mathrm{bdd}(u_E) = \mathrm{bdd}(u_F) = \mathrm{acl}(u_F)$. Moreover, by the definition of F and our assumption, u_F is definable over $\mathrm{acl}(u_E)$. Hence, $\mathrm{bdd}(u_E) = \mathrm{acl}(u_E)$, and by 5.1.18 again, Ltp $\equiv$ stp over u_E, as wanted. ■

Question Does the converse of (1) hold (for simple T)? Is the above (2)(a)$\Rightarrow$(b) true for all T?

5.2 Lascar Types Are Strong Types in Low Theories

In this section we show for low theories that Ltp $\equiv$ stp, a weaker condition than elimination of hyperimaginaries. According to 5.1.19 and 3.2.11(3) it suffices to show that Ltp $\equiv$ stp for finite real tuples over a finite real set A. For convenience we suppress A to $\emptyset$. We shall use Proposition 2.5.5.

In the rest of this section T is low, and tuples are finite. Let $r(x,y)$ be a type over $\emptyset$, closed under conjunctions, defining equality of Lascar strong types $\mathrm{Ltp}(x) = \mathrm{Ltp}(y)$. Fix $p(x) \in S_n(\emptyset)$ and $\varphi(x,y) \in r$. Define

$$
\begin{aligned}
R(u,v) \equiv\ & p(u) \wedge p(v) \text{ and there is } v' \mathop{\smile\!\!\!\!\!\!\mid}\, u \text{ and } r(v,v') \text{ such that} \\
& \varphi(x,u) \wedge \varphi(x,v') \text{ does not fork.} \\
S(u,v) \equiv\ & p(u) \wedge p(v) \text{ and there is } v' \mathop{\smile\!\!\!\!\!\!\mid}\, u \text{ and } r(v,v') \text{ such that} \\
& \varphi(x,u) \wedge \varphi(x,v') \text{ forks.}
\end{aligned}
$$

For $a \models p$, $R(a,a)$ holds (3.2.1). Note that if $u \mathop{\smile\!\!\!\!\!\!\mid}\, v'$, then $I = \langle u_i v'_i | i < \omega \rangle$ is a Morely sequence of $\mathrm{tp}(uv')$ iff I is indiscernible in $\mathrm{tp}(uv')$ and $\{u_i, v'_i | \, i < \omega\}$ is independent. Hence, due to 2.2.6, 2.2.9, 2.5.5 both R and S are type-definable relations depending on φ. (This is the place where lowness of T is used.) Clearly,

$$R(u,v) \wedge r(v,w) \models R(u,w), \text{ and } S(u,v) \wedge r(v,w) \models S(u,w). \quad (*)$$

Moreover, for $u, v \models p(x)$, $\neg R(u,v)$ iff $S(u,v)$. Since there always is $v' \mathop{\smile\!\!\!\!\!\!\mid}\, u$ with $r(v,v')$, either $R(u,v)$ or $S(u,v)$ holds. But Proposition 3.2.7 says it is not the case that $R(u,v)$ and $S(u,v)$ hold at the same time.

Therefore, as type-definable R and S partition $p \times p$, by compactness, both turn out to be definable. Namely, there is a formula $\sigma(x,y) (= \sigma_\varphi(x,y))$ such that

$$p(u) \wedge p(v) \models R(u,v) \leftrightarrow \sigma(u,v).$$

Then, according to (*) and compactness again, there is a formula $\delta(x) \in p(x)$ such that

$$\delta(u) \wedge \delta(v) \wedge \sigma(u,v) \wedge r(v,w) \models \sigma(u,w). \quad (**)$$

Now let $E_\varphi(x,y)$ be a definable equivalence relation such as

$$E_\varphi(x,y) = \big[\neg\delta(x) \wedge \neg\delta(y)\big] \vee \big[\delta(x) \wedge \delta(y) \wedge \forall z(\delta(z) \rightarrow (\sigma(z,x) \leftrightarrow \sigma(z,y))\big].$$

Since $r(x,y)$ implies that $\mathrm{tp}(x) = \mathrm{tp}(y)$, from (**), $\models r(x,y) \rightarrow E_\varphi(x,y)$. Hence, each E_φ is a finite equivalence relation as the number of classes of $r(x,y)$ is bounded. Now, for $a, b \models p$, assume that $\neg r(a,b)$. Then there must exist $\psi(x,y) \in r$ such that $\neg\psi(a,b)$ (†). Also, since r is transitive, there, too, exists by compactness $\phi(x,y) \in r$ such that

$$\exists x x' (\phi(x,y) \wedge \phi(x,x') \wedge \phi(x',z)) \models \psi(y,z). \quad (***)$$

Then $\neg E_\phi(a,b)$. Otherwise, since $\sigma_\phi(a,a)$, we have $\sigma_\phi(a,b)$. Hence, for some c with $r(c,b)$, we have $\phi(x,a) \wedge \phi(x,c)$ does not fork. Thus, $\exists x, c(\phi(x,a) \wedge \phi(x,c) \wedge \phi(c,b))$, and from (***), $\models \psi(a,b)$, which contradicts (+).

Therefore, $p(x) \wedge p(y) \models r(x,y) \leftrightarrow \bigwedge_{\varphi \in r} E_\varphi(x,y)$. In conclusion, Ltp $\equiv$ stp as desired.

5.3 Elimination of Finitary Hyperimaginaries for Small Theories

This section aims to show that in a small theory any type-definable equivalence relation of a finite arity over a finite set is a conjunction of definable ones. In this section, tuples and sets are *all real* and $x, y, \ldots$ are *finite* real variables; T is arbitrary.

For the rest, given a type-definable equivalence relation $E(x,y)$ on $\mathcal{M}^n$ over A, we let

$$E^*(x,y) \equiv \exists z(E(x,z) \wedge \mathrm{tp}(z/A) = \mathrm{tp}(y/A)).$$

As similarly in 5.1.9(3)(b), it follows that E^* again is a type-definable equivalence relation over A, and is the *finest* among those *coarser* than both E and equality of n-types over A. By symmetry, $\models E^*(x,y)$ iff $\exists z(\mathrm{tp}(z/A) = \mathrm{tp}(x/A)) \wedge E(z,y)$, and, hence, for each $p \in S_n(A)$, $\exists y(p(y) \wedge E(x,y))$ type-defines exactly the E^* class of any $c \models p$.

Proposition 5.3.1 *Given a type-definable equivalence relation $\bar{E}(x,y)$ on $\mathcal{M}^n$ over some A is equivalent to an infinite conjunction of definable equivalence relations over A if and only if, so is $\bar{E}^*(x,y)$ and so is each restriction of $\bar{E}(x,y)$ to each $p(x) \in S_n(A)$.*

Proof ($\Rightarrow$) Assume left. Hence, $\bar{E}(x,y) \leftrightarrow \bigwedge_i E_i(x,y)$ for some set $\{E_i(x,y) \mid i \in I\}$ of definable equivalence relations over A closed under conjunctions. Now, for each $i \in I$, let Ψ_i be a collection of formulas $\psi_i(x)$ over A such that $\exists x(\psi_i(x) \wedge E_i(x,y))$ implies that $\psi_i(y)$. Then clearly $\psi_i(x) \leftrightarrow \psi_i(y)$ is a definable equivalence relation coarser than E_i. Let $\Psi(x) = \bigcup_i \Psi_i$. Define

$$\bar{E}'(x,y) \text{ iff } \bigwedge \{\psi \in \Psi \mid \psi(x) \leftrightarrow \psi(y)\}.$$

It suffices to show $\bar{E}'$ and $\bar{E}^*$ are equivalent. Since $\bar{E}'$ is coarser than $\bar{E}$ and equality of types, we have $\bar{E}^*(x,y) \models \bar{E}'(x,y)$. Conversely, assume that $\neg\bar{E}^*(a,b)$. Then by compactness, there must be say E_0 $(0 \in I)$ and $\sigma \in \mathrm{tp}(b/A)$ such that $\{E_0(x,a), \sigma(x)\}$ is inconsistent. Then let

$$\psi_0(x) = \exists z(E_0(x,z) \wedge \sigma(z)).$$

Clearly, $\psi_0(x) \in \Psi_0$, and $\neg\psi_0(a)$, $\psi_0(b)$ hold. Thus, as desired $\neg\bar{E}'(a,b)$.

($\Leftarrow$) Assume the right-hand side. Fix $\epsilon(x,y) \in \bar{E}(x,y)$ and $p(x) \in S_n(A)$. Now, there are sets $\{E_j^*(x,y) \mid j \in I\}$, $\{E_i^p(x,y) \mid i \in I\}$ of definable equivalence relations over A closed under conjunctions such that

$$\models \bar{E}^*(x,y) \leftrightarrow \bigwedge_j E_j^*(x,y) \text{ and } p(x) \wedge p(y) \models \bar{E}(x,y) \leftrightarrow \bigwedge_i E_i^p(x,y)$$

hold. Note that since $\bar{E}^*$ has boundedly many classes, each E_j^* is a finite equivalence relation. Now, by compactness there are $\psi_{p,\epsilon}(x) \in p(x)$, $i = i(p,\epsilon) \in I$, and $\epsilon'(x,y) \in \bar{E}$ implying $\epsilon(x,y)$ (all depending on p, ϵ) such that

$$\psi_{p,\epsilon}(x) \wedge \psi_{p,\epsilon}(y) \wedge E_i^p(x,y) \models \epsilon(x,y) \tag{†}$$

and

$$\psi_{p,\epsilon}(x) \wedge \psi_{p,\epsilon}(y) \wedge \epsilon'(x,y) \models E_i^p(x,y) \tag{‡}$$

There also is $\epsilon'' \in \bar{E}$ implying ϵ' such that

$$\exists x'y'((\epsilon''(x,x') \wedge \epsilon''(x',y') \wedge \epsilon''(y',y)) \models \epsilon'(x,y). \tag{$\star$}$$

Claim 1 There are $p_k \in S_n(A)$, formulas $F_{k,\epsilon}(x)$ over A $(k = 1, \ldots, m)$, and $E_{j_0}^*$ (all only depending on ϵ) such that

(i) $\{F_{1,\epsilon}(x), \ldots, F_{m,\epsilon}(x)\}$ partitions $\mathcal{M}^n$,
(ii) $F_{k,\epsilon}(x) \models \exists y(\psi_{p_k,\epsilon}(y) \wedge \epsilon''(x,y))$, and
(iii) Each set defined by $F_{k,\epsilon}(x)$ is a finite union of $E_{j_0}^*$ classes:

By the remark preceding 5.3.1, $\bar{E}^*(x,c) \models \exists y(\psi_{p,\epsilon}(y) \wedge \epsilon''(x,y))$ for any $c \models p$. Hence, for some $j = j(p,\epsilon)$, $E_j^*(x,c) \models \exists y(\psi_{p,\epsilon}(y) \wedge \epsilon''(x,y))$. Note that $E_j^*(x,c)$ is indeed equivalent to a formula $F'_{p,\epsilon}(x)$ in $p(x)$ since the solution set is A-invariant. Hence, by compactness on all $p \in S_n(A)$, there is a finite set $\{p_1, \ldots, p_m\} \subseteq S_n(A)$ (only depending on ϵ) such that $\models \forall x(F'_{p_1,\epsilon}(x) \vee \ldots \vee F'_{p_m,\epsilon}(x))$ and each $F'_{p_k,\epsilon} \wedge (\neg F'_{p_1,\epsilon} \wedge \ldots \wedge \neg F'_{p_{k-1},\epsilon})$ (which we put $F_{k,\epsilon}$) is consistent. Clearly, we can find $E_{j_0}^*$, too, satisfying Claim 1.

We now let $E_\epsilon(x,y)$ be

$$\bigvee_{k=1}^{m} F_{k,\epsilon}(x) \wedge F_{k,\epsilon}(y) \wedge (\forall x'y' \left(\psi_{p_k,\epsilon}(x') \wedge \psi_{p_k,\epsilon}(y') \wedge \epsilon''(x,x') \wedge \epsilon''(y,y') \rightarrow E^{p_k}_{i(p_k,\epsilon)}(x',y')\right).$$

According to Claim 1(i),(ii), it forms an A-definable equivalence relation.

Claim 2 $\{\epsilon \in \bar{E} | E_\epsilon(x,y)\}$ type-defines $\bar{E}$: If $\bar{E}(a,b)$ holds, and a realizes say $F_{0,\epsilon}$, then since $\bar{E}^*(a,b)$, we have $E_{j_0}^*(a,b)$ and thus $b \models F_{0,\epsilon}$, too, by Claim 1(iii). If $a', b' \models \psi_{p_0,\epsilon}$ and $\epsilon''(a,a'), \epsilon''(b,b')$, then due to (‡)($\star$), we have $E^{p_0}_{i(p_0,\epsilon)}(a',b')$. Hence, $E_\epsilon(a,b)$ holds.

Conversely, if $E_\epsilon(a,b)$, and say $a, b \models F_{0,\epsilon}$ then by Claim 1(ii), there are $a', b' \models \psi_{p_0,\epsilon}$ such that $\epsilon''(a,a') \wedge \epsilon''(b,b')$, and so $E^{p_0}_{i(p_0,\epsilon)}(a',b')$ holds. Thus, from (†),

$\epsilon(a', b')$, and hence, we have $\epsilon(a, a') \wedge \epsilon(a', b') \wedge \epsilon(b', b)$. Therefore, it follows that $\{\epsilon \in \bar{E} | E_\epsilon(x, y)\} \models \bar{E}(x, y)$. ■

Lemma 5.3.2 *Let $c \models p(x) \in S_n(A)$, and let $\bar{E}(x, y)$ be an equivalence relation on $\mathcal{M}^n$ over A. Then*

$$\bar{F}(x, y) := \exists z\big(\text{tp}(cx/A) = \text{tp}(zy/A) \wedge \bar{E}(c, z)\big)$$

is an equivalence relation over cA, coarser than equality of n-types over Ac. If $\bar{F}$ is equivalent to some conjunction of definable equivalence relations over Ac, then the restriction of $\bar{E}$ to $p(x)$ is equivalent to a conjunction of definable equivalence relations over A.

Proof Since $\bar{E}$ is A-invariant, it is not hard to see that $\bar{F}$ forms an equivalence relation. Moreover, if $\text{tp}(d/Ac) = tp(e/Ac)$ for n-tuples d, e, then, obviously, $\bar{F}(d, e)$ holds. Hence, $\bar{F}$ is coarser than equality of n-types over Ac. Assume now that

$$\bar{F}(x, y) \leftrightarrow \bigwedge_i E_i(x, y; c) \qquad (*)$$

for some formulas $E_i(x, y; c)$ over Ac defining equivalence relations. For each i, we let

$$F_i(x, y) \equiv \forall u(E_i(u, x, x) \leftrightarrow E_i(u, y, y)).$$

We claim that $p(x) \wedge p(y) \models \bar{E}(x, y) \leftrightarrow \bigwedge_i F_i(x, y)$: It suffices to show that $p(y) \models \bar{E}(c, y) \leftrightarrow \bigwedge_i F_i(c, y)$. Let $a \models p$ be given. Firstly, assume that $\bar{E}(c, a)$. Let u be an arbitrary n-tuple. Now, by (*) and the definition of $\bar{F}$, for some u' with $\text{tp}(au/A) = tp(cu'/A)$, we have $E_i(u, u'; c)$ for each i. Then as $E_i(x, y; c)$ is an equivalence relation, for each i, $E_i(u, c, c) \leftrightarrow E_i(u', c, c) \leftrightarrow E_i(u, a, a)$. Therefore, $F_i(c, a)$ holds for each i. Conversely, if $F_i(c, a)$ for each i, then as $E_i(a, a; a)$, we have $E_i(a, c, c)$ for each i. Thus, $\bar{F}(a, c), \bar{F}(c, a)$ and $\bar{E}(c, a)$ hold. ■

We apply the above propositions to small (countable) theories (e.g., ω-categorical theories).

Proposition 5.3.3 *Let T be small. Let $\bar{E}(x, y)$ be a type-definable equivalence relation over a finite set A coarser than equality of n-types over A. Then, there are formulas $E_i(x, y)$ $(i \in \omega)$ over A, each of which defines an equivalence relation, such that*

$$\bar{E}(x, y) \leftrightarrow \bigwedge\nolimits_{i \in \omega} E_i(x, y).$$

Proof Since T being small is preserved under naming finitely many constants, we assume $A = \emptyset$. Suppose now that $\neg\bar{E}(a, b)$ (i.e., $\{\bar{E}(x, a), \bar{E}(x, b)\}$ is inconsistent), and $p(x) = tp(a)$, $q(y) = tp(b)$. As $\bar{E}$ is coarser than equality of types, $p(x) \cup q(y) \models \neg\bar{E}(x, y)$. Hence, we can choose $\varphi(x) \in p(x)$, $\psi(y) \in q(y)$ among those formulas witnessing

$\varphi(x) \wedge \psi(y) \vdash \neg\bar{E}(x,y)$ such that the set X, defined by $\neg(\varphi(y) \vee \psi(y))$, has the minimum Cantor–Bendixson rank and degree.

We claim $X = \emptyset$: If not, then $CB(X) = (\alpha, n)$. Now there is $p'(y) \in S(\emptyset)$ containing $\neg(\varphi(y) \vee \psi(y))$ such that the CB-rank of $p'(y)$ is α. Since it is not possible that both $\varphi(x) \wedge p'(y) \wedge \bar{E}(x,y)$ and $p'(x) \wedge \psi(y) \wedge \bar{E}(x,y)$ are consistent, we may assume that there is $\sigma(y) \in p'(y)$ such that $\varphi(x) \wedge \sigma(y) \vdash \neg\bar{E}(x,y)$. Hence, for $\psi'(y) := \psi(y) \vee \sigma(y) \in q(y)$, we have $\varphi(x) \wedge \psi'(y) \vdash \neg\bar{E}(x,y)$ as well. Now since the set X' defined by $\neg(\varphi(y) \vee \psi'(y))$ is contained in X and $CB(\sigma(y)) \geq (\alpha, 1)$, we have $CB(X') < (\alpha, n)$. This contradicts the choice of φ and ψ. Hence, the claim follows.

Now since $\{\varphi(x), \psi(x)\}$ cannot be consistent, by the above claim, it partitions the universe, and hence $\bar{E}(x,y) \wedge \varphi(x) \models \varphi(y)$. Let us rewrite $\varphi(x)$ as $\varphi_{p,q}(x)$. It follows that $\bar{E}(x,y)$ is a conjunction of formulas $\varphi_{p',q'}(x) \leftrightarrow \varphi_{p',q'}(y)$, for all $p', q' \in S(\emptyset)$ witnessing $p'(x) \wedge q'(y) \models \neg\bar{E}(x,y)$. ■

By 5.3.1 and 5.3.2, the previous result is extended to any finitary type-definable equivalence relations over a finite set.

Corollary 5.3.4 (T small) *Any type-definable equivalence relation on $\mathcal{M}^n$ over finite A is an infinite conjunction of A-definable equivalence relations.*

Based on 3.2.11 and 5.1.19 we have the following:

Corollary 5.3.5 *Assume T is small. Then T has elimination of finitary hyperimaginaries. If, in addition, T is simple, then* Ltp $\equiv$ *stp in a hyperimaginary context.*

Corollary 5.3.4 is stronger than elimination of finitary hyperimaginaries.

Example 5.3.6 Let M be a model consisting of the universe $\mathbb{Q}$, and unary predicates $U_a = \{x \in \mathbb{Q} | x \leq a\}$ $(a \in \mathbb{Q})$. For each $a \in \mathbb{Q}$, let us consider the following types (over the empty set):

(1) type a^+ determined by $\{U_b(x) | a < b \in \mathbb{Q}\} \cup \{\neg U_b(x) | a \not< b \in \mathbb{Q}\}$,

(2) type a^- determined by $\{U_b(x) | a \leq b \in \mathbb{Q}\} \cup \{\neg U_b(x) | a \not\leq b \in \mathbb{Q}\}$.

Now let E be the equivalence relation type-defined by the conjunction of the formulas $(U_c(x) \rightarrow U_d(y)) \wedge (U_c(y) \rightarrow U_d(x))$ for $c < d$. It is easy to see, for each $a \in \mathbb{Q}$, (type a^+)$\cup$(type a^-) consists of an E class.

Now, if E were a conjunction of (finite) definable equivalence relations E_i over $\emptyset$, then, for every i, each of (type a^+)$\cup$(type a^-) should be contained in one of the E_i classes. Quantifier elimination shows, then, E_i must be trivial. (Note that E_i cannot be of the form $U_a(x) \leftrightarrow U_a(y)$.) Therefore, E is not a conjunction of definable equivalence relations. But Th(M) is stable, so it has e.h.i.

Question Does any small simple theory have e.h.i?

If T is not simple, the answer is no. Any theory having the strict order property has an infinitary non-eliminable hyperimaginary [2].

5.4 Elimination of Hyperimaginaries for Supersimple Theories

In this section, we shall show supersimple T has elimination of hyperimaginaries (Theorem 5.4.9). Before we start, we should point out a direct consequence from 2.5.15, which will be used freely.

Remark 5.4.1 In supersimple T, given a real tuple $\bar{b}$ possibly infinite, $SU(\bar{b}) < \infty$ iff there is some finite real subtuple b of $\bar{b}$ such that $\bar{b} \in \operatorname{acl}(b)$.

We will also freely use that in simple T, if $e \in \operatorname{bdd}(ab)$ and $SU(a/b) < \infty$, then $SU(e/b) < \infty$ (4.2.13(2)). Results in Section 4.4.2 will be referred to frequently.

In the rest of this section, T **is supersimple**; $a, b, \ldots, x, y, \ldots$ are finite imaginary tuples or variables; $\bar{a}, \hat{a}, \ldots, \bar{x}, \hat{y}, \ldots$ are possibly infinite imaginary tuples or variables; and hyperimaginaries are written in boldface letters $\boldsymbol{a}, \boldsymbol{b}, \ldots$.

Lemma 5.4.2 *Let $E(\bar{x}, \bar{y})$ be a set of formulas over $\emptyset$ type-defining an equivalence relation on $\operatorname{tp}(\bar{a})$ with $SU(\bar{a}) < \infty$. Then there are finite $a \subseteq \bar{a}$ with $\bar{a} \in \operatorname{acl}(a)$, and $\psi(x, y) \in E$ such that for any $\bar{b}, \bar{c} \equiv \bar{a}$, if $\psi(b, c)$ (where b, c are corresponding subtuples of $\bar{b}, \bar{c}$ for x, y) then $\bar{b}_E, \bar{c}_E$ are interbounded.*

Proof Recall our convention in 4.1.2(3). In particular, E is closed under the projections of formulas. Now, since $SU(\bar{a}) < \infty$, there is finite $a_0 \subseteq \bar{a}$ such that $\bar{a} \in \operatorname{acl}(a_0)$. Let $x_0 \subseteq \bar{x}$ be the corresponding variable for a_0 so that $\bar{x} \in \operatorname{acl}(x_0)$. Select $\psi_0(x_0, a_0) \in E(\bar{x}, \bar{a})$ having the least D-rank. For the rest, for a formula $\alpha(x', y') \in E$, $\alpha(\bar{x}, \bar{y})$ means $\alpha(x', y') \wedge \bar{x} = \bar{x} \wedge \bar{y} = \bar{y}$, equivalently on $\operatorname{tp}(\bar{a})$, $\alpha(x', y') \wedge x' \in \operatorname{acl}(x_0) \wedge y' \in \operatorname{acl}(y_0)$. Now there are $\psi_1, \psi_2 \in E$ such that

$$\models \psi_{k+1}(\bar{x}, \bar{y}) \wedge \psi_{k+1}(\bar{y}, \bar{z}) \rightarrow \psi_k(\bar{x}, \bar{z}) \text{ for } k = 0, 1.$$

We claim that $\psi(x, y) = \psi_2(x, y)$ $(x_0 \subseteq x \subseteq \bar{x})$ is a desired formula. So assume that $\bar{b} \equiv \bar{a}$ and $\psi(b, a)$, and we want to show that $\bar{a}_E$ and $\bar{b}_E$ are interbounded. Suppose not; say $\bar{b}_E \notin \operatorname{bdd}(\bar{a}_E)$. Then we can take an indiscernible sequence $I = \langle \bar{b}_i | i < \omega \rangle$ with $\bar{b}_0 = \bar{b}$ in $\operatorname{tp}(\bar{b}/\operatorname{bdd}(\bar{a}_E))$ such that $\not\models E(\bar{b}_i, \bar{b}_j)$ for $i \neq j$. Now, $\psi(\bar{b}_i, \bar{a}_i)$ holds for some $\bar{a}_i$ with $E(\bar{a}_i, \bar{a})$ (*). As in 4.2.3(1), I is some $\bar{a}'$-indiscernible with $E(\bar{a}', \bar{a})$. By moving $\bar{a}'$ to $\bar{a}$ we can assume I is $\bar{a}$-indiscernible while keeping (*). So for each i, $\psi_1(\bar{b}_i, \bar{a})$ since $\psi(\bar{b}_i, \bar{a}_i) \wedge \psi(\bar{a}_i, \bar{a})$ holds. Moreover, there are $\chi, \varphi \in E$ such that $\not\models \chi(\bar{b}_i, \bar{b}_j)$ for $i \neq j$, and $\models \varphi(\bar{x}, \bar{y}) \rightarrow \psi_1(\bar{x}, \bar{y})$, $\models \varphi(\bar{x}, \bar{y}) \wedge \varphi(\bar{y}, \bar{z}) \rightarrow \chi(\bar{x}, \bar{z})$. Then it follows that $\varphi(\bar{x}, \bar{b}_0) \rightarrow \psi_0(x_0, a_0)$ and for $i \neq j < \omega$, $\varphi(\bar{x}, \bar{b}_i) \wedge \varphi(\bar{x}, \bar{b}_j)$ is inconsistent. Indeed, we have $\exists(\bar{x} \setminus x_0)\varphi(\bar{x}, \bar{b}_0) \rightarrow \psi_0(x_0, a_0)$, and $\exists(\bar{x} \setminus x_0)\varphi(\bar{x}, \bar{b}_0)$ divides over a_0 since $\bar{x} \in \operatorname{acl}(x_0)$. Thus, $D(\exists(\bar{x} \setminus x_0)\varphi(\bar{x}, \bar{b}_0)) < D(\psi_0(x_0, a_0))$, contradicting the choice of ψ_0. We have proved the claim and lemma. ∎

Showing the following proposition implies elimination of finitary hyperimaginaries is our major task in this section.

Proposition 5.4.3 *Let $\bar{b}$ be a sequence of imaginaries, and let* $\boldsymbol{a}$ *be a hyperimaginary. Suppose that:*

(1) $\bar{b} \subseteq \mathrm{acl}(b)$ *for some finite subtuple* b *of* $\bar{b}$ *(equivalently* $SU(\bar{b}) < \infty$*), and*
(2) $\boldsymbol{a} \in \mathrm{dcl}(\bar{b})$.

Then $\boldsymbol{a}$ *is equivalent to a sequence of imaginaries.*

Proof We shall prove the following lemma.

Lemma 5.4.4 *Suppose, first, that Proposition 5.4.3 holds for any* $\bar{b}', \boldsymbol{a}'$ *satisfying (1)(2) with* $R(\bar{b}'/\boldsymbol{a}') < \alpha$ $(\alpha > 0)$*. Suppose, second, that given* $\bar{b}, \boldsymbol{a}$ *satisfying (1)(2),* $\mathrm{tp}(\bar{b}/\boldsymbol{a})$ *is internal in the family of formulas of D-rank* α*, and is orthogonal to all types of D-rank* $< \alpha$*. Then* $\boldsymbol{a}$ *is equivalent to some imaginary tuple.*

We, first, point out that under the assumption of Lemma 5.4.4, Proposition 5.4.3 can be proved using the induction argument on $R(\bar{b}/\boldsymbol{a})$. Note that $R(\bar{b}/\boldsymbol{a})$ is defined since $SU(\bar{b}) < \infty$. If $R(\bar{b}/\boldsymbol{a}) = 0$, then $\bar{b} \in \mathrm{bdd}(\boldsymbol{a})$ and Proposition 5.4.3 follows from 4.1.5. Now suppose that 5.4.3 holds for all $\bar{b}', \boldsymbol{a}'$ fulfilling (1),(2) with $R(\bar{b}'/\boldsymbol{a}') < \alpha$ (the induction hypothesis). We now let $R(\bar{b}/\boldsymbol{a}) = \alpha$. Then from 4.4.16, there are hyperimaginaries $\boldsymbol{a} = \boldsymbol{a}_0 \subseteq \boldsymbol{a}_1 \subseteq \ldots \subseteq \boldsymbol{a}_n \subseteq \bar{b}$ and $\alpha_0, \ldots, \alpha_{n-1} \leq \alpha$ such that $\bar{b} \in \mathrm{bdd}(\boldsymbol{a}_n)$, and for $i < n$, $\mathrm{tp}(\boldsymbol{a}_{i+1}/\boldsymbol{a}_i)$ is internal in the family of formulas of D-rank α_i, and $\mathrm{tp}(\bar{b}/\boldsymbol{a}_i)$ is orthogonal to any type of D-rank $< \alpha_i$. It again follows from 4.1.5, $\boldsymbol{a}_n$ is interdefinable with an imaginary tuple $\bar{a}_n$. We can further show that so is $\boldsymbol{a}_{n-1}$: Note that the sequence $\boldsymbol{a}_0' = \boldsymbol{a}_{n-1}, \boldsymbol{a}_1' = \bar{a}_n (\subseteq \bar{b})$, and $\alpha_0' = \alpha_{n-1}$ witness $R(\bar{a}_n/\boldsymbol{a}_{n-1}) \leq \alpha_{n-1}$. Thus, if $\alpha_{n-1} < \alpha$, then apply the induction hypothesis. If $\alpha_{n-1} = \alpha$, then use Lemma 5.4.4 with $\bar{a}_n, \boldsymbol{a}_{n-1}$.

Hence, we replace $\boldsymbol{a}_{n-1}$ by a sequence $\bar{a}_{n-1}$ of imaginaries. The same argument with $\bar{a}_{n-1}, \boldsymbol{a}_{n-2}$ shows that $\boldsymbol{a}_{n-2}$, too, is equivalent to an imaginary tuple. By continuing this, we conclude that so is $\boldsymbol{a}_0 = \boldsymbol{a}$. Therefore, Proposition 5.4.3 is proved (under 5.4.4).

In the rest, we devote ourselves to show Lemma 5.4.4. Hence, suppose the two suppositions in the lemma hold. Now according to 4.1.5 and 4.1.2(2), we can write $\boldsymbol{a} = \bar{b}/E_0 = \bar{b}_{E_0}$ for some type-definable equivalence relation E_0 finer than equality of types. Note that $\mathrm{tp}(\bar{b}/\bar{b}_{E_0}) \equiv E_0(\bar{x}, \bar{b})$. As in 4.3.9(7), for $E(\bar{x}, \bar{y}) := E_0(\bar{x}, \bar{y}) \wedge \bar{x} \equiv^L \bar{y}$, we have $\mathrm{Ltp}(\bar{b}/\bar{b}_{E_0}) \equiv E(\bar{x}, \bar{b}) \equiv \mathrm{tp}(\bar{b}/\bar{b}_E) \equiv \mathrm{Ltp}(\bar{b}/\bar{b}_E)$, and $\bar{b}_E \in \mathrm{bdd}(\bar{b}_{E_0})$, $\bar{b}_{E_0} \in \mathrm{dcl}(\bar{b}_E)$. We shall show that $\bar{b}_E$ is interdefinable with a sequence of imaginaries. Then from 4.1.5, so is $\bar{b}_{E_0}$, and the proof is finished. Now as $\bar{b}$ and $\bar{b}_E$ also satisfy the second supposition in the lemma (since $\bar{b}, \bar{b}_{E_0}$ do), we reset $\boldsymbol{a} = \bar{b}_E$.

We list several ingredients necessary for showing the forthcoming main goal. When $\bar{x}$ is a variable tuple for $\bar{b}$, then $x \subseteq \bar{x}$ corresponds to b. Put $\bar{q} = \mathrm{tp}(\bar{b})$.

(i) We take notational convention in the proof of 5.4.2. In particular, given a formula $\psi(x', y') \in E$, $\psi(\bar{x}, \bar{y})$ can be identified as a single formula $\psi(x', y') \wedge x' \in \mathrm{acl}(x) \wedge y' \in \mathrm{acl}(y)$.

(ii) Now due to 5.4.2, there is $\psi_0 \in E(\bar{x},\bar{y})$ such that for $\bar{a},\bar{c} \models \bar{q}$, $\psi_0(\bar{a},\bar{c})$ implies that $\bar{a}_E, \bar{c}_E$ are interbounded. There, too, is $\psi_1 \in E$ such that $\exists \bar{w}(\psi_1(\bar{w},\bar{x}) \wedge \psi_1(\bar{w},\bar{y})) \models \psi_0(\bar{x},\bar{y})$. For the rest, we reset $\{\varphi \wedge \psi_1 | \varphi \in E(\bar{x},\bar{y})\}$ for E.

(iii) Note that $\mathrm{tp}(b/\bar{b}_E)$ is internal in the family of formulas of D-rank α. Thus we can find an imaginary tuple $\hat{b} \supseteq \bar{b}$, and formulas $\theta_i(z_i)$ of D-rank α over $\hat{b}'(\equiv_{\bar{b}_E} \hat{b}$, so $\hat{b}' \supseteq \bar{b}_E)$ with $c_i \models \theta_i$ $(i = 1, \ldots, n)$, such that $b \perp_{\bar{b}_E} \hat{b}'$; and $b = f(c_1 \ldots c_n)$, where f is some $\hat{b}'$-definable function. Let $\chi(x, \hat{b}')$ be the formula $\exists z_1 \ldots z_n(\theta_1(z_1) \wedge \ldots \wedge \theta_n(z_n) \wedge x = f(z_1 \ldots z_n))$, realized by b. We shall use the number n in the proof of Lemma 5.4.7 below. Clearly, there is no harm to assume that $\hat{b} \setminus \bar{b}$ is finite, so that $SU(\hat{b}) < \infty$.

(iv) Let $\hat{q} := \mathrm{tp}(\hat{b}') = \mathrm{tp}(\hat{b})$. From now on, given $\hat{d} \models \hat{q}$, $\bar{d}$ and d represent tuples such that $d\bar{d}\hat{d} \equiv b\bar{b}\hat{b}$. Similarly, if $\hat{x}$ is a variable tuple for $\hat{d}$ then $x \subseteq \bar{x}(\subseteq \hat{x})$ are variables for $d, \bar{d}$.

(v) By 3.1.5, there is a type-definable F on $\hat{q}$ such that $\hat{b}/F$ is equivalent to $\langle \mathbf{e} \in \mathrm{dcl}(\hat{b}) | R(\mathbf{e}/\bar{b}_E) < \alpha \rangle$. Then the following hold:

- (v1) $R(\hat{b}_F/\bar{b}_E) < \alpha$. It follows from 4.4.20.
- (v2) $\mathrm{tp}(\hat{b}/\hat{b}_F) \perp$ any type of D-rank $< \alpha$; again, due to 4.4.20.
- (v3) $\bar{b}_E \in \mathrm{dcl}(\hat{b}_F)$: Clear by the definition.
- (v4) $\mathrm{tp}(\hat{b}/\hat{b}_F) = F(\hat{x}, \hat{b})$ is an amalgamation base: If F' is finer than F but $\hat{b}_{F'} \in \mathrm{bdd}(\hat{b}_F)$, then $R(\hat{b}_{F'}/\bar{b}_E) < \alpha$ as well by 4.4.19(2). Hence, F' is equivalent to F, and (v4) follows as in 4.3.9(7).
- (v5) For any $\hat{d} \equiv \hat{b}$, $\mathrm{tp}(\bar{b}/\bar{b}_E) \perp \mathrm{tp}(\hat{d}_F/\bar{d}_E)$: Since $\mathrm{tp}(\bar{b}/\bar{b}_E)$ is orthogonal to all types of D-rank $< \alpha$ (the second supposition of 5.4.4), this comes from 4.4.18(3) and (v1).

In the rest, we shall show the following main goal:

Goal There is $\tilde{b}(\in \mathrm{dcl}(\hat{b}))$, a sequence of imaginaries, such that $\tilde{b} \in \mathrm{dcl}(\hat{b}_F)$ and $\bar{b}_E \in dcl(\tilde{b})$.

Note that once the goal is achieved, then since $R(\tilde{b}/\bar{b}_E) \leq R(\hat{b}_F/\bar{b}_E) < \alpha$ (4.4.19(2)), from the first supposition in Lemma 5.4.4, we can conclude that $\bar{b}_E$ is equivalent to an imaginary tuple. Thus, the proof of the lemma is finished. For this, we introduce the following.

Definition 5.4.5 Let $\psi \in E$. Define

$$R_\psi(\hat{u}, \bar{v}) \equiv \hat{q}(\hat{u}) \wedge \bar{q}(\bar{v}) \wedge \text{ there exists } \hat{w} \models \hat{q} \text{ such that:}$$

(a) $F(\hat{w}, \hat{u})$ and $\hat{w} \perp_{\hat{u}_F} \bar{v}$;

(b) $\psi_0(\bar{w}, \bar{v})$ and $\chi(x, \hat{w}) \wedge \psi(\bar{x}, \bar{w}) \wedge \psi(\bar{x}, \bar{v})$ does not fork over $\hat{u}_F$.

$$S_\psi(\hat{u}, \bar{v}) \equiv \hat{q}(\hat{u}) \wedge \bar{q}(\bar{v}) \wedge \text{ there is } \hat{w} \models \hat{q} \text{ such that:}$$

(c) $F(\hat{w}, \hat{u})$ and $\hat{w} \mathop{\smile\!\!\!\!\!\!|}_{\hat{u}_F} \bar{v}$;

(d) $\neg\psi_0(\bar{w}, \bar{v})$ or $\chi(x, \hat{w}) \wedge \psi(\bar{x}, \bar{w}) \wedge \psi(\bar{x}, \bar{v})$ forks over $\hat{u}_F$.

Lemma 5.4.6

(1) R_ψ is type-definable over $\emptyset$.

(2) $R_\psi(\hat{b}, \bar{b})$ holds.

(3) If $R_\psi(\hat{a}, \bar{b})$ holds for all $\psi \in E$, then $\bar{a}_E = \bar{b}_E$.

Proof (1) According to 4.2.13(1), (a) is type-expressible. It remains to see that $\langle \hat{w}_i\bar{v}_i | i < \omega\rangle$ is Morley in $\mathrm{tp}(\hat{w}\bar{v}/\hat{u}_F)$ is type-expressible. Note that due to (a) and (v3) above, $\bar{u}_E = \bar{w}_E$. So by the choice of ψ_0, $\bar{v}_E$ and $\bar{u}_E$ are interbounded. Then from (v5), we have $\bar{v} \mathop{\smile\!\!\!\!\!\!|}_{\bar{u}_E} \hat{u}_F$. Therefore, again from (a) and (v3), it follows that $I = \langle \hat{w}_i\bar{v}_i | i < \omega\rangle$ is Morley in $\mathrm{tp}(\hat{w}\bar{v}/\hat{u}_F)$ iff I is $\hat{u}_F$-indiscernible in $\mathrm{tp}(\hat{w}\bar{v}/\hat{u}_F)$; for each φ, k and each $i < \omega$, $D(\mathrm{tp}(\hat{w}_i/\hat{u}_F\hat{w}_j\bar{v}_j : j < i), \varphi, k) \geq D(\mathrm{tp}(\hat{w}/\hat{u}_F), \varphi, k) = D(\mathrm{tp}(\hat{u}/\hat{u}_F), \varphi, k)$; and $D(\mathrm{tp}(\bar{v}_i/\hat{u}_F\hat{w}_i\hat{w}_j\bar{v}_j : j < i), \varphi, k) \geq D(\mathrm{tp}(\bar{v}/\hat{u}_F), \varphi, k) = D(\mathrm{tp}(\bar{v}/\bar{u}_E), \varphi, k) = D(\mathrm{tp}(\bar{v}/\bar{v}_E), \varphi, k)$. Then again as in 4.2.13(1), the latter is type-expressible.

(2) In (iii) above, we have $\bar{b} \mathop{\smile\!\!\!\!\!\!|}_{\bar{b}_E} \hat{b}'$ and $\hat{b}' \equiv_{\bar{b}_E} \hat{b}$. Hence, there is $\bar{d}$ such that $\bar{d}\hat{b} \equiv_{\bar{b}_E} \bar{b}\hat{b}'$ and $\bar{d} \mathop{\smile\!\!\!\!\!\!|}_{\bar{b}_E} \hat{b}$; so $\bar{d} \mathop{\smile\!\!\!\!\!\!|}_{\hat{b}_F} \hat{b}$ by (v3). Choose $\bar{d}\hat{b}'' \equiv_{\hat{b}_F} \bar{d}\hat{b}$ such that $\bar{d}'\hat{b}'' \mathop{\smile\!\!\!\!\!\!|}_{\hat{b}_F} \hat{b}$; so $\bar{d}' \mathop{\smile\!\!\!\!\!\!|}_{\hat{b}_F} \hat{b}''\hat{b}$. Then by taking $\hat{b}''$ for $\hat{w}$ in (a), it is not hard to see that $R_\psi(\hat{b}, \bar{b})$ follows.

(3) A direct consequence of (a), (v3), and (b). ■

We remark that there do not exist $e_1 \ldots e_k \not\mathop{\smile\!\!\!\!\!\!|}_c a_j$ $(j = 1, \ldots, k+1)$ such that $D(e_i/c) \leq \alpha$ $(i = 1, \ldots, k)$, $\{a_1, \ldots, a_{k+1}\}$ is c-independent and each $\mathrm{tp}(a_j/c)$ is orthogonal to any type of D-rank $< \alpha$. (As usual e_i is finite real. This will crucially be used in bounding "forking numbers" to show the type-definability of S_ψ.): It is easy to see when $k = 1$. Assume the remark holds for $1, \ldots, k-1$. To lead a contradiction, suppose that there exist such $e_1 \ldots e_k \not\mathop{\smile\!\!\!\!\!\!|}_c a_j$ $(j = 1, \ldots, k+1)$. Put $A_j = \{a_1, \ldots, a_{k+1}\} \setminus \{a_j\}$ and $A_{ij} = \{a_1, \ldots, a_{k+1}\} \setminus \{a_i, a_j\}$, and so on. Now $e_k \mathop{\smile\!\!\!\!\!\!|}_c a_j$ since, otherwise, due to $k = 1$ case, $e_k \mathop{\smile\!\!\!\!\!\!|}_c A_j$ ($\mathrm{tp}(A_j/c) \perp$ any type of D-rank $< \alpha$, by 4.4.2(4)), so A_j is ce_k-independent, and $e_1 \ldots e_{k-1} \not\mathop{\smile\!\!\!\!\!\!|}_{e_kc} a_\ell$ for $a_\ell \in A_j$ contradicting the induction hypothesis. Similarly, we have $e_k \mathop{\smile\!\!\!\!\!\!|}_{ca_i} a_j$ $(i \neq j)$ since, otherwise, again $e_k \mathop{\smile\!\!\!\!\!\!|}_{ca_i} A_{ij}$ so $e_k \mathop{\smile\!\!\!\!\!\!|}_c A_j$, leading again to a contradiction. Iterating this, by forking transitivity we have $e_k \mathop{\smile\!\!\!\!\!\!|}_c a_1 \ldots a_{k+1}$, again reaching a contradiction.

Lemma 5.4.7 *S_ψ is type-definable over $\emptyset$.*

Proof It suffices to prove the following claim as the rest of the argument is the same as that of 5.4.6(1).

Claim In 5.4.5(d), assume that (c) and $\psi_0(\bar{w}, \bar{v})$ hold. Thus, $\hat{u}_F = \hat{w}_F$ and $\bar{w}_E (= \bar{u}_E)$ is interbounded with $\bar{v}_E$. Now then $\chi(x, \hat{w}) \wedge \psi(\bar{x}, \bar{w}) \wedge \psi(\bar{x}, \bar{v})$ forks over $\hat{u}_F$ iff there is a sequence $\langle \hat{w}_i\bar{v}_i | i < \omega\rangle$ Morley in $\mathrm{tp}(\hat{w}\bar{v}/\hat{u}_F)$ such that $\{\chi(x, \hat{w}_i) \wedge \psi(\bar{x}, \bar{w}_i) \wedge \psi(\bar{x}, \bar{v}_i) | i < n+2\}$ is inconsistent (n is the number in (iii) above): Assume the left-hand side. We shall prove the right-hand side. (The other direction is obvious.) Suppose not; say there is a Morley sequence $\langle \hat{w}_i \bar{v}_i |\ i < \omega \rangle$ in $\mathrm{tp}(\hat{w}\bar{v}/\hat{u}_F)$ such that $\{\chi(x, \hat{w}_i) \wedge \psi(\bar{x}, \bar{w}_i) \wedge \psi(\bar{x}, \bar{v}_i) |\ i < n+2\}$ is realized by some (finite) tuple $\bar{a}$. Hence, it follows that $\bar{a} \not\perp_{\hat{w}_0} \hat{w}_i \bar{v}_i$ for $i = 1, \ldots, n+1$. Now since $\chi(\bar{a}, \hat{w}_0)$ holds, there are e_j $(j = 1, \ldots, n)$ such that $\bar{a} \in \mathrm{acl}(\hat{w}_0, e_1 \ldots e_n)$ (see our notational convention in (i) above), and $D(e_j/\hat{w}_0) \leq \alpha$. Thus, $e_1 \ldots e_n \not\perp_{\hat{w}_0} \hat{w}_i \bar{v}_i$ for $i = 1, \ldots, n+1$. Then the following subclaim and the remark before this lemma induce a contradiction, and hence the claim is proved.

Sub Claim For $i > 0$, $\mathrm{tp}(\hat{w}_i \bar{v}_i/\hat{w}_0)$ is orthogonal to any type of D-rank $< \alpha$. Since $\hat{w}_i \perp_{\hat{w}_0} \bar{v}_i$, it suffices to show that each of $\mathrm{tp}(\hat{w}_i/\hat{w}_0)$ and $\mathrm{tp}(\bar{v}_i/\hat{w}_0)$ is orthogonal to any such types (4.4.2(4)). From $\hat{w}_i \perp_{\hat{w}_i/F} \hat{w}_0$ and (v2) above, the former follows. For the latter, note, first, that $(\bar{v}_i)_E$ and $\bar{u}_E$ are interbounded. Hence, from (v5) above, we have $\bar{v}_i \perp_{\bar{u}_E} \hat{u}_F$. Now since $\bar{v}_i \perp_{\hat{u}_F} \hat{w}_0$, we have that $\bar{v}_i \perp_{(\bar{v}_i)_E} \hat{w}_0$. Then according to our second supposition of Lemma 5.4.4, the latter follows. ■

Lemma 5.4.8 *In $\hat{q}(\hat{u}) \wedge \bar{q}(\bar{v})$, $\neg R_\psi(\hat{u}, \bar{v})$ iff $S_\psi(\hat{u}, \bar{v})$.*

Proof It is easy to see the left implies the right. For the right to imply the left, assume $R_\psi(\hat{u}, \bar{v})$ and $S_\psi(\hat{u}, \bar{v})$ lead to a contradiction. Hence, there are $\hat{w}, \hat{w}' \models F(\hat{x}, \hat{u})$ such that $\hat{w}$ witnesses 5.4.5(a),(b), and $\hat{w}'$ witnesses (c),(d). By extension, we additionally assume that $\hat{w}\bar{v} \perp_{\hat{u}_F} \hat{w}'$. Then there are $\bar{a} \perp_{\hat{u}_F} \hat{w}\bar{v}$ realizing $\chi(x, \hat{w}) \wedge \psi(\bar{x}, \bar{w}) \wedge \psi(\bar{x}, \bar{v})$, and $\bar{a}'\hat{w}' \equiv^L_{\hat{u}_F} \bar{a}\hat{w}$ (v4). Thus, it follows from the independence theorem, $\chi(x, \hat{w}') \wedge \psi(\bar{x}, \bar{w}') \wedge \psi(\bar{x}, \bar{v})$ does not fork over $\hat{u}_F$ either. In particular, by (ii) above, $\exists \bar{x}(\psi_1(\bar{x}, \bar{w}') \wedge \psi_1(\bar{x}, \bar{v}))$, and thus $\psi_0(\bar{w}', \bar{v})$ hold, a contradiction. ■

From Lemma 5.4.8, for each $\psi \in E$, there is a formula $\rho_\psi(\hat{x}, \bar{y})$ defining R_ψ on $\hat{q}(\hat{x}) \wedge \bar{q}(\bar{y})$. By the definition of R_ψ, we have that

$$F(\hat{x}, \hat{y}) \wedge \bar{q}(\bar{z}) \models \rho_\psi(\hat{x}, \bar{z}) \leftrightarrow \rho_\psi(\hat{y}, \bar{z}).$$

Hence, there is $\delta_\psi(\bar{z}) \in \bar{q}$ such that

$$F(\hat{x}, \hat{y}) \models E_\psi(\hat{x}, \hat{y}) (\equiv \forall \bar{z}\, (\delta_\psi(\bar{z}) \rightarrow (\rho_\psi(\hat{x}, \bar{z}) \leftrightarrow \rho_\psi(\hat{y}, \bar{z})))).$$

Hence, if we set $\tilde{b} = \langle \hat{b}_{E_\psi} \mid \psi \in E \rangle$, then $\tilde{b} \in \mathrm{dcl}(\hat{b}_F)$. Moreover, $\bar{b}_E \in \mathrm{dcl}(\tilde{b})$, since if we have $E_\psi(\hat{a}, \hat{b})$ for all $\psi \in E$, then $R_\psi(\hat{a}, \bar{b})$ holds for each ψ (5.4.6(2)), and $\bar{a}_E = \bar{b}_E$ (4.4.6(3)). Therefore, $\tilde{b}$ is the sequence of imaginaries fulfilling our Goal. The proof of Lemma 5.4.4, and so that of Proposition 5.4.3, are completed. ■

Theorem 5.4.9 *Supersimple theories have elimination of hyperimaginaries.*

Proof Proposition 5.4.3 says elimination of finitary hyperimaginaries holds in supersimple theories. Now the theorem follows from Corollary 5.0.4. ■

5.5 Definability

In this section, we fix **simple** T, and variables $x, y, \ldots$ are finite *imaginary* tuples unless stated otherwise; so T in $\mathcal{L}$ below, indeed, is T^{eq} in $\mathcal{L}^{\text{eq}}$.

Definition 5.5.1

(1) Let a formula $\varphi(x,y) \in \mathcal{L}$ be given. Recall from Chapter 1 that for an imaginary tuple b, we say $\varphi(x,b)$ is an *instance* of φ. By a *φ-formula,* we mean a formula over a parameter equivalent to a finite Boolean combination of instances of φ. For example, both $x = x$ and $c \neq c$ are φ-formulas. For a complete type $p(x)$ of an imaginary tuple over an imaginary set, and the formula $\varphi(x,y)$, we define

$$p\lceil\varphi := \{\psi(x) \mid \psi \text{ is a } \varphi\text{-formula in } p\}.$$

(2) Let $p(x)$ be an amalgamation base of an imaginary tuple. We say $\varphi(x,y) \in \mathcal{L}$ is *p-stable* if for any set of imaginaries B (e.g., a model), whenever $q_1(x)$ and $q_2(x) \in S(B)$ are both in $\mathcal{P}_p$, then $q_1\lceil\varphi = q_2\lceil\varphi$.

Clearly, p-stability of a formula only depends on $\mathcal{P}_p$. It is easy to see that the set of all p-stable formulas is closed under a Boolean combination. For convenience, we write $\bigcup\mathcal{P}$, a class of formulas, for the union of types in $\mathcal{P}$ over imaginary parameters (equivalently over models).

Lemma 5.5.2 *Suppose that $p(x)$ is an amalgamation base. Let $e = \text{Cb}(p)$, a hyperimaginary, and let $\varphi(x,y)$ be an $\mathcal{L}$-formula.*

(1) *For any imaginary tuple d, we have $\varphi(x,d) \in \bigcup\mathcal{P}_p$ if and only if $p\lceil e \wedge \varphi(x,d)$ does not fork over e. Moreover, there is a partial type $\Delta_\varphi(y)$ over e such that, for tuple d, $\varphi(x,d) \in \bigcup\mathcal{P}$ if and only if $\models \Delta_\varphi(d)$.*

(2) *If $\varphi(x,y)$ is $p(x)$-stable, then there is a formula $\delta(y)$ over e such that for any imaginary tuple d,*

$$\varphi(x,d) \in \bigcup\mathcal{P} \text{ if and only if } \models \delta(d).$$

We call $\delta(y)$, the $\varphi(x,y)$-definition of p (or $\mathcal{P}_p$).

(3) *The formula $\varphi(x,y)$ is not $p(x)$-stable iff there is some d such that both*

$$p\lceil e \wedge \varphi(x,d) \text{ and } p\lceil e \wedge \neg\varphi(x,d)$$

do not fork over e.

Proof (1) Recall that for an amalgamation base q,

$$q \in \mathcal{P} \text{ iff } q \text{ is a non-forking extension of } p\lceil e \text{ (4.3.4(2))}.$$

Hence, if $\varphi(x,d) \in \bigcup \mathcal{P}$ then $p\lceil e \wedge \varphi(x,d)$ does not fork over e. The converse easily follows, too, by taking a non-forking extension of $p\lceil e \wedge \varphi(x,d)$ over a model M such that $d \in M$ and $e \in \mathrm{dcl}(M)$. The moreover clause follows from 4.2.13(1).

(2) Suppose that $\varphi(x,y) \in \mathcal{L}$ is $p(x)$-stable. By compactness, it suffices to show that $\{\Delta_\varphi(y), \Delta_{\neg\varphi}(y)\}$ partitions the universe: Suppose that $\Delta_\varphi(d)$ and $\Delta_{\neg\varphi}(d)$. Then as in the proof of (1), there are a model M containing d and amalgamation bases $q_1, q_2 \in S(M)$ in $\mathcal{P}$ such that $\varphi(x,d) \in q_1$ and $\neg\varphi(x,d) \in q_2$. This contradicts p-stability of $\varphi(x,y)$. On the other hand, by the extension axiom, given d, either $p(x)\lceil e \wedge \varphi(x,d)$ does not fork over e or $p(x)\lceil e \wedge \neg\varphi(x,d)$ does not fork over e. Hence, either $\Delta_\varphi(d)$ or $\Delta_{\neg\varphi}(d)$.

(3) Right to left follows easily from (1). For the converse, assume φ is not p-stable. Then there is $\psi(x,b)$ equivalent to a Boolean combination of the instances of φ such that both $p\lceil e \wedge \psi(x,b)$ and $p\lceil e \wedge \neg\psi(x,b)$ do not fork over e. Now by induction on Boolean combination, it is routine to show that there must be some d such that $p\lceil e \wedge \varphi(x,d)$ and $p\lceil e \wedge \neg\varphi(x,d)$ do not fork over e. ■

Proposition 5.5.3 *An $\mathcal{L}$-formula $\varphi(x,y)$ is stable if and only if for every amalgamation base $p(x)$, $\varphi(x,y)$ is $p(x)$-stable.*

Proof Suppose that $\varphi(x,y)$ is not $p(x)$-stable.

Claim There is an indiscernible sequence $\langle a_i b_i | \ i \in \mathbb{Z}\rangle$ over $e(= \mathrm{Cb}(p))$ such that, for $i \leq j < k \in \mathbb{Z}$, $a_j \models p\lceil e \wedge \varphi(x,b_k) \wedge \neg\varphi(x,b_i)$, and $a_j \downarrow_e b_k$, $a_j \downarrow_e b_i$. (Hence, $\varphi(x,y)$ is not stable.) Since $\varphi(x,y)$ is not p-stable, there is some d and c_0, c_1 such that $c_0 \models p\lceil e \wedge \varphi(x,d)$, $c_1 \models p\lceil e \wedge \neg\varphi(x,d)$, and $c_0 \downarrow_e d$, $c_1 \downarrow_e d$. We may further assume that $c_0 \downarrow_{de} c_1$. Now as $p\lceil e$ is an amalgamation base, there is an indiscernible sequence $I = \langle c_i | i \in \mathbb{Z}\rangle$ over e (4.2.19). As $c_0c_1 \downarrow_e d$, we can assume that $I = \langle c_ic_{i+1} | i \in 2\mathbb{Z}\rangle$ is an indiscernible sequence over de in $\mathrm{tp}(c_0c_1/de)$. Let $J_0 = \langle c_{2n} | n \leq 0\rangle$, $J_1 = \langle c_{2n+1} | n \geq 0\rangle$. For $c_i \in J_0$, $c_i \models p\lceil e \wedge \varphi(x,d)$, and $c_i \downarrow_e d$. For $c_j \in J_1$, $c_j \models p\lceil e \wedge \neg\varphi(c_i,d)$, and $c_j \downarrow_e d$. Moreover, $J = J_0 \frown J_1$ is an indiscernible sequence over e. Using automorphisms over e preserving the order of J, and compactness with the Ramsey argument, we can find the desired sequence in the claim.

Conversely, suppose that $\varphi(x,y)$ is not stable. Then there is an indiscernible sequence $J_0 \frown J_1 \frown J_2$ over $\emptyset$ and c such that $\varphi(c,d)$ for each $d \in J_1$, and $\neg\varphi(c,d')$ for each $d' \in J_2$ (*). Moreover, we may assume that $J_1 \frown J_2$ is a Morley sequence over J_0 (3.2.3). There is a model M containing J_0 which is independent from $J_1 \frown J_2$ over J_0. Then $J_1 \frown J_2$ is an independent sequence over M. By 3.2.4, we can further assume that $J_1 \frown J_2$ is a Morley sequence of $\mathrm{tp}(d/M)$ for $d \in J_1$ (**). Let $q(x) = \mathrm{tp}(c/M)$. Then, by (*) and (**), both $q(x) \wedge \varphi(x,d)$ and $q(x) \wedge \neg\varphi(x,d)$ do not fork over M. Hence, $\varphi(x,y)$ is not q-stable. ■

Based on the claim in the proof of the previous proposition, it holds.

Corollary 5.5.4 *A formula $\varphi(x, y)$ is $p(x)$-stable iff there does not exist an $e(= \mathrm{Cb}(p))$-indiscernible sequence $I = \langle a_i | i \in \mathbb{Z} \rangle$ of realizations of $p\lceil e$ and d such that $a_i \perp_e d$ for all $i \in \mathbb{Z}$, and $\varphi(a_i, d)$ iff $0 < i$.*

Until Lemma 5.5.7, $p(x)$ is an amalgamation base, and a hyperimaginary $e := \mathrm{Cb}(p)$.

Notation 5.5.5

$$p(x)\text{- SCb} := \{\text{the name of } \varphi(x,y)\text{-definition of } \mathcal{P}_p \,:\, \varphi \text{ is a } p\text{-stable formula}\}.$$

$$\mathrm{SCb}(p(x)) := \{\text{the name of } \varphi(x,y)\text{-definition of } \mathcal{P}_p \,:\, \varphi \text{ stable}\}.$$

Remark 5.5.6 Note that $\mathrm{SCb}(p) \subseteq p\text{-SCb} \subseteq \mathcal{M}^{\mathrm{eq}}$. Indeed, $p\text{-SCb} \subseteq \mathrm{dcl}(e)$.

Proof Let f be an automorphism fixing the hyperimaginary e. We want to see that f fixes p-SCb pointwise, too. Given p-stable $\varphi(x, y)$, let $\psi(y)$ be a φ-definition. Now for imaginary d, $\psi(d)$ holds iff $\varphi(x, d) \in \bigcup \mathcal{P}$ iff $\varphi(x, f(d)) \in \bigcup \mathcal{P}$ (f fixes $\mathcal{P}$ setwise) iff $\psi(f(d))$. Hence, the set p-SCb is fixed by f. ■

From Lemma 5.5.2 we make an observation:

Proposition 5.5.7

(1) *Suppose that there is a family $\{\varphi_i(x, a_i) \in \mathcal{L} |\; i \in I\}$ of the instances of p-stable formulas closed under conjunctions, such that*

$$\models p\lceil e \leftrightarrow \bigwedge_{i \in I} \varphi_i(x, a_i).$$

Then e is interdefinable with the sequence of the names of the $\varphi_i(x, y_i)$ definitions of p $(i \in I)$. In particular, e and p-SCb are interdefinable.

(2) *If the canonical type $p\lceil e$ is axiomatized by the instances of stable formulas, then e is $\mathrm{SCb}(p)$.*

(3) *Thus, either (1) or (2) holds for any amalgamation base p, then T eliminates hyperimaginaries.*

Proof (1),(2) We will prove (1), and the proof of (2) will be similar. Let $B(\subseteq \mathcal{M}^{\mathrm{eq}})$ be the set of names of φ_i-definitions $(i \in I)$. Let f be an automorphism of $\mathcal{M}$ fixing B pointwise. By 4.3.9(1), it suffices to prove that $p\lceil e \wedge f(p\lceil e)$ does not fork over e. So due to the assumption, it is enough to show that, for $\varphi_i(x, a_i)$, $p\lceil e \wedge \varphi_i(x, f(a_i))$ does not fork over e. Now, there is a $\varphi_i(x, y_i)$-definition $\psi_i(y_i)$. From 5.5.2, we have $\psi(a_i)$. Since f fixes B pointwise, $\psi_i(f(a_i))$ holds. Hence, $p\lceil e \wedge \varphi_i(x, f(a_i))$ does not fork over e. This shows the interdefinability between e and p-SCb.

(3) comes from Proposition 5.0.3. ■

Fact 5.5.8 (T any) *Let a partial type $\Gamma(\bar{x})$ ($\bar{x}$ possibly infinite) be the disjoint union of $\Gamma_0(\bar{x})$ and $\Gamma_1(\bar{x})$. Suppose that $\Gamma(\bar{x})$ is not equivalent to $\Gamma_0(\bar{x})$. Then there is $\psi(x) \in \Gamma_1$ such that for any $\varphi(x') \in \Gamma(\bar{x})$, if $\models \varphi(x') \to \psi(x)$ then $\varphi(x') \in \Gamma_1(\bar{x})$.*

Proof If $\Gamma(\bar{x})$ is not equivalent to $\Gamma_0(\bar{x})$, then there must exist $\bar{d}$ realizing $\Gamma_0(\bar{x})$, while not realizing $\Gamma(\bar{x})$. Now choose $\psi(x) \in \Gamma$ such that $d(\subseteq \bar{d}) \not\models \psi(x)$. Then, clearly $\psi(x) \in \Gamma_1$ and satisfies the fact. ■

We state important consequences of elimination of hyperimaginaries for supersimple theories (Theorem 5.4.9). Because of e.h.i., we work in $\mathcal{M}^{\mathrm{eq}}$.

Theorem 5.5.9 *Assume T is supersimple. Let $p(x) = \mathrm{tp}(b/B)$ be an amalgamation base, and let $C = \mathrm{Cb}(p)$. Then there are p-stable formulas $\varphi_i(x, y_i)$ such that*

$$\models p\lceil C \leftrightarrow \bigwedge_i \varphi_i(x, b_i).\ (b_i \in B)$$

Proof Let $r(x, \bar{y}) = \mathrm{tp}(b, C)$, where $\bar{y}$ is possibly infinite. Then $r(x, \bar{y}) = r_s(x, \bar{y}) \cup r_u(x, \bar{y})$, where $r_s(x, \bar{y})$ is the set of p-stable formulas in r, and $r_u = r - r_s$.

Claim r and r_s are equivalent: By supersimplicity there is a finite tuple $c \subseteq C$ such that $b \mathop{\smile\!\!\!\!\!\!|}_c C$. Hence, C and c are interalgebraic. Suppose now the claim does not hold. Then by Fact 5.5.8, r and r_u are equivalent and we can assume that r_u is closed under the conjunctions with the formulas in r.

Sub Claim Let c_1 be a finite tuple from C, and let $\varphi(x, yy_1) \in r_u$ be given where yy_1 is a variable tuple for cc_1. Then there is $c'c'_1 \equiv cc_1$ such that $c \neq c'$ and $r(x, C)(= p\lceil C) \wedge \varphi(x, c'c'_1)$ does not fork over C: Since $c_1 \in \mathrm{acl}(c)$, given a formula $\sigma(yy_1) \in \mathrm{tp}(cc_1)$ we can assume $\sigma(yy_1)$ implies that $y_1 \in \mathrm{acl}(y)$. Moreover, we have $\varphi(x, yy_1) \wedge \sigma(yy_1) \in r_u$. Hence, by 5.5.2(3) there is dd_1 such that both $r(x, C) \wedge \varphi(x, dd_1) \wedge \sigma(dd_1)$ and $r(x, C) \wedge \neg\varphi(x, dd_1) \wedge \sigma(dd_1)$ do not fork over C. Then since $r(x, C)$ has a unique extension over $\mathrm{acl}(C)$ (Remark 4.3.2), $dd_1 \notin \mathrm{acl}(C) = \mathrm{acl}(c)$. It follows that $d \neq c$. Thus, by compactness with 4.2.13(1), the subclaim follows.

Now then, again by compactness with the subclaim, we can define $C' \equiv C$ such that the corresponding $c'(\in C') \neq c$ and $r(x, C) \wedge r(x, C')$ do not fork over C. This violates 4.3.9(1), a contradiction. Thus, the claim is proved.

Therefore, Theorem 5.5.9 holds with $b_i \in C$. Note that $C \subseteq \mathrm{dcl}(B)$, so $b_i \in \mathrm{dcl}(b')$ for finite $b' \in B$ witnessed by a functional formula $f(b', z_i)$. Clearly, $\exists z_i(\varphi_i(x, z_i) \wedge f(b', z_i))$ is equivalent to $\varphi_i(x, b_i)$, so it is p-stable. This finishes the proof of Theorem 5.5.9. ■

From Proposition 5.5.7 and Theorem 5.5.9, the following is obtained.

Corollary 5.5.10 *Assume that T is supersimple. For an amalgamation base $p(x)$,* Cb(p) *is (interdefinable with) p-*SCb.

In stable theories, the above definability result comes first, then using the definability of types, notions of canonical base, and elimination of hyperimaginaries follow later. In the context of simple theories matters are much more complicated as we have seen and the order of the development is also reversed. Note that in the proof of Theorem 5.5.9, supersimplicity ($x \neq y$ is not even type-expressible if x, y are infinite) and elimination of hyperimaginaries are crucially used.

Now we turn our attention to so-called *stable forking*, which is to do with the local stability and definability we have just seen. In the rest of this section, sets and tuples are imaginary (except possibly canonical bases), and formulas are from $\mathcal{L} = \mathcal{L}^{eq}$. We restate some in Fact 2.2.4. Recall that T is assumed to be simple.

Fact 5.5.11 *Let $\delta(x, y)$ be stable.*

(1) *Let $A \subseteq B \subseteq C$ and let $q(x) \in S(C)$. Then $q\lceil\delta$ does not fork over A iff $q\lceil\delta$ does not fork over B and $(q\lceil B)\lceil\delta = (q\lceil\delta)\lceil B$ does not fork over A.*

(2) *Let $A \subseteq M$ and let $p(x) \in S(M)$. Then $p\lceil\delta$ does not fork over A iff the δ-definition of p is over $\mathrm{acl}(A)$.*

(3) *Let $p(x)$ be a complete type. For any $\varphi(x, b) \in p\lceil\delta$, there is $\phi(y) \in \mathrm{tp}(b)$ such that $\varphi(x, y) \wedge \phi(y)$ is stable.*

Definition 5.5.12 We say an arbitrary simple T has *stable forking* if for any imaginary sets $A \subseteq B$ and $q(x) \in S(B)$, whenever $q(x)$ forks over A, then there is an instance of a stable formula $\psi(x, b) \in q(x)$ such that $\psi(x, b)$ forks over A. We say T has *strong stable forking* if it satisfies the same definition of stable forking without the condition $A \subseteq B$. Hence, strong stable forking implies stable forking.

Proposition 5.5.13

(1) *The following are equivalent.*

 (a) *T has stable forking.*

 (b) *For any $A \subseteq M$ (a model) and $q(x) \in S(M)$, whenever $q(x)$ forks over A, then there is an instance of a stable formula $\psi(x, b) \in q(x)$ such that $\psi(x, b)$ forks over A.*

 (c) *Any amalgamation base $p(x)$ does not fork over $\mathrm{SCb}(p)$, i.e., $\mathrm{Cb}(p) \subseteq \mathrm{bdd}(\mathrm{SCb}(p))$.*

(2) *Suppose that $\mathrm{Ltp} \equiv \mathrm{stp}$ in real context, and stable forking holds. Then T eliminates hyperimaginaries.*

Proof (1) (a)$\Rightarrow$(b) Obvious.

(b)$\Rightarrow$(a) Assume (b). Suppose $q(x) \in S(B)$ forks over $A(\subseteq B)$. Let M be a model containing B and let $q'(x) \in S(M)$ be a non-forking extension of $q(x)$. So $q'(x)$ forks over A. According to (b), there is a stable formula $\phi(x, y)$ such that $q'\lceil\phi$ forks over A. Then by Fact 5.5.11(1), $q\lceil\phi$ forks over A, and again by 5.5.11(3), this is witnessed by an instance of a stable formula in $q\lceil\phi \subseteq q(x)$.

(b)⇒(c) Assume (b). Let $p'(x)$ be some non-forking extension of $p(x)$ to a model $M \supseteq \mathrm{Cb}(p)$. It suffices to show that $p'(x)$ does not fork over $C := \mathrm{SCb}(p) \subseteq M = M^{eq}$. So by (b) it is enough to prove that, for each stable $\phi(x,y)$, $p'(x)\lceil\phi(x,y)$ does not fork over C. Now this follows from Fact 5.5.11(2) and the definition of $\mathrm{SCb}(p)$.

(c)⇒(b) Assume (c). Let $q(x) \in S(M)$. So $q(x)$ is an amalgamation base. Suppose $q(x)$ forks over $A \subseteq M$. Then $\mathrm{Cb}(p)$ is not bounded over A, and thus by (c), $\mathrm{SCb}(p) \nsubseteq \mathrm{acl}(A)$. Hence, from 5.5.11(2), for some stable $\varphi(x,y)$, we have that $q\lceil\varphi$ forks over A. So as in 5.5.11(3), some instance of a stable formula in $q\lceil\varphi$ forks over A. We have proved (b).

(2) By (1)(c), for any amalgamation base $p(x)$, $\mathrm{Cb}(p) \subseteq \mathrm{bdd}(\mathrm{SCb}(p))$. But since $\mathrm{Ltp} \equiv \mathrm{stp}$, we have $\mathrm{SCb}(p) \subseteq \mathrm{Cb}(p) \subseteq \mathrm{acl}(\mathrm{SCb}(p))$ (5.1.18). Now e.h.i. follows from 4.1.5(1) and Proposition 5.0.3. ■

Indeed, more is shown in [76] that if the assumption of (2) above holds, then for an amalgamation base $p(x)$, p-SCb is $\mathrm{Cb}(p)$.

Question Is there a simple theory where stable forking does not hold?

But there are examples of supersimple theories where strong stable forking fails to hold, and $\mathrm{Cb}(p)$ is *not* equivalent to $\mathrm{SCb}(p)$ [76],[73].

Definition 5.5.14 We say simple T is 1-*based* or *modular* if for any hyperimaginaries b and A, $\mathrm{Cb}(b/A) \subseteq \mathrm{bdd}(b)$.

The class of 1-based theories is another important subclass of simple theories. 1-Based theories will be studied more intensively in later chapters.

Proposition 5.5.15 *Assume that T is 1-based having elimination of hyperimaginaries.*

(1) *Any canonical type $p(x)$ is axiomatized by the instance of stable formulas. Thus,* $\mathrm{Cb}(p)$ *is interdefinable with* $\mathrm{SCb}(p)$.
(2) *T has strong stable forking.*

Proof (1) For $a \models p(x)$, we have imaginary $C = \mathrm{Cb}(p) \subseteq \mathrm{acl}(a)$. Hence, $p(x)$ ($= \mathrm{tp}(a/C)$) is axiomatized by the instances of stable formulas of the form $\psi(x,y)$ implying $y \in \mathrm{acl}(x)$. The rest follows from 5.5.7.

(2) Assume that $p(x) = \mathrm{tp}(c/B)$ forks over A. We want to find a stable instance in p which forks over A. Let $C = \mathrm{acl}(c) \cap B'$ where $B' = \mathrm{acl}(B)$. So $\mathrm{tp}(c/B')$ does not fork over C. As in the proof of (1), $\mathrm{tp}(c/C)$ is axiomatized by the instances (over B') of stable formulas.

Claim $\mathrm{tp}(c/C)$ forks over A: Otherwise there is $c' \equiv_C c$ such that $c' \perp_A C$. Hence, $C \subseteq \mathrm{acl}(A)$, since $C \subseteq \mathrm{acl}(c') \subseteq \mathrm{acl}(c'A)$. Now by transitivity, any non-forking extension $q(x)$ of $\mathrm{tp}(c/B')$ over $B' \cup A$ does not fork over C. So $q(x)$ does not fork over A, contradicting $p(x)(\subseteq q(x))$ forks over A. The claim is proved.

Hence there is an instance $(c \models)\varphi(x, b')$ $(b' \in B')$ of a stable formula that forks over A. Thus, $c \models \exists z(\varphi(x, z) \wedge \psi(z, b))$, where $\psi(b', b)$ with $b \in B$ and $\psi(z, y)$ implies $z \in \mathrm{acl}(y)$. Clearly, $\exists z(\varphi(x, z) \wedge \psi(z, b))$ is a desired stable instance in $p(x)$. ■

5.6 Bibliographical Remarks

Proposition 5.0.3 is from [85]. Section 5.1 is from [81],[85],[17],[67]. Corollary 5.1.8 is observed in [18]. Proposition 5.1.17 and 5.1.19(2) are in [71]. A good summary on the Lascar group is also written in [125],[120]. Proposition 5.1.19 and the proof of 5.1.3 are possibly new. Lascar's original definition of the topology for $\mathrm{Gal}_L(T)$ uses ultrafilters. The Lascar group and related subjects are being studied by many authors ([56],[94],[40],[18], and more). Section 5.2 is due to S. Buechler [12]. Section 5.3 is from [67] and [102]. The celebrated result of elimination of hyperimaginaries for supersimple theories is due to Buechler, Pillay, and Wagner [13]. Section 5.5 is from [69] and [76].

6

Constructing Simple Structures

The main theme of this chapter is to introduce/summarize several ways to construct simple structures (= structures having simple theories). In the meantime we turn our attention to a couple of important subclasses of simple structures, namely 1-*based* (or *modular*) and *CM-trivial* structures.

Let us temporarily recall a brief history of geometric stability theory. As mentioned in Chapter 1, in the proof of Morley's categoricity theorem, the notion of strongly minimal definable sets played a crucial role. In Example 1.2.8(1), typical examples of strongly minimal structures are listed. Zilber's trichotomy or principle says those three are essentially all the examples of strongly minimal structures. In other words, a strongly minimal set X is either *trivial* or *modular* (= 1-*based*) or an *infinite field is definable* in X^{eq}. As the triviality of the strongly minimal set implies modularity, trichotomy is sometimes called dichotomy, and by a result of Macintyre [87] the field must be algebraically closed. A more detailed description is provided in Chapter 7. Zilber himself worked on this conjecture in ω-categorical context. He proved that any ω-categorical strongly minimal structure is modular [126], and if it is non-trivial then a canonical infinite vector pace over a finite field is interpretable. (The field case is devoid.) Later the result was extended by Cherlin et al. to the class of ω-categorical superstable (= ω-categorical ω-stable) structures [29]. The proof uses a fact they showed ahead that any such structure has finite SU-rank. (On the other hand, as in Remark 6.1.3, it is easy to see that any modular ω-categorical T is supersimple having finite SU-rank.)

But in non-ω-categorical context, Hrushovski constructed a counter-example. Namely, he built a non-modular strongly minimal structure in which no infinite group is interpretable [52]. By a similar method he constructed a strictly stable ω-categorical non-modular structure [118].

Now it is natural to ask whether the mentioned results are extended in the context of simple theories. More precisely, let T be ω-categorical supersimple: Is T modular? Does T have finite SU-rank? Surprisingly, Hrushovski finds a counter-example of the first question by generalizing his construction method (Claim 6.2.13). The second question is still open. But there are partial answers. F. O. Wagner answers the question positively when

T is the theory of a group with extra relations (Theorem 7.5.12). Evans and Wagner give a positive answer when T is in addition CM-trivial (Theorem 6.1.5). Indeed, all of Hrushovski's mentioned (stable/simple) counter-examples are CM-trivial. So if the ω-categorical supersimple T is always CM-trivial (not yet known though), then the second question is positively answered.

In Section 6.3 we study several construction methods obtaining simple structures from stable/simple structures by adding predicates or automorphisms.

6.1 Modularity and CM-Triviality

In this section T **is simple**, and we work with hyperimaginaries until Definition 6.1.4. Let X be an e-invariant set of hyperimaginaries. Recall that X is said to be 1-*based* (or *modular*) if $\mathrm{Cb}(A/Be) \subseteq \mathrm{bdd}(Ae)$ for any small sets $A(\subseteq X)$ and B. According to Remark 4.3.9(6), X is 1-based iff for any n-tuple $a \in X^n$ and B, $\mathrm{Cb}(a/Be) \subseteq \mathrm{bdd}(ae)$.

The set X being 1-based is independent from the choice of a parameter. Namely, if modular X is e'- invariant as well, then for $A \subset X$ and B, by extension, we can assume that $AB \perp_{e'} e$. Then $c := \mathrm{Cb}(A/Be') = \mathrm{Cb}(A/Be'e) \subseteq \mathrm{bdd}(Ae) \subseteq \mathrm{bdd}(Aee')$. Moreover, $c \in \mathrm{bdd}(Be')$. Now, $Be' \perp_{Ae'} e$, so $c \perp_{Ae'} e$ and $c \in \mathrm{bdd}(Ae')$.

We call T 1-based if $\mathcal{M}$ is so (Definition 5.5.14). According to Remark 4.3.9(6) this definition is compatible when T is stable (see Chapter 1). Note that for the big set X, $\mathrm{dcl}(X)$ and $\mathrm{bdd}(X)$ are defined as the big union of $\mathrm{dcl}(A)$ and $\mathrm{bdd}(A)$, respectively, where A ranges over small subsets of X. We put $X^{\mathrm{heq}} := \mathrm{dcl}(Xe)$. We point out another folklore of modularity.

Fact 6.1.1 *For an e-invariant set X, the following are equivalent:*

(1) *X is modular,*

(2) $\mathrm{bdd}(eX)$ *is modular, and*

(3) *Some e-invariant Y with $X \subseteq Y \subseteq \mathrm{bdd}(eX)$ is modular.*

Proof It suffices to show (1)$\Rightarrow$(2) since the other direction is obvious and the equivalence of (3) comes from those. Let $A \subseteq \mathrm{bdd}(Ce)$ with small $C \subseteq X$. Given B, we want to see $A \perp_{\mathrm{bdd}(Ae)\cap\mathrm{bdd}(Be)} B$ (*). By extension we can assume that $C \perp_{Ae} B$. Hence, $\mathrm{bdd}(Ce) \cap \mathrm{bdd}(Be) \subseteq \mathrm{bdd}(Ae)$ (**). Now, based on (1), $C \perp_{\mathrm{bdd}(Ce)\cap\mathrm{bdd}(Be)} B$, so we have $A \perp_{\mathrm{bdd}(Ce)\cap\mathrm{bdd}(Be)} B$. Then (*) follows from (**) and transitivity. ■

We now state several conditions equivalent to 1-basedness.

Fact 6.1.2 *For an e-invariant set X, the following are equivalent:*

(1) *X is 1-based.*

(2) *For (small) $A, B \subseteq X$, $\mathrm{Cb}(A/Be) \subseteq \mathrm{bdd}(Ae)$.*

(3) *For* $A, B \subseteq X$,

$$A \mathop{\downarrow}_{\mathrm{bdd}(Ae)\cap\mathrm{bdd}(Be)\cap\mathrm{dcl}(Xe)} B.$$

(4) *For* $A, B \subseteq X$, *we have*

$$A \mathop{\downarrow}_{\mathrm{bdd}(Ae)\cap\mathrm{bdd}(Be)} B.$$

(5) *For* $A, B \subseteq X$ *and* C, *we have that* $A \downarrow_{Ce} B$ *if and only if*

$$\mathrm{bdd}(ACe) \cap \mathrm{bdd}(BCe) = \mathrm{bdd}(Ce).$$

(6) *For any e-indiscernible* $I = \langle a_i | i \in \omega\rangle$ *with* $a_i \subseteq X$, *it follows that* $I - \{a_0\}$ *is Morley over* ea_0.

Proof (1)⇒(2) Clear.

(2)⇒(3) If $c = \mathrm{Cb}(A/Be)$, then $A \downarrow_{ce} Be$. By (2), we have $ce \subseteq \mathrm{bdd}(Ae)$. Moreover, by 4.3.9(5), $c \subseteq \mathrm{dcl}(A_i | i < \omega)$, where $\langle A_i | i < \omega\rangle (\subseteq \mathrm{dcl}(X))$ is a Morley sequence of $\mathrm{Ltp}(A/Be)$. Hence, $ce \subseteq \mathrm{dcl}(Xe)$, and by transitivity, (3) follows.

(3)⇒(4) By transitivity.

(4)⇒(1) Let $I_\omega = \langle A_i | i \leq \omega\rangle$ be a Morley sequence of $\mathrm{tp}(A/Be)$ for some $A = A_\omega \subseteq X$ and B. Then by 2.4.2, we know that $c = \mathrm{Cb}(A/Be) = \mathrm{Cb}(A/Ie)$, where $I = I_\omega - \{A\} \subseteq X$. Then from (4), $c \in \mathrm{bdd}(Ae)$.

(4)⇔(5) In (5), the implication from left to right is true for *any* simple theory, and it is easy to see that the reverse implication is equivalent to (4).

(6)⇒(1) Again by 2.4.2, for $A_0 \subseteq X$, we have

$$c = \mathrm{Cb}(A_0/Be) = \mathrm{Cb}(A_\omega/Ie),$$

where $I = \langle A_i | i < \omega\rangle$, and $I^\frown\langle A_\omega\rangle$ forms a Morley sequence of $\mathrm{tp}(A_0/Be)$. Then based on (6), we have that $A_\omega \downarrow_{A_0e} I$ and $c \in \mathrm{bdd}(A_0e)$.

(4)⇒(6) Note first that (4) implies that $a_0 \downarrow_A a_1a_2\ldots$, where

$$A = \mathrm{bdd}(ea_0) \cap \mathrm{bdd}(ea_1a_2\ldots).$$

Now as pointed out in 4.2.3(1), any indiscernible sequence over B is indiscernible over $\mathrm{bdd}(B)$ (*), so $\mathrm{tp}(a_1a_2\ldots/\mathrm{bdd}(ea_0)) = \mathrm{tp}(a_2a_3\ldots/\mathrm{bdd}(ea_0))$. Hence $A = \mathrm{bdd}(ea_0) \cap \mathrm{bdd}(ea_2a_3\ldots)$. Again by (*), $\mathrm{tp}(a_0/\mathrm{bdd}(ea_2a_3\ldots)) = \mathrm{tp}(a_1/\mathrm{bdd}(ea_2a_3\ldots))$, and thus

$$A = \mathrm{bdd}(ea_1) \cap \mathrm{bdd}(ea_2a_3\ldots).$$

Then, due to (4) and transitivity, we have $a_1 \downarrow_A a_2a_3\ldots$, and $a_1 \downarrow_{a_0A} a_2a_3\ldots$. Since Aa_0 and ea_0 are interbounded, it follows that $a_1 \downarrow_{ea_0} a_2a_3\ldots$, and we obtain (6). ■

Remark 6.1.3

(1) The reader can note that the above (3),(4),(5) are equivalent to the cases where B need not be a subset of X. The name "1-baseness" comes from (6), and "modularity" comes from (4) as the independence behaves similarly to that in modules (see Chapter 8).

(2) Assume T is 1-based. Then due to (6) above and 2.4.2, 2.3.9, for any A and a, we can find $a' \equiv_A a$ (a, a' are components in a Morley sequence) such that for all φ, k, $D(\text{tp}(a/A), \varphi, k) = D(\text{tp}(a/a'), \varphi, k)$. Thus, if additionally T is ω-categorical, then T is supersimple since every forking chain of real types of finite tuples must have finite length. Moreover, since $SU(a/A) = SU(a/a')$ and the SU-rank is connected, the 1-based ω-categorical T has *finite SU-rank* (i.e., any real complete n-type in T has finite SU-rank; equivalently, because of the Lascar inequalities, every real complete 1-type in T has finite SU-rank).

(3) Clearly, if T is 1-based then so is any invariant set of T. Conversely, if T has finite SU-rank and any complete type of SU-rank 1 is 1-based then T is 1-based [45] (however, there is even a non-1-based stable theory where any complete type of SU-rank 1 is 1-based [99, 4.6.1]). Therefore if T is ω-categorical, then T is 1-based iff T has finite SU-rank and any complete type of SU-rank 1 is 1-based.

We now introduce the notion of CM-triviality. (CM stands for Cohen–Macaulay.)

Definition 6.1.4 Simple T is said to be *CM-trivial* if for any hyperimaginaries c and $A \subseteq B$, whenever $\text{bdd}(cA) \cap \text{bdd}(B) = \text{bdd}(A)$, then $\text{Cb}(c/A) \subseteq \text{bdd}(\text{Cb}(c/B))$.

Note that if T is 1-based then it is CM-trivial, since by 6.1.2(5), for $A \subseteq B$, $\text{bdd}(cA) \cap \text{bdd}(B) = \text{bdd}(A)$ implies that $c \perp_A B$. Now we aim to show that any ω-categorical supersimple CM-trivial theory has finite SU-rank. So this extends the finiteness of SU-rank result in Remark 6.1.3(2) to the context of ω-categorical supersimple CM-trivial theories. (But as mentioned in Chapter 1, CM-trivial ω-categorical stable T need not even be superstable.)

In the rest of the section, we work in $\mathcal{M}^{\text{eq}}$. Assume now that a real type $p(x) \in S_n(\emptyset)$ is regular in simple ω-categorical T. Let D be its solution set. As pointed out in Remark 4.4.4, (D, cl) forms a pregeometry, where $a \in \text{cl}(A)$ iff $a \not\perp A$. We put its canonical geometry as $(D' = D/E, \text{cl}')$, where $E(x,y) := x \not\perp y$ is an $\emptyset$-definable equivalence relation on D. It easily follows that cl' is again forking dependence; and D', a sort in $D^{\text{eq}} = \text{dcl}^{\text{eq}}(D)$, is also the solution set of some regular $p' \in S(\emptyset)$. Clearly, p, p' are non-orthogonal.

Theorem 6.1.5 *Let T be ω-categorical and CM-trivial.*

(1) *A real regular type $p \in S_n(\emptyset)$ is given, and as above, regular $p' \in S(\emptyset)$ represents the canonical geometry of p. Then $SU(p') = 1$.*

(2) *If T is supersimple, then T has finite SU-rank.*

Proof (1) We use the same terminology explained before 6.1.5. To lead a contradiction, assume that $SU(p') \neq 1$ (*).

Claim 1 (D', cl') is not locally finite; i.e., there is finite $J \subseteq D'$ such that $\mathrm{cl}'(J)$ is infinite. Note that D' is infinite, since we can find $x \equiv^L y \models p$ with $x \perp y$. Then due to (*), $SU(p') > 1$. Hence, there are $a \in D'$ and a set B such that $a \not\perp B$ and $a \notin \mathrm{acl}(B)$. Let $I(= \langle a_i | i < \omega \rangle)a$ be an infinite Morley sequence in $\mathrm{tp}(a/B)$. By 2.4.2, $a \perp_I B$ and so $a \not\perp I$. Hence, there is a finite initial segment $J \subseteq I$ such that $a \not\perp J$. By indiscernibility, $Ia \subseteq \mathrm{cl}'(J)$ and $\mathrm{cl}'(J)$ is infinite. Claim 1 is proved.

Hence, there is the least $n > 0$ such that some subset $F = \{c_0, \ldots, c_n\} \subseteq D'$ has infinite $\mathrm{cl}'(F)$. Name $A = \{c_2, \ldots, c_n\}$ to the language, and consider the localized pregeometry (D'_A, cl'_A), where $D'_A = D' \setminus \mathrm{cl}'(A)$. There, according to ω-categoricity, if we let P_k be the set of (the codes of) k-dimensional closed sets, then it is $\emptyset$-definable in $(D'_A)^{\mathrm{eq}}$. We call an element of P_k, a *point, line, plane*, respectively, for $k = 1, 2, 3$. For $u \in P_k, v \in P_{k'}$ with $k < k'$, by uLv we mean that the corresponding closed set of u is contained in that of v. By the choice of n, any point is a finite subset of D'_A. We let $v \in P_2$ be the code of the infinite line $\mathrm{cl}'_A(c_0c_1)$. Note that P_1 forms a *geometry* with forking dependence over $\emptyset$.

Claim 2 There are $u \in P_1, w \in P_3$ with $uLvLw$ such that $u \notin \mathrm{acl}(vw)$, $w \notin \mathrm{acl}(v)$; $v \notin \mathrm{acl}(uw)$; and $u \perp_v w$: Since v is an infinite line, there is a point $u_0 Lv$ such that $u_0 \notin \mathrm{acl}(v)$ (†). Now due to ω-categoricity, we have m such that there are at most m points in the algebraic closure of arbitrary two lines. Let $I' = \langle u_0, \ldots, u_m \rangle$ be an initial segment of an infinite Morley sequence of $\mathrm{tp}(u_0/v)$. There are $u'_1 (\equiv_{u_0} u_1) \perp_{u_0} I'$, and $v'u'_1 \equiv_{u_0} u_1 v$ (‡). Since $\{u'_1, u_0, u_1\}$ is independent and v' is a line containing $u_0 u'_1$, we have $v \neq v'$, and u_0 is the unique point in the intersection of lines v and v', so $u_0 \in \mathrm{dcl}(vv')$ (**). By the choice of m, we have some $0 < i \leq m$ with $u = u_i \notin \mathrm{acl}(vv')$ (***). Let w be the unique plane containing $\{u'_1, u_0, u_1\}$. All the mentioned points and lines lie on w. Hence, $w \in \mathrm{dcl}(vv')$ and from (***), we have *firstly* $u \notin \mathrm{acl}(vw)$. By taking an infinite Morley sequence of $\mathrm{tp}(u'_1/u_0u_1)$, we easily see that $w \notin \mathrm{acl}(u_0u_1)$. As $v \in \mathrm{dcl}(u_0u_1)$, we, too, have $w \notin \mathrm{acl}(v)$. *Secondly*, since $v' \in \mathrm{dcl}(u'_1u_0)$ and $u_0 \perp u$, we have from (‡), $v' \perp_{u_0} u$ and $u_0 \perp_v u$. Then, based on (†), we have $u_0 \notin \mathrm{acl}(uv')$. Thus, from (**), we get $v \notin \mathrm{acl}(uv')$, and $v \notin \mathrm{acl}(uw)$ since $w \in \mathrm{dcl}(uv')$. *Lastly*, from (‡), it follows that $u'_1 \perp u_0 uv$, $u'_1 \perp_v u$, and $w \perp_v u$. We have proved Claim 2.

Claim 3 Let $C = \mathrm{acl}(uw) \cap \mathrm{acl}(vw)$. Then $\mathrm{Cb}(u/C) \not\subseteq \mathrm{acl}(v)$. Let $B = \mathrm{acl}(\mathrm{Cb}(u/C))$. Hence, $B \subseteq \mathrm{acl}(C) = C$ and $u \perp_B C$. When $w \in B$, there is nothing to prove since $w \notin \mathrm{acl}(v)$ by Claim 2. Hence, we can assume $w \notin B$, and there is $\langle w = w_0, w_1, w_2 \rangle$, an initial segment of an infinite Morley sequence of $\mathrm{tp}(w/B)$. Since $w \in C$, by type amalgamation, there is $u'(\perp_B w_0w_1w_2)$ such that $u'w_i \equiv_B uw$ for $i < 3$ (⋆). In particular $u'Lw_i$. Now by Claim 2, we have $u \notin B \subseteq \mathrm{acl}(vw)$, so from (⋆), it follows that $u' \notin B$ and $u' \notin \mathrm{acl}(w_0w_1w_2)$. Hence, the three planes intersect on infinitely many points, in particular, a line say v_1. Indeed for $i \neq j < 3$, by dimension counting, clearly v_1 is the unique line that lies on both w_i and w_j, i.e., $v_1 \in \mathrm{dcl}(w_iw_j)$. Then since $\{w_0, w_1, w_2\}$ is B-independent, we have $w_1 \perp_B v_1$; $v_1 w_0 \perp_{w_1 B} w_2 v_1$; and $v_1 \perp_B v_1$. Thus, $v_1 \in B$. By Claim 2, $v \notin B \subseteq \mathrm{acl}(uw)$. Hence, $v \neq v_1$, and $w \in \mathrm{dcl}(vv_1)$, because both

$v, v_1 Lw(= w_0)$. Therefore, as $w \notin \mathrm{acl}(v)$ (Claim 2), we get $v_1 \in B - \mathrm{acl}(v)$. We have shown Claim 3.

Now $C = \mathrm{acl}(uC) \cap \mathrm{acl}(vC)$. By Claim 2, we have $u \perp_v w$ and

$$\mathrm{Cb}(u/Cv) = \mathrm{Cb}(u/vw) = \mathrm{Cb}(u/v) \subseteq \mathrm{acl}(v).$$

Thus, by Claim 3, $\mathrm{Cb}(u/C) \nsubseteq \mathrm{Cb}(u/Cv)$. It contradicts CM-triviality of T.

(2) If a real type has an infinite SU-rank, then due to the connectedness of the SU-rank and supersimplicity with Remark 4.4.2(1), there is a regular $p \in S_n(b)$ (b finite real) having SU-rank ω. Since $T(b)$ still is ω-categorical, (1) and Remark 4.4.2(1) are contradictory. ■

6.2 Hrushovski Construction

This section is devoted to constructing an ω-categorical supersimple non-1-based theory. The method is called the Hrushovski construction. Instead of unfolding full generalizations we will stick to a case for our goal.

Assume that R is a ternary relation, and $\mathcal{L} = \{R\}$. In this section, all the $\mathcal{L}$-structures we refer to are irreflexive (i.e., $R(a, b, c)$ implies that a, b, c are distinct) and symmetric (i.e., $R(a_0, a_1, a_2)$ implies that $R(a_i, a_j, a_k)$ for $\{i, j, k\} = \{0, 1, 2\}$). We allow $\emptyset$ as an $\mathcal{L}$-structure. Now for an $\mathcal{L}$-structure A, R^A denotes the collection of 3-subsets of A realizing R, so among permutations of a 3-tuple realizing R, we pick only once.

For finite A, we let

$$\delta(A) := |A| - |R^A|,$$

and let $\mathcal{C}$ be the class of every finite $\mathcal{L}$-structure A such that $\delta(A') > 0$ for all $(\emptyset \neq) A' \subseteq A$. (As we indeed are interested in isomorphism types of the structures, we may choose $\mathcal{C}$ to be a countably infinite set of all the representatives for the isomorphism types, and complete everything in this section *up to isomorphisms.*)

Definition 6.2.1

(1) Let A_1, A_2 be $\mathcal{L}$-structures, and $A_0 = A_1 \cap A_2$. We say A_1, A_2 are *freely amalgamated over* A_0 if any tuple from $A_0 \cup A_1$ realizing R is already in A_0 or A_1; i.e., $A_1 \cup A_2$ carries no other R-relations than those in A_1 or A_2. In that case, $A_1 \amalg_{A_0} A_2$ denotes the $\mathcal{L}$-structure having the universe $A_1 \cup A_2$, which we call the *free amalgam* of A_1, A_2 over A_0. (In the rest, for $A_0' \subseteq A_0$, $A_1 \amalg_{A_0'} A_2$ indeed means $A_1 \amalg_{A_0'} A_2'$, where $A_2' \cong A_2$ and $A_1 \cap A_2' = A_0'$. There is such A_2' and as we are interested in the isomorphism types, this notation will not give any harm.)

(2) Let $A \subseteq B$ be finite $\mathcal{L}$-structures. We say A is *closed* in B, written $A \leq B$, if $\delta(A) < \delta(A')$ for all A' with $A \subset A' \subseteq B$. An embedding $\sigma : C \to D$ is *closed over* $C'(\subseteq C \cap D)$ if $\sigma(C) \leq D$ and σ fixes C' pointwise.

Lemma 6.2.2 *Let A, B, C be finite $\mathcal{L}$-structures.*

(1) *We have $A \leq A$, and $\mathcal{C} = \{B \mid \emptyset \leq B\}$. For $B \in \mathcal{C}$, if $A \subseteq B$ then $A \in \mathcal{C}$; and $\delta(B) = 0$ iff $B = \emptyset$.*

(2) (Submodularity)

$$\delta(A \cup B) \leq \delta(A) + \delta(B) - \delta(A \cap B),$$

and equality holds iff A, B are freely amalgamated over $A \cap B$.

(3) *Let $A \leq C$ and $B \subseteq C$. Then $B \cap A \leq B$. In particular if $A \subseteq B$ then $A \leq B$.*

(4) (Transitivity) *Let $A \leq B \leq C$. Then $A \leq C$.*

(5) *Let $C \in \mathcal{C}$ and $A = B \cap C \leq B \in \mathcal{C}$. Then $F := B \amalg_A C \in \mathcal{C}$ and $C \leq F$.*

(6) *If $A \subseteq B$, then there is the unique minimal closed superstructure of A, denoted by $\mathrm{cl}_B(A)$, in B.*

Proof (1) Clear.

(2) Note that $|A \cup B| = |A| + |B| - |A \cap B|$. Moreover, $(R^A \cup R^B) \setminus R^{A \cap B} \subseteq R^{A \cup B}$, and equality holds iff A, B are freely amalgamated over $A \cap B$. Hence, (2) follows.

(3) For any A' with $A \cap B \subset A' \subseteq B$, we have $A \subset A \cup A' \subseteq C$. Hence, $\delta(A) < \delta(A \cup A')$ (*). Moreover, $A \cap B = A \cap A'$. Hence, by (2) and (*),

$$\delta(A \cap B) = \delta(A \cap A') \leq \delta(A) + \delta(A') - \delta(A \cup A') < \delta(A').$$

(4) Let B' be given such that $A \subset B' \subseteq C$ and let $A' := B' \cap B$. So $A' \subseteq B'$, and by (3) with $B \leq C$, we have $A' \leq B'$ (**). If $B' = A' \subseteq B$, then $\delta(A) < \delta(B')$ since $A \leq B$, and we are done. So we can assume $A' \subset B'$, and it follows that $\delta(A) \leq \delta(A') < \delta(B')$ by (**) and that $A \leq B$.

(5) Let F' be such that $C \subset F' \subseteq F$, and let $B' := F' \cap B$. Then $A \subset B' \subseteq B$, so $\delta(A) < \delta(B')$. Note that $F' = B' \amalg_A C$ as an $\mathcal{L}$-structure. So by (2), $\delta(F') = \delta(B') + \delta(C) - \delta(A) > \delta(C)$. Thus $C \leq F$. It remains to see $F \in \mathcal{C}$. Consider $(\emptyset \neq) F_0 \subseteq F$. If already $F_0 \subseteq B$, then $\delta(F_0) > 0$ since $B \in \mathcal{C}$. Otherwise, $\emptyset \neq F_0 \cap C \leq F_0$ by (3) and that $C \leq F$. Thus, $0 < \delta(F_0 \cap C) \leq \delta(F_0)$ since $C \in \mathcal{C}$.

(6) Let $n := \min\{\delta(C) \mid A \subseteq C \subseteq B\}$. Now there is a maximal sequence $A \subseteq C_0 \subset \ldots \subset C_m \subseteq B$ such that $\delta(C_i) = n$ for all $i \leq m$. Based on our choices, $C_m \leq B$, and is *minimal* in the sense that any proper subset of C_m containing A cannot be closed in B. Now assume $C' \supseteq A$ is another minimal closed set in B. Then $C_m \leq C_m \cup C'$. Hence, by (3), $C_m \cap C' \leq C'$, and by (4), $C_m \cap C' \leq B$. Thus, due to minimality, $C_m \cap C' = C' = C_m$. ■

Definition 6.2.3 Let $\mathcal{B} \subseteq \mathcal{C}$.

(1) We say $\mathcal{B}$ is *closed under free amalgamation* if for any $A_0 \leq A_i \in \mathcal{B}$ $(i = 1, 2)$ with $A_0 = A_1 \cap A_2$, we have $A_1 \amalg_{A_0} A_2 \in \mathcal{B}$.

(2) Tentatively, we call $\mathcal{B}$ *closed* if $\mathcal{B}$ is closed under substructures (i.e., $B \in \mathcal{B}$ and $A \subseteq B$, then $A \in \mathcal{B}$, too), isomorphisms, and free amalgamation. Call $\mathcal{B}$ *non-trivial* if there are ω-many isomorphism types in $\mathcal{B}$. Call $\mathcal{B}$ *regular* if $\mathcal{B}$ is non-trivial closed, and there is no countable chain $A_0 \subset A_1 \subset \ldots$. with $A_i \in \mathcal{B}$ such that $\delta(A_i)$ are all the same.

(3) By a *generic* model for $(\mathcal{B}, \leq)$, we mean a countably infinite $\mathcal{L}$-model $M_{\mathcal{B}}$ such that every finite substructure of $M_{\mathcal{B}}$ is in $\mathcal{B}$; and for any finite $A \leq M_{\mathcal{B}}$ (it means that $A \leq A'$ for any finite substructure $(A \subseteq)A'$ of $M_{\mathcal{B}}$), and $A \leq B \in \mathcal{B}$, there is a closed embedding $\sigma : B \to M_{\mathcal{B}}$ over A.

Remark 6.2.4

(1) By Lemma 6.2.2, $\mathcal{C}$ itself is non-trivial and closed.

(2) If A_i $(i = 1, 2)$ are freely amalgamated over $A_0 = A_1 \cap A_2$, then any $C(\subseteq A_1 \cup A_2)$ is a free amalgam of $C \cap A_i$ over $C \cap A_0$. In addition, if $A_0 \leq A_i$, then by 6.2.2(3), $C \cap A_0 \leq C \cap A_i$.

Proposition 6.2.5 *Let $\mathcal{B}(\subseteq \mathcal{C})$ be non-trivial and closed. Then there is a generic model $M_{\mathcal{B}}$ for $\mathcal{B}$.*

Proof Let $A_0, A_1, \ldots$ be an enumeration of all the representatives for the isomorphism types in $\mathcal{B}$. We shall construct $M_0 \leq M_1 \leq \ldots$, a chain of finite structures $M_i \in \mathcal{B}$ such that (*) for any $i, j < n$, if $A_i \cong B \leq C \cong A_j$ and $B \leq M_{n-1}$ then there is a closed embedding $\sigma : C \to M_n$ over B. Note now that due to closedness, $\emptyset \in \mathcal{B}$ and put $M_0 = \emptyset$. For induction, assume that $M_0 \leq \ldots \leq M_n$ from $\mathcal{B}$ satisfying (*) are constructed. Let D_n be the set of all the representatives for the pairs (B, C) such that $B \leq M_n$ and $A_i \cong B \leq C \cong A_j$ for some $i, j \leq n$. Clearly, $D_n(\neq \emptyset)$ is finite and write $D_n = \{(B_k, C_k) | \, k < m\}$. Now define inductively $X_0 = M_n$ and $X_{k+1} := C_k \amalg_{B_k} X_k$ (see 6.2.1(1)). Then by 6.2.2(5), we have $B_k \leq C_k, X_k \leq X_{k+1}$. Put $M_{n+1} := X_m$. By transitivity, $M_n \leq M_{n+1}$. Moreover, by closedness, $M_{n+1} \in \mathcal{B}$ and M_{n+1} satisfies (*) for $n + 1$.

We claim that $M_{\mathcal{B}} := \bigcup_{n<\omega} M_n$ is a desired generic model. Clearly, $M_{\mathcal{B}}$ is countably infinite (every A_i is embedded into $M_{\mathcal{B}}$). If A is a finite substructure of $M_{\mathcal{B}}$ then $A \subseteq M_n \in \mathcal{B}$ for some $n < \omega$. So $A \in \mathcal{B}$. Now if finite $A \leq M_{\mathcal{B}}$ and $B \geq A$, then $A \leq M_n$ for some n, so we can assume $B \leq M_{n+1} \subseteq M_{\mathcal{B}}$. Then, for any finite $(B \subseteq)C \subseteq M_{\mathcal{B}}, C \subseteq M_m \geq M_{n+1}$. So by 6.2.2(3), $B \leq C$; and $B \leq M_{\mathcal{B}}$. ■

In the rest of this section, for $A \subset M$, acl(A) only takes in the home-sort.

Proposition 6.2.6 *Let $\mathcal{B}(\subseteq \mathcal{C})$ be regular with its generic model $M_{\mathcal{B}}$.*

(1) Let $A \subseteq M_{\mathcal{B}}$ be finite. Then there is the unique finite smallest $C \supseteq A$ such that $C \leq M_{\mathcal{B}}$. Write C as $\mathrm{cl}_{M_{\mathcal{B}}}(A)$, or simply $\mathrm{cl}_{\mathcal{B}}(A)$, even $\mathrm{cl}(A)$ when the underlying generic model is clear. $\mathrm{cl}(\emptyset) = \emptyset$, *and* $\delta(\mathrm{cl}(A)) = \min\{\delta(B) | \, A \subseteq$ *finite* $B \subseteq M_{\mathcal{B}}\}$.

(2) *Any two generic models $M_{\mathcal{B}}, M_{\mathcal{B}}'$ of $\mathcal{B}$ are isomorphic. More generally, if $A \leq M_{\mathcal{B}}$ and $A' \leq M_{\mathcal{B}}'$, and $\sigma : A \to A'$ is an isomorphism, then σ is extended to an isomorphism between $M_{\mathcal{B}}$ and $M_{\mathcal{B}}'$.*

(3) *Assume that given n there is $m \geq n$ such that for all finite $A \subseteq M_{\mathcal{B}}$ with $|A| \leq n$, $A \leq M_{\mathcal{B}}$ hold if for all $(A \subseteq)B \subseteq M_{\mathcal{B}}$ with $|B| \leq m$, $A \leq B$ hold. Then the theory $T_{\mathcal{B}} := \mathrm{Th}(M_{\mathcal{B}})$ is ω-categorical, and for finite $A \subset M_{\mathcal{B}}$, $\mathrm{acl}(A) = \mathrm{cl}_{\mathcal{B}}(A)$.*

Proof (1) The proof will be similar to that of Lemma 6.2.2(6). Replace B there by $M_{\mathcal{B}}$ and use the regularity of $\mathcal{B}$.

(2) We use the standard back-and-forth argument. Enumerate $M_{\mathcal{B}} = \{m_0, m_1, \ldots\}$, $M_{\mathcal{B}}' = \{m_0', m_1', \ldots\}$. Let $B_0 := \mathrm{cl}_{M_{\mathcal{B}}}(Am_0)$. Then $A \leq B_0 \leq M_{\mathcal{B}}$. Since $M_{\mathcal{B}}'$ is generic, there is $(A' \leq)B_0' \leq M_{\mathcal{B}}'$ and an isomorphism $\sigma_0 : B_0 \to B_0'$ extending σ. Let $B_1' := \mathrm{cl}_{M_{\mathcal{B}}'}(B_0'm_0')$. Then similarly we can find $(B_0 \leq)B_1$ closed in $M_{\mathcal{B}}$, and an isomorphism $\sigma_1 : B_1 \to B_1'$ extending σ_0. By iterating this, we can find a desired isomorphism between $M_{\mathcal{B}}$ and $M_{\mathcal{B}}'$.

(3) Note that as there are only finitely many isomorphism types of a given size, for $n_0 \leq n_1$, there is an $\mathcal{L}$-formula $\varphi(\bar{x}_0, \bar{x}_1)$ with $|\bar{x}_i| = n_i$ such that for A_i with $|A_i| = n_i$, $(A_0, A_1) \models \varphi$ iff $A_0 \leq A_1$. Now the assumption says even the condition $A \leq M_{\mathcal{B}}$ for arbitrary A with $|A| = n$ can also be described by an $\mathcal{L}$-formula in variables $x_1, \ldots, x_n$. Hence, there is an $\mathcal{L}$-theory $(M_{\mathcal{B}} \models)T_0$ saying that any (countably infinite) model of T_0 is generic. Thus, by (2), $T_{M_{\mathcal{B}}}$ is ω-categorical.

Now for finite $A \subset M_{\mathcal{B}}$, $\mathrm{cl}(A) \subseteq \mathrm{acl}(A)$ since $\mathrm{cl}(A)$ is finite and unique over A, and $M_{\mathcal{B}}$ saturated. Conversely, if $a \in M_{\mathcal{B}} \setminus \mathrm{cl}(A)$, then for $B := \mathrm{cl}(Aa)$, the free amalgam B_n of n copies of B over $\mathrm{cl}(A)$ is embedded closedly into $M_{\mathcal{B}}$ over $\mathrm{cl}(A)$. By (2), this means a has n-many $\mathrm{cl}(A)$-automorphism conjugates for each $n < \omega$, so $a \notin \mathrm{acl}(A)$. ■

We now look for a suitable regular subclass of $\mathcal{C}$ whose generic model is supersimple ω-categorical. Fix a function $f : \mathbb{R}^{>0} \to \mathbb{R}^{>0}$ (while $f(0) = 0$), which is unbounded, concave, $0 \leq f(1)$ but $f(n) \leq n$ for all $n < \omega$, and $0 < f'(x) \leq \frac{1}{x}$ for all $x > 0$. (For example, $f(x) = \ln x$.) Let

$$\mathcal{C}_f := \{A \mid A \text{ is a finite } \mathcal{L}\text{-structure such that } \delta(|B|) \geq f(|B|) \text{ for all } B \subseteq A\}.$$

Claim 6.2.7

(1) $\mathcal{C}_f$ is a regular subclass of $\mathcal{C}$.

(2) The theory T_f of the generic model M_f for $\mathcal{C}_f$ is ω-categorical.

Proof (1) We can think of the (x, y) plane as the $(|A|, \delta(|A|))$ plane. If $A \in \mathcal{C}_f$ then $A \in \mathcal{C}$, since $f(1) \geq 0$ and f is strictly increasing. The condition "$f(n) \leq n$ for all $n < \omega$" implies that any finite A with $|A| = n$ and $R^A = \emptyset$ is in $\mathcal{C}_f$, so $\mathcal{C}_f$ is not trivial. Clearly, $\mathcal{C}_f$ is closed under substructures. Since f is strictly increasing, we have checked all the regularity conditions for $\mathcal{C}_f$ except that it is closed under free amalgamation.

Let $A_0 \leq A_i \in \mathcal{C}_f$ and $A_0 = A_1 \cap A_2$, and let $F := A_1 \amalg_{A_0} A_2$. We claim, first, that $f(|F|) \leq \delta(|F|)$: We can assume that $|A_0| < |A_1| \leq |A_2|$. Now it follows, in the $(|A|, \delta(|A|))$ plane, that the four points $a := (|A_0|, \delta(|A_0|)$, $b := (|A_1|, \delta(|A_1|)$, $c := (|A_2|, \delta(|A_2|)$, $d := (|F|, \delta(|F|)$ form vertexes of a parallelogram. Since f is concave (its slope is decreasing), to show the claim it suffices to see that

$$f'(|A_2|) \leq \text{the slope of } \overline{cd} = \frac{\delta(F) - \delta(A_2)}{|F| - |A_2|} = \frac{\delta(A_1) - \delta(A_0)}{|A_1| - |A_0|}.$$

But since $A_0 \leq A_1$,

$$\frac{\delta(A_1) - \delta(A_0)}{|A_1| - |A_0|} \geq \frac{1}{|A_1|} \geq \frac{1}{|A_2|} \geq f'(|A_2|).$$

Hence, the claim follows.

Now due to 6.2.4(2) and the claim again, $F \in \mathcal{C}_f$.

(2) The assumption in Proposition 6.2.6(3) is satisfied: Given n, since f is unbounded, we can take $m \geq f^{-1}(\delta(|A|))$, where $\delta(|A|)$ is maximal among $|A| \leq n$. ■

We keep working with $\mathcal{C}_f$ and the theory $T_f = \mathrm{Th}(M_f)$. As usual let $\mathcal{M}$ be a large saturated model of T_f. In the rest $a, b, c, \ldots$ are *finite* tuples from $\mathcal{M}$ and $A, B, C, \ldots$ are sets from $\mathcal{M}$. Given a, A from $\mathcal{M}$, we put $d(a) := \delta(\mathrm{acl}(a))$ and $d(a/A) := \min\{d(aA') - d(A') | A'(\subseteq A) \text{ finite}\}$. Since M_f is saturated this definition makes sense.

Claim 6.2.8

(1) If $A \subset B$ finite, then

$$d(A) = d(\mathrm{cl}(A)) = \min\{\delta(C) | A \subseteq \text{ finite } C\},$$

and $d(A) \leq d(B)$.

(2) (Submodularity) Assume A, B are finite. Then

$$d(AB) \leq d(A) + d(B) - d(\mathrm{cl}(A) \cap \mathrm{cl}(B)).$$

(3) If $A \subseteq B$, then $d(a/A) \geq d(a/B) = d(a/B_0)$ for some finite $B_0 \subseteq B$. If B is finite, then $d(a/B) = d(aB) - d(B)$.

(4) $d(a/A) \geq 0$, and equality holds iff $a \in \mathrm{acl}(A)$. $d(a/\emptyset) = d(a)$.

(5) $d(a/A) = d(a/\mathrm{acl}(A)) = d(\mathrm{acl}(a)/A)$, and $d(ab/A) = d(a/bA) + d(b/A)$.

Proof (1) Clear from 6.2.6(1).

(2) Immediate from 6.2.2(2) and 6.2.6(1).

(3) Let $A \subseteq B$ be finite. Then by (1),(2),

$$d(aA) + d(B) \geq d(aAB) + d(\mathrm{cl}(aA) \cap \mathrm{cl}(B)) \geq d(aB) + d(A).$$

This inequality shows (3).

(4) Clear.

(5) follows from the definition. ■

Now for sets A, B, C from $\mathcal{M}$ we define an independence relation $A \mathop{\smile\!\!\!\!\!\mid}^f_B C$ as the following: For each finite tuple $a \in A$, $d(a/B) = d(a/BC)$.

Claim 6.2.9

(1) $\mathop{\smile\!\!\!\!\!\mid}^f$ is automorphism invariant, and satisfies finite character, (super) local character, symmetry, and transitivity.

(2) $A \mathop{\smile\!\!\!\!\!\mid}^f_B C$ iff $A \mathop{\smile\!\!\!\!\!\mid}^f_B BC$ iff $\mathrm{acl}(A) \mathop{\smile\!\!\!\!\!\mid}^f_{\mathrm{acl}(B)} \mathrm{acl}(BC)$.

(3) Suppose a and $B \subseteq C$ are given. Put $\bar{B} := \mathrm{acl}(aB)$.

  (a) Let B, C be finite. Then $a \mathop{\smile\!\!\!\!\!\mid}^f_B C$ (i.e., $d(a/B) = d(a/C)$) iff $\bar{B}$ and $\mathrm{acl}(C)$ are freely amalgamated over $\mathrm{acl}(B)$ and $\delta(\mathrm{acl}(\bar{B}C)) = \delta(\bar{B}C)$.

  (b) In general, if $a \mathop{\smile\!\!\!\!\!\mid}^f_B C$ then still $\bar{B}$ and $\mathrm{acl}(C)$ are freely amalgamated over $\mathrm{acl}(B)$ (in particular, $\bar{B} \cap \mathrm{acl}(C) = \mathrm{acl}(B)$).

(4) $\mathop{\smile\!\!\!\!\!\mid}^f$ satisfies the extension axiom too.

Proof (1) We show symmetry (others come directly from the definition and 6.2.8(3)): Based on finite character, it suffices to check $d(a/Bc) = d(a/B)$ implies that $d(c/Ba) = d(c/B)$. Let $n := d(a/Bc) = d(a/B)$. Now, there is finite $b \in B$ such that $n = d(ab) - d(b)$. Then by the definition and 6.2.8(3), for any finite $(b \subseteq)b' \in B$, $n = d(ab') - d(b') = d(ab'c) - d(b'c)$. So similarly, $d(c/B) = d(c/Ba)$.

(2) follows by 6.2.8(5).

(3) (a) There is no harm to assume $B = \mathrm{acl}(B)$ and $\mathrm{acl}(C) = C$. So $B \leq \bar{B} \cap C$. According to 6.2.6(1) and 6.2.2(2),

$$\begin{aligned} d(a/C) &= \delta(\mathrm{acl}(aC)) - \delta(C) \leq \delta(\bar{B}C) - \delta(C) \leq \delta(\bar{B}) - \delta(\bar{B} \cap C) \\ &\leq \delta(\bar{B}) - \delta(B) = d(a/B). \end{aligned}$$

Now $a \mathop{\smile\!\!\!\!\!\mid}_B C$ iff above inequalities are all equal iff $\bar{B}$ and C are freely amalgamated over $\bar{B} \cap C$; $\delta(B) = \delta(\bar{B} \cap C)$, so $B = \bar{B} \cap C$; and $\delta(\mathrm{acl}(aC)) = \delta(\mathrm{acl}(\bar{B}C)) = \delta(\bar{B}C)$.

(b) This clearly holds even if B, C are infinite, according to 6.2.8(3) and (a).

(4) We want to show that for $B \subseteq C$ and a, there is $a' \equiv_B a$ such that $a' \mathop{\smile\!\!\!\!\!\mid}^f_B C$. Without loss of generality, we assume B, C are closed. Let us first prove it when B, C are finite. Let $F := \mathrm{acl}(aB) \amalg_B C$. By 6.2.2(5), $C \leq F$. Then by 6.2.3(3), F is embedded into $\mathcal{M}$ and its image $F' = \mathrm{acl}(a'B) \amalg_B C$ over C is closed, so by (3) $a' \mathop{\smile\!\!\!\!\!\mid}^f_B C$. According to 6.2.6(2), $\mathrm{tp}(\mathrm{acl}(a'B)) = \mathrm{tp}(\mathrm{acl}(aB))$, in particular, $a \equiv_B a'$, as desired.

Now, in general (B, C can be infinite), there is finite $B_0 \subseteq B$ such that $a \mathop{\smile\!\!\!\!\!\mid}^f_{B_0} B$. Moreover, due to ω-categoricity of T_f, for finite $B_0 \subseteq B'(\subseteq B)$ and finite $B' \subseteq B''(\subseteq C)$, the condition that $x \mathop{\smile\!\!\!\!\!\!|}^f_{B'} B''$ and $x \equiv_{B'} a$ is consistent and definable over B''. Then by compactness while varying finite $B' \subseteq B''$, and transitivity, the extension axiom follows. ■

We suggest a further condition producing the independence theorem over algebraically closed sets.

Definition 6.2.10 Let $\{i, j, k\} = \{0, 1, 2\}$. We say $\mathcal{C}_f$ has 3-*closed amalgamation* if whenever there are a family $\{A_i, B_i \in \mathcal{C}_f \mid i < 3\}$ of sets $\geq A \in \mathcal{C}_f$, and closed A-embeddings $\sigma_{ij} : A_i \to B_j$ such that $B_j = \mathrm{cl}_{B_j}(A_{ij}A_{kj})$ $(A_{ij} := \sigma_{ij}(A_i) \leq B_j)$ and

$$\delta(B_j) = \delta(A_{ij}A_{kj}) = \delta(A_{ij}) + \delta(A_{kj}) - \delta(A),$$

then there is $B \in \mathcal{C}_f$ and closed A-embeddings $\tau_i : B_i \to B$ such that $\tau_j.\sigma_{ij} = \tau_k.\sigma_{ik}$, $B = \mathrm{cl}_B(A'_iA'_jA'_k)$ $(A'_i := \tau_j.\sigma_{ij}(A_i) \leq B)$ and

$$\delta(B) = \delta(A'_iA'_jA'_k) = \delta(A'_i) + \delta(A'_j) + \delta(A'_k) - 2\delta(A).$$

The notion of closed 3-amalgamation is analogous to that of n-amalgamation introduced in Chapter 9. The above equalities of the δ-values are given so in order to accord with 6.2.9(3)(a). We shall see for some f, $\mathcal{C}_f$ has closed 3-amalgamation.

Claim 6.2.11 If $\mathcal{C}_f$ has closed 3-amalgamation, then $\mathop{\smile\!\!\!\!\!\!|}^f$ satisfies the independence theorem over any algebraically closed set (in the home-sort $\mathcal{M}$).

Proof Let $A = \mathrm{acl}(A) \subseteq A_i$ $(i = 1, 2)$, $A_1 \mathop{\smile\!\!\!\!\!\!|}^f_A A_2$, $a_0 \mathop{\smile\!\!\!\!\!\!|}^f_A A_2$, $a''_0 \mathop{\smile\!\!\!\!\!\!|}^f_A A_1$, and $a_0 \equiv_A a''_0$. We want to find a' such that $a_0 \equiv_{A_2} a' \equiv_{A_1} a''_0$ and $a' \mathop{\smile\!\!\!\!\!\!|}^f_A A_1A_2$. We can clearly assume A_1, A_2 are also closed. We shall show the independence theorem over A when A, A_i are all finite. Then compactness yields the general case similarly to the proof of Claim 6.2.9(4).

We use closed 3-amalgamation and match up with the notations in Definition 6.2.10. Let $A_0 := \mathrm{acl}(a_0A)$ and $A''_0 := \mathrm{acl}(a''_0A)$, $B_0 := \mathrm{acl}(A_1A_2)$, $B_1 := \mathrm{acl}(a_0A_2) = \mathrm{acl}(A_0A_2)$, $B_2 := \mathrm{acl}(a''_0A_1) = \mathrm{acl}(A''_0A_1)$, and let $\sigma_{20} = \sigma_{21} := \mathrm{id}_{A_2}$, $\sigma_{10} = \sigma_{12} := \mathrm{id}_{A_1}$, and $\sigma_{01} = \mathrm{id}_{A_0}$. Lastly, let σ_{02} be an A-automorphism sending $a_0 \in A_0$ to $a''_0 \in A''_0 \subseteq B_2$. Now based on Claim 6.2.9(3)(a), $\{A_i, B_i\}$ satisfies the δ-value equalities in 6.2.10. Hence, by 3-closed amalgamation there are $B \leq \mathcal{M}$ and embeddings $\tau_j : B_j \to B$ $(j < 3)$ satisfying the δ-value equalities. We can clearly assume τ_0 is an identity map. We let $A'_0 := \tau_1(A_0) = \tau_2(A''_0)$. Since each embedding cab be extended to an automorphism, we have $A''_0 \equiv_{A_1} A'_0 \equiv_{A_2} A_0$. Moreover, again due to Claim 6.2.9(3)(a), the equalities say $A'_0 \mathop{\smile\!\!\!\!\!\!|}^f_A A_1A_2$. This completes the proof. ■

Therefore, by Theorem 3.3.1 and Proposition 3.2.12, 5.1.18, if $\mathcal{C}_f$ has closed 3-amalgamation, then T_f is supersimple ω-categorical, and $\mathop{\smile\!\!\!\!\!\!|}^f$ coincides with non-forking independence; for any algebraically closed set A in the home-sort, A and $\mathrm{bdd}(A)$ are interdefinable. Moreover, it also is immediate to show that T_f has weak elimination of imaginaries (see, for example, [124]).

We now find a function f such that $\mathcal{C}_f$ has closed 3-amalgamation. Let $f_0 := \log_3 x + 1$. Then clearly f_0 satisfies all the conditions stated before Claim 6.2.7; thus $\mathcal{C}_0 := \mathcal{C}_{f_0}$ is regular.

Claim 6.2.12 $\mathcal{C}_0$ has closed 3-amalgamation.

Proof Suppose that a family $\{A_i, B_i \in \mathcal{C}_0 | \ i < 3\}$ of sets $\geq A \in \mathcal{C}_0$, and closed A-embeddings $\sigma_{ij} : A_i \to B_j$ are given as in 6.2.10. We use the same notation. We can assume that $\sigma_{01} = \sigma_{02} = \mathrm{id}_{A_0}$ by identifying $A_{01} = A_{02} = A_0$. Let $F_{12} := B_1 \amalg_{A_0} B_2$. Note that the first equalities in 5.2.10 say $A_{10}A_{20} = A_{10} \amalg_A A_{20} \subseteq B_0$. Hence, as structures $A_{10}A_{20} \cong A_1 \amalg_A A_2 \cong A_{12} \amalg_A A_{21} \subseteq F_{12}$. So in order to ease our notation, we identify $A_{10}A_{20}$ and $A_{12}A_{21}$ as A_1A_2 over A, and take $B := F_{12} \amalg_{A_1A_2} B_0$. It easily follows that $A_1 \cup A_2 \leq F_{12}$ by comparing δ-values of A_1A_2 and any superset of A_1A_2 in F_{12}. Hence, by 6.2.2(5), $B_0 \leq B$. Indeed, $F_{12} \in \mathcal{C}_0$ and $B \in \mathcal{C}$, by 6.2.2.(5). Now the second equalities in 6.2.10 easily follow from the first and our choice of B. The second equalities conversely say $B = (B_0 \amalg_{A_1} B_2) \amalg_{A_0A_2} B_1 = B_0B_1 \amalg_{A_0A_1} B_2$ (with our ease of notation). Thus, similarly both B_1, B_2 are closed in B, and take each τ_i as an identity map. It remains to see that $B \in \mathcal{C}_0$.

Let $C \subseteq B$. We shall show that $f_0(|C|) \leq \delta(C)$, equivalently that $|C| \leq 3^{\delta(C)-1}$. Let $C_i' := C \cap B_i \in \mathcal{C}_0$. Then $C_i' \leq C$. Now let, say, $\delta(C_0') \geq \delta(C_1'), \delta(C_2')$. If $\delta(C_0') = \delta(C)$, then $C_0' = C \in \mathcal{C}_0$ and we are done. Thus, we can assume $\delta(C_i') < \delta(C)$. Then

$$|C| \leq \sum_{i=0}^{2} |C_i'| \leq \sum_{i=0}^{2} 3^{\delta(C_i')-1} \leq 3 \cdot 3^{\delta(C_0')-1} \leq 3^{\delta(C)-1}.$$

■

Hence, the theory T_0 of the generic model of $\mathcal{C}_0$ is supersimple ω-categorical. Note that any singleton a is closed, so $\mathcal{M}$ itself is the solution set of the unique 1-type $p(x) \in S_1(\emptyset)$. Moreover, $d(a/A) \leq d(a) = 1$. Therefore, $SU(p) = 1$ and $(\mathcal{M}, \mathrm{acl}(-))$ forms a geometry (see Chapter 8). Note also that any $\emptyset \leq A \in \mathcal{C}_0$ can be closedly embedded into $\mathcal{M}$, so there are two 2-types realized, say, by ab and cd such that $\mathrm{acl}(ab) = \{a, b\}$, whereas $\mathrm{acl}(cd)$ contains a third point $e \in \mathcal{M}$ with $R(a, c, e)$.

Claim 6.2.13 T_0 is not 1-based.

Proof We borrow a technique from Chapter 8. By the independence theorem over $\emptyset$, there is a non-modular triple (ca_1a_2) in the home-sort such that $a_1a_2 \equiv ab, ca_i \equiv cd$ above. Hence, the geometry $(\mathcal{M}, \mathrm{acl}(-))$ is not modular (Remark 8.1.2). To lead a contradiction assume T_0 is 1-based. Then $(G(\mathcal{M}), \mathrm{bdd}(-))$ is modular due to Theorem 8.3.6, so there must be $u \in \mathrm{bdd}(a_1a_2) = \mathrm{dcl}(a_1a_2)$ and $u \in G(\mathcal{M})$ such that $\{a_1, a_2, u\}$ is pairwise independent. Now due to Lemma 4.1.5(3) and ω-categoricity, $u \in \mathcal{M}^{\mathrm{eq}}$. Thus, by weak elimination of imaginaries, there is a tuple $a' \in \mathrm{acl}(a_1a_2)$ in the home-sort interalgebraic with u, in particular $a' \not\perp a_i$, a contradiction. ■

6.3 Generic Predicates/Automorphisms and Lovely Pairs

In Example 2.6.10, we briefly explain methods of constructing simple structures from stable/simple structures by adding generic predicates or automorphisms. This section summarizes the construction techniques more in detail. We also describe the notion of a lovely pair of simple structures introduced in [7]. Given a pair of simple structures $N \prec M$, one may ask when (M, P), where P is a unary predicate and $P(M) = N$, is again a simple structure. The lovely pair notion gives some sufficient condition for the positive answer of the question. This notion also has an application in partially solving a deep conjecture that a unidimensional theory is supersimple (Fact 6.3.15).

In this section sets are real or imaginary.

6.3.1 *Generic Predicates*

While a pure infinite set is a prototypical trivial stable structure, the random graph is the basic unstable simple structure obtained from a pure set by adding a *generic* binary relation. One may wonder whether the similar process of adding a random predicate can be given to an arbitrary stable structures making it an unstable simple structure. This issue is dealt with in this subsection.

Throughout, we assume our complete theory T in $\mathcal{L}$ has quantifier elimination. Let P be a new unary predicate and $\mathcal{L}_P := \mathcal{L} \cup \{P\}$. For a notational distinction, let T_0 be (incomplete) T in $\mathcal{L}_P$. As stated, we shall summarize the facts proved in [27]. Recall that T has $\exists^\infty$*-elimination* if for any $\varphi(x, y) \in \mathcal{L}$ there is $\psi(y) \in \mathcal{L}$ such that $\varphi(x, b)$ defines an infinite set iff $\psi(b)$ holds. Recall that T_0 has a *model companion* T_P (in $\mathcal{L}_P$) if $(T_0 \subseteq) T_P$ is model complete and any model of T_0 has an extension which is a model of T_P. The model companion is unique (up to implication).

Fact 6.3.1

(1) *T_0 has model companion (in $\mathcal{L}_p$) iff T eliminates $\exists^\infty$ (in $\mathcal{L}$).*

(2) *If T has $\exists^\infty$-elimination, then for any $\varphi(\bar{x}, \bar{z}) \in \mathcal{L}$ there is $\phi(\bar{z}) \in \mathcal{L}$ such that*

$$\phi(\bar{z}) \equiv \exists \bar{x}(\varphi(\bar{x}, \bar{z}) \wedge \bar{x} \cap \operatorname{acl}(\bar{z}) = \emptyset).$$

(3) *If T_0 has the model companion T_P then T_P is axiomatized by the following sentences:*

  (a) *T together with;*

  (b) *For each $\varphi(\bar{x}, \bar{z}) \in \mathcal{L}$ with $\bar{x} = \{x_0, \ldots, x_{n-1}\}$ and each subset $I \subseteq n$,*

$$\forall \bar{z}\ [\ \exists \bar{x}(\varphi(\bar{x}, \bar{z}) \wedge (\bar{x} \cap \operatorname{acl}_T(\bar{z}) = \emptyset) \wedge \textstyle\bigwedge_{i \neq j < n} x_i \neq x_j)) \\ \rightarrow \exists \bar{y}(\varphi(\bar{y}, \bar{z}) \wedge \textstyle\bigwedge_{i \in I} P(y_i) \wedge \textstyle\bigwedge_{j \notin I} \neg P(y_j))].$$

  where $\operatorname{acl}_T(\bar{z})$ is an algebraic closure of $\bar{z}$ in T.

Now we assume that T eliminates $\exists^\infty$.

Fact 6.3.2

(1) *Let* $(M, P), (N, Q)$ *be models of* T_P *and let* $A \subseteq M \cap N$. *Then* $(M, P)_A \equiv (N, Q)_A$ *iff* $(\mathrm{acl}_T(A) \cap M, P \cap \mathrm{acl}_T(A))$ *is A-isomorphic to* $(\mathrm{acl}_T(A) \cap N, P \cap \mathrm{acl}_T(A))$.

(2) *A completion of* T_P *is obtained by describing* $P \cap \mathrm{acl}_T(\emptyset)$.

(3) $\mathrm{acl}_T(A) = \mathrm{acl}_{T_P}(A)$ *in any completion of* T_P.

Fact 6.3.3 *Suppose that T is simple and* T'_P *is a completion of* T_P.

(1) *If T is (super)simple, then so is* T'_P.

(2) *For some tuples a, b, if* $\mathrm{acl}(a, b) \not\subseteq \mathrm{acl}(a) \cup \mathrm{acl}(b)$ *(in T or* T'_P*), then* T'_P *is not stable.*

(3) *For sets, A is independent from B over C in T iff the same holds in* T'_P.

(4) *Assume T is stable. Then* T'_P *has the independence theorem over any algebraically closed set in* T^{eq}. *If T has (weak) elimination of imaginaries, so is* T'_P.

Contrary to (3) above, it is possible that $\mathrm{tp}(a/b)$ forks over c in T'_P while $\mathrm{tp}(a/b)$ does not fork over c in T (so here $c \not\subseteq b$) [73].

One can do the same thing for a sort in $\mathcal{M}^{\mathrm{eq}}$; in particular, P can be an n-ary predicate by taking the sort $\mathcal{M}^n$. If a sort is the set of 2-element subsets of $\mathcal{M}$ where $\mathcal{M}$ is an infinite set in an empty language, the resulting structure is unstable: $\{b, c\} \in \mathrm{acl}(\{a, b\}\{a, c\}) \setminus (\mathrm{acl}(\{a, b\}) \cup \mathrm{acl}(\{a, c\}))$. Indeed, that is the random graph.

We finish this subsection by supplying a promised (Remark 2.5.13) example where the D-rank does not reflect forking.

Example 6.3.4 Let T be a theory in $\mathcal{L} := \{Q_i, E_i | \ i < \omega\}$. T says that unary predicates Q_i are disjoint and each $E_i(x, y)$ defines an equivalence relation only on Q_i having only infinitely many infinite classes. T is complete and stable. Now we add a generic predicate $R(x, y)$ on the sort of 2-element sets of each Q_i. Let T_R be the resulting complete simple theory. So in each Q_i, R supplies the random graph relation. On the other hand, there is no R-relation between elements in Q_i and Q_j $(i \neq j < \omega)$. By compactness, on $p(x) \in S_1(T_R)$ axiomatized by $\{\neg Q_i(x) | \ i < \omega\}$, R also gives the random graph relation. Now choose $a, b \models p$ and $a \neq b$, which are independent in T_R. Note that $D(p) = D(a) = 2$ in T and T_R. But $D(a/b) = 1$ in T_R $(= 2$ in $T)$ since $\mathrm{tp}(a/b)$ is isolated by $R(x, b)$ and $D(R(x, b)) = 1$.

6.3.2 Generic Automorphisms

We now consider an issue of adding an automorphism to make a stable structure simple. Assume the underlying complete $T = T^{\mathrm{eq}}$ in $\mathcal{L} = \mathcal{L}^{\mathrm{eq}}$ has quantifier elimination, and $\mathrm{acl}(-)$ below is $\mathrm{acl}^{\mathrm{eq}}(-)$. For a new unary function symbol σ we let $\mathcal{L}_\sigma := \mathcal{L} \cup \{\sigma\}$. Let $T^0_\sigma := \mathrm{Th}(\{(M, \sigma) | \ M \models T, \ \sigma \in \mathrm{Aut}(M)\})$, a theory in $\mathcal{L}_\sigma$. There are no known practical criteria determining when T^0_σ has the model companion T_σ even if T is stable, but it is known that if T has the strict order property then T_σ does not exist [64]. In the rest, T is stable. At least we have the following fact.

Definition 6.3.5 Let $(M, \sigma_0) \models T^0_\sigma$. We say σ_0 is $|T|^+$-*generic* if for any algebraically closed substructure $A \subseteq M$ (in T) with $|A| \leq |T|$ and $\sigma_A := \sigma_0 \lceil A \in \mathrm{Aut}(A)$, whenever $f : (A, \sigma_A) \to (N, \sigma_1) \models T^0_\sigma$ is an $\mathcal{L}_\sigma$-embedding, then there is an $\mathcal{L}_\sigma$-embedding $g : f(A, \sigma_A) \to (M, \sigma_1)$ such that $g.f$ is the identity map on A.

Fact 6.3.6

(1) *If T^0_σ has the model companion T_σ then $T' := \mathrm{Th}(\{(M, \sigma_M) \models T^0_\sigma : \sigma_M\ |T|^+$ – generic$\})$ is equal to T_σ. Conversely, if T' is model-complete then it is the model companion of T^0_σ.*

(2) *Assume T_σ exists. If $(M_1, \sigma_1), (M_2, \sigma_2) \models T_\sigma$ and $A = \mathrm{acl}(A) \subseteq M_i$ (in T) with $\sigma_A := \sigma_i \lceil A \in \mathrm{Aut}(A)$, then $(M_1, \sigma_1)_A \equiv (M_2, \sigma_2)_A$.*

For the rest we assume that the model companion T_σ of T^0_σ exists. As usual, fix a completion T'_σ of T_σ and a large saturated model $(\mathcal{M}, \sigma) \models T'_\sigma$. For a set A, $\mathrm{acl}_\sigma(A) := \mathrm{acl}(\bigcup_{n\in\mathbb{Z}} \sigma^n(A))$ in T. Note that the completion T'_σ is obtained from T_σ together with a description of $(\mathrm{acl}_\sigma(A), \sigma)$.

The following results are independent from the choice of the completion.

Fact 6.3.7

(1) *$\mathrm{acl}_\sigma(A)$ is the algebraic closure of A in T'_σ.*

(2) *T'_σ is simple. If T is superstable then T'_σ is supersimple.*

(3) *$A \perp_B C$ in T'_σ iff $\mathrm{acl}_\sigma(AB) \perp_{\mathrm{acl}_\sigma(B)} \mathrm{acl}_\sigma(BC)$ in T.*

(4) *If there are a model $M \models T$ and tuples a, b such that $\mathrm{acl}(Mab) \neq \mathrm{dcl}(\mathrm{acl}(Ma)\ \mathrm{acl}(Mb))$ in T then T'_σ is unstable (simple).*

We refer the reader to Examples 2.6.9 and 2.6.10 and [27] for specific examples of T_σ.

6.3.3 *Lovely Pairs*

The notion of lovely pairs generalizes Poizat's notion of beautiful pairs for stable structures [106], and Vassiliev's notion of generic pairs for SU-rank 1 structures [116].

Throughout T is complete simple in $\mathcal{L}$. Given a new unary predicate P, $\mathcal{L}_P := \mathcal{L} \cup \{P\}$. As usual, given a, A from $(\mathcal{M}, P)$ with $\mathcal{M} \models T$, $\mathrm{tp}_\mathcal{L}(a/A)$ is an $\mathcal{L}$-type and $\mathrm{tp}_{\mathcal{L}_P}(a/A)$ is an $\mathcal{L}_P$-type, and when we say forking or non-forking of $\mathcal{L}$-types, it is computed in T (i.e., in $\mathcal{M}$).

Definition 6.3.8 Let $\kappa > |T|$. By a κ-*lovely pair* of models of T, we mean an $\mathcal{L}_p$-structure (M, P) with $M \models T$ such that for any $A \subseteq M$ with $|A| < \kappa$, and $p \in S_n(A)$ in T,

(1) There is $a(\in M) \models p$ such that $\mathrm{tp}_\mathcal{L}(a/A \cup P(M))$ does not fork over A in T; and

(2) If p does not fork over $P(A)(:= P(M) \cap A)$, then p is realized in $P(M)$ (so it follows in $\mathcal{L}$, $P(M) \prec M$ and both are κ-saturated).

By a *lovely pair* of models of T, we mean $|T|^+$-lovely pair. Let $\mathcal{C}_{T,P}$ denote the class of lovely pairs, and let $T_P := \mathrm{Th}(\mathcal{C}_{T,P})$.

Fact 6.3.9

(1) *For each* $\kappa > |T|$, κ*-lovely pairs exist.*
(2) *All lovely pairs are elementarily equivalent so* T_P *is complete.*

Definition 6.3.10

(1) We say $C_{T,P}$ is *first order* if any $|T|^+$-saturated model of T_P is a lovely pair.
(2) Given $\varphi(x,y), \psi(y,z) \in \mathcal{L}$, we let

$$Q_{\varphi,\psi}(u) := \forall b\big(b \models \psi(b,u) \rightarrow \varphi(x,b) \text{ does not divide over } u\big).$$

(3) We say the rank $D_{\varphi(x,y)}(-)$ is *definable* if for each $\phi(x,b)$ and α, there is $\delta_\alpha(y) \in \mathcal{L}$ such that $D_\varphi(\phi(x,b)) = \alpha$ iff $\delta_\alpha(b)$.

Fact 6.3.11 *The following are equivalent:*

(1) $C_{T,P}$ *is first-order.*
(2) *T is low and,* $Q_{\varphi,\phi}$ *is type-definable in T for each* $\varphi(x,y), \psi(y,z) \in \mathcal{L}$.
(3) *T is low and* $Q_{\varphi,\phi}$ *is definable in T for each* $\varphi, \psi \in \mathcal{L}$.
(4) *For each* $\varphi \in \mathcal{L}$, *the* D_φ*-rank is finite and definable in T.*

If T is low then for $\varphi(x,u) \in \mathcal{L}$, that "$\varphi(x,u)$ divides over v" is $\emptyset$-type-definable on (u,v), say by $D_{\varphi(x,u),v}(u,v)$.

In Facts 6.3.12–6.3.14 below, $C_{T,P}$ is first order.

Fact 6.3.12 T_P *is axiomatized by:*

$$\Big\{\forall v\big[\neg Q_{\varphi,\psi}(v) \rightarrow \exists u(\psi(u,v) \wedge \forall x(P(x) \rightarrow \neg\phi(x,u)))\big], \forall uv\big[P(v) \wedge \neg\phi(u,v) \rightarrow \exists x(P(x) \wedge \varphi(x,u))\big] : \varphi(x,u), \psi(u,v) \in \mathcal{L}, \phi \in D_{\varphi,v}(u,v)\Big\}.$$

Temporarily for a set A, let $\widehat{A} := A \cup \mathrm{Cb}(a/P(\mathcal{M}))$ in T.

Fact 6.3.13

(1) T_P *is simple. If T is stable/supersimple/1-based, so is* T_P, *respectively.*
(2) *For a tuple a, and* $B \subseteq C$, *the following are equivalent:*
   (a) $a \mathop{\smile\!\!\!\!\!\mid}_B C$ *in* T_P.
   (b) *In T,* $a \mathop{\smile\!\!\!\!\!\mid}_{B \cup P(\mathcal{M})} C \cup P(\mathcal{M})$ *and* $\widehat{aB} \mathop{\smile\!\!\!\!\!\mid}_{\widehat{B}} \widehat{C}$.

There are a couple of applications of the notion of lovely pairs to geometric stability/simplicity theory.

Fact 6.3.14 *Let T be stable (so* T_P *stable too). The following are equivalent:*

(1) T_P *eliminates imaginaries.*
(2) *Any imaginary in* T_P *is interdefinable (in* T_P*) with an imaginary in T.*

(3) *T_P has weak elimination of imaginaries.*

(4) *Any imaginary in T_P is interalgebraic (in T_P) with an imaginary in T.*

(5) *No infinite group is definable in T.*

Fact 6.3.15 *Assume T is countable simple unidimensional and eliminates hyperimaginaries. Then T is supersimple if $C_{T,P}$ is first order.*

Later, Shami [111] removed the condition that $C_{T,P}$ is first order from Fact 6.3.15, using different and deep arguments. Any stable unidimensional theory is superstable due to Hrushovski [50].

6.4 Bibliographical Remarks

Theorem 6.1.5 is from [39]. For more canonical hierarchy beyond 1-basedness and CM-triviality, we refer to [100],[37]. Section 6.2 is due to Hrushovski [56] and the account here follows from [120]. More on the Hrushovski construction is available as lecture notes or slides on the Internet homepages of J. T. Baldwin, F. O. Wagner, D. M. Evans, and others. Sections 6.3.1 and 6.3.2 are from [27], and 6.3.3 is from [7]. Fact 6.3.14 is in [105], and Fact 6.3.15 is in [101].

7

Groups

In this chapter we study groups (type-)definable or hyperdefinable in structures having simple theories. Groups are basic building blocks constituting further complicated algebraic objects. Our knowledge of hyperdefinable groups initiates deeper understanding on algebraic behaviours of given abstract simple structures. The notion of generic elements will play a crucial role in understanding such groups.

Throughout this chapter our theory T is **simple** unless said otherwise. We assume some familiarity with the theory of stable groups in [107], [99], or [119]. Given two properties P and Q, we call a group *P-by-Q* if the group has a normal subgroup satisfying the property P with its quotient holding the property Q. Recall from Chapter 1 that, unless mentioned otherwise, an *automorphism* is that of the ambient first-order structure. Note that even if the structure itself is a group with extra relations, a group automorphism need not be a structure automorphism.

7.1 Type-Definable Groups

We study type-definable groups in $\mathcal{M} \models T$. In this section, the arities of variables are all finite *real*. Before we start, let's state some basic folklores (true in *any* theory) on type-definable groups.

Suppose that $(G(x), \cdot(x,y,z), 1_G)$ (x, y, z finite) is a type-definable group G over say $\emptyset$; i.e., $G(x)$ $\emptyset$-type-defines the universe of the group G, and $\cdot(x,y,z)(\models G(x) \wedge G(y) \wedge G(z))$ type-defines the graph of the group operation. We claim that there is $\sigma_0(x) \in G(x)$ and $\varphi(x,y,z) \in \cdot(x,y,z)$ such that

$$\sigma_0(x) \wedge \sigma_0(y) \models \varphi(x, 1_G, x) \wedge \varphi(1_G, y, y) \wedge \exists! z\, \varphi(x,y,z))(^*).$$

Note first that $G(x) \wedge G(y) \wedge \cdot(x,y,z_1) \wedge \cdot(x,y,z_2) \wedge z_1 \neq z_2$ is inconsistent. Hence, by compactness there are $\sigma_0(x) \in G(x)$ and $\varphi(x,y,z) \in \cdot(x,y,z)$ such that

$$\sigma_0(x) \wedge \sigma_0(y) \models \varphi(x,y,z_1) \wedge \varphi(x,y,z_2) \rightarrow z_1 = z_2.$$

Moreover, $G(x) \wedge G(y) \models \exists z\varphi(x,y,z) \wedge \varphi(x, 1_G, x) \wedge \varphi(1_G, y, y)$. Hence, by refining σ_0, we can further assume that (*) holds with σ_0 and φ.

Now we write the unique z in (*) as $x.y$. Indeed, we can enforce that . is a binary operation on the whole universe whose restriction on G is the group operation (by, for example, putting $x.y = x$ for not both $x, y \models \sigma_0$).

We further claim that for any $\sigma'' \in G(x)$, there is $\sigma'(x) \in G(x)$ implying that $\sigma''(x) \wedge \sigma_0(x)$ such that $1_G \models \sigma'$; associativity holds inside σ' (i.e., for any $x, y, z \models \sigma'$, we have $(x.y).z = x.(y.z)$, *but* $x.y$ *need not satisfy* σ'); any $x \models \sigma'$ has the unique inverse $y \models \sigma'$ such that $x.y = y.x = 1_G$ (**). For a fixed n, by compactness, clearly we can find $\sigma_0'(\in G(x)) \models \sigma'' \wedge \sigma_0$ such that any n elements in σ_0' satisfy all the possible associativity. Now, by compactness there is $\sigma_1'(x)(\models \sigma_0' \in G(x)) \models \exists y(\sigma_0'(y) \wedge x.y = y.x = 1)$. Moreover, by associativity with (*), the inverse y of $x \models \sigma_1'$ must be unique. Now, by taking σ' as $\sigma_1'(x) \vee \sigma_1'(x^{-1})(\models \sigma_0')$, (**) follows.

In conclusion, we can assume that σ_0 in (*) also satisfies the properties for σ' in (**). Note that the definable structure $(\sigma_0, ., 1)$ is almost like a group except that for $a, b \models \sigma_0$, $a.b$ *need not* realize σ_0.

Throughout this section, we fix the $\emptyset$-type-definable group G, and $\sigma_0(x)$. We fix some more notation. For a formula $\varphi(x; y)$, $\varphi(z.x; y)(= z^{-1}.\varphi(x; y))$ and $\varphi(x^{-1}; y)$ mean, respectively,

$$\exists w(\varphi(w,y) \wedge w = z.x \wedge \sigma_0(z) \wedge \sigma_0(x) \wedge \sigma_0(w)),$$

and

$$\exists w\big(\varphi(w,y) \wedge \sigma_0(x) \wedge \sigma_0(w) \wedge w = x^{-1}\big).$$

Note that for some $g_0, g_1 \in G$, $g_0.\psi(x, a)$ is consistent while $g_1.\psi(x, a)$ *need not* be, if for all $c \models \sigma_0(x) \wedge \psi(x, a)$, $g_1.c \not\models \sigma_0$.

For a type $p(x)$ over A, and $g \in G$,

$$g.p := \{g.\varphi(x; a) |\ \varphi(x, a) \in p(x)\}, \text{ and } p(x^{-1}) = \{\psi(x^{-1}) |\ \psi(x) \in p\}.$$

Similarly, we define $\varphi(x.z; y) = \varphi(x, y).z^{-1}$ and $p(x).g$. Note that if H is a type-definable subgroup of G then $g.H$ ($H.g$, resp.) is a left (right, resp.) coset, so the notation is compatible. In addition, $S_G(A) := \{p(x) \in S(A) |\ p(x) \models G(x)\}$ and $S_G^L(A) := S_G(\text{bdd}(A))$. As usual $|G : H|$ denotes the index of a subgroup H in G.

We remark that for $c, g \models G$, if $c \in \text{dcl}(A)$ then $c.\text{tp}(g/A) \equiv \text{tp}(c.g/A)(\equiv \text{tp}(c.g/Ac) \equiv \text{tp}(c.g/Ac^{-1}))$ since $g \equiv_A g'$ implies that $c.g \equiv_A c.g'$. We will freely use this fact throughout. In general, $c.\text{tp}(g/A)$ merely is a partial type over Ac^{-1} realized by $c.g$.

We now introduce the notion of *generic* elements or types, which is a central concept in our theoretic development. The notion of generic types comes from stability theory [119]. Namely, a type $p \in S_G(\emptyset)$ is said to be *generic* if $g.p$ does not fork over $\emptyset$ for all $g \in G$. The genericity of a type is preserved under automorphisms and (group) translations. Here we define the *genericity* of a type a little differently as it is initially more convenient to work with. Then later, it will be proved that the two notions coincide (Theorem 7.1.7).

Definition 7.1.1 Let $g \in G$. We say g is *left-generic* in (or for) G over A, if for any $a \in G$ with $g \downarrow_A a$, we have $a.g \downarrow Aa$. It follows, by putting $a = 1_G$, that $g \downarrow A$. We say g is *right-generic* over A if the same but $g.a \downarrow Aa$ holds. A (partial) type $p(x)$ over A implying $G(x)$ is said to be *(left/right-)generic* if some generic g over A realizes $p(x)$.

Lemma 7.1.2

(1) *Suppose that $g, a \in G$ and $\text{tp}(g/Aa)$ is left-generic. Then so is $\text{tp}(a.g/Aa)$.*

(2) *Let $A \subseteq B$, and let $p(x)(\in S_G(A)) \subseteq q(x) \in S_G(B)$. Then $q(x)$ is left-generic iff $q(x)$ is a non-forking extension of left-generic $p(x)$.*

(3) *If $g(\in G)$ is left-generic over A and $g \downarrow_A a(\in G)$, then $a.g$ is left-generic over Aa.*

(4) *If $g \in G$ is left-generic over A, then so is g^{-1}.*

(5) *Suppose that $I = \langle a_i | \ i < \alpha \rangle$ is some A-independent sequence from G such that each a_i is left-generic over A. Then I is left-generic for the group G^α over A.*

Proof (1) Let $a.g \downarrow_{Aa} b(\in G)$. Hence, $a^{-1}.a.g \downarrow_{Aa} b.a$; i.e., $g \downarrow_{Aa} b.a$. Then since $\text{tp}(g/Aa)$ is left-generic, we have $b.a.g \downarrow Aa(b.a)$. It follows that $b.a.g \downarrow Aa(a^{-1}.b^{-1})$, $b.a.g \downarrow Aab^{-1}$, and $b.a.g \downarrow Aab$. Therefore, $\text{tp}(a.g/Aa)$ is left-generic.

(2) Let $g \models q(x)$. ($\Rightarrow$) Since q does not fork over $\emptyset$, it is a non-forking extension of p. To show p is left-generic, choose $a(\in G) \downarrow_{Ag} (\models p)$. Now, we have $B'(\equiv_{Ag} B) \downarrow_{Ag} a$. Hence, $a \downarrow_A B'g$ and $a \downarrow_{B'} g$ hold. Since $\text{tp}(g/B')$ is left-generic, we have $a.g \downarrow B'a$, and $a.g \downarrow Aa$.

($\Leftarrow$) Choose $b(\in G) \downarrow_B g$. Thus, $bB \downarrow_A g$. Since $\text{tp}(g/A)$ is left-generic, we have $b.g \downarrow Ab$. Moreover, $g \downarrow_{Ab} B$ and $b.g \downarrow_{Ab} B$ hold. Therefore, $b.g \downarrow Bb$.

(3) comes from (1),(2).

(4) Choose $h \equiv_A g$, and $g \downarrow_A h$. By genericity, $h.g \downarrow_A h$, and from (2), both $\text{tp}(h/A(h.g))$ and $\text{tp}(h/A(g^{-1}.h^{-1}))$ are left-generic. Then by (1), $\text{tp}(g^{-1}/A(g^{-1}.h^{-1}))$ is left-generic, and so $g^{-1} \downarrow A(g^{-1}.h^{-1})$ holds. Thus, again from (2), $\text{tp}(g^{-1}/A)$ is generic.

(5) This easily comes from induction and forking calculation. ■

Corollary 7.1.3 *$g \in G$ is left-generic over A iff it is right-generic over A.*

Proof ($\Rightarrow$) Let $g \downarrow_A a(\in G)$. Then g^{-1} is left-generic and $g^{-1} \downarrow_A a^{-1}$. It follows that $a^{-1}.g^{-1} \downarrow Aa^{-1}$, and $g.a \downarrow Aa$. Hence, g is right-generic.

($\Leftarrow$) Similar to ($\Rightarrow$). Note that the right-generic version of Lemma 7.1.2 holds too. ■

Hence, from now on, we take out left- and right-, and simply say *generic* elements or types. Now we aim to prove the existence of the generic types. For this, we introduce the $D_G(p(x), \varphi(x,y), k)$-rank (a variation of the $D(p(x) \wedge G(x), \varphi(x,y), k)$-rank) for a type $p(x)$ and $k \geq 2$. We let $D_G(p(x), \varphi(x,y), k) \geq 0$ if $\exists x(p(x) \wedge G(x))$; and $D_G(p(x), \varphi(x,y), k) \geq n+1$ if there are $g \in G$ and b_i $(i < \omega)$ such that $\{\varphi(x, b_i) | \ i < \omega\}$

is k-inconsistent and $D_G(g.p \cup \{\varphi(x,b_i)\},\varphi,k) \geq n$ for each i. We write $D_G(a/A,\varphi,k)$ for $D_G(\mathrm{tp}(a/A),\varphi,k)$.

Remark 7.1.4 As we want the $D_G(-,\varphi,k)$-rank to be invariant under any group translation (7.1.5(2)), we required translating p by some g, instead of $\{\varphi(x,b_i)\}_i$. Also this definition is compatible with the $D_G(-,\varphi,\epsilon,k)$-rank in Section 7.2 for hyperdefinable groups. Note now, by compactness and induction, that $D_G(p,\varphi,k) \geq n+1$ is a type-definable condition. Hence, in the definition, we can replace ω by any infinite κ.

Lemma 7.1.5

(1) *$D_G(p(x),\varphi,k) = D_G(p(x) \wedge G(x),\varphi,k)$.*

(2) *For a type $p(x) \models G(x)$, and $h \in G$, $D_G(p,\varphi,k) = D_G(h.p,\varphi,k)$. Hence, $D_G(a/A,\varphi,k) = D_G(h.a/A,\varphi,k)$, for $a \in G$ and $h \in \mathrm{dcl}(A) \cap G$.*

(3) *$D_G(p_1(x),\varphi,k_1) \leq D_G(p_2(x),\varphi,k_2)$ if $p_1 \vdash p_2$ and $k_1 \leq k_2$.*
For every $p(x)$, $\varphi(x,y)$, k, there is a finite subset $q(x)$ of p such that $D_G(p,\varphi,k) = D_G(q,\varphi,k)$;
$D_G(p \cup \bigvee_{i<n} q_i,\psi,k) = \max\{D_G(p \cup q_i),\psi,k) : i < n\}$.

(4) *Let $\varphi(x,y)$, k be given. For a type $p(x)$ over A, there is a completion $p'(x) \in S(A)$ of p such that $D_G(p,\psi(x,y),k) = D_G(p',\psi(x,y),k)$.*

(5) *$D_G(G(x),\varphi,k)$ is finite.*

(6) *Let $q(x) \in S_G(B)$. Then $q(x)$ is a non-forking extension of $p(x) \in S_G(A)$ with $A \subseteq B$ iff $D_G(p,\varphi,k) = D_G(q,\varphi,k)$ for any φ,k. In particular, $D_G(a/A,\varphi,k) = D_G(a/\mathrm{bdd}(A),\varphi,k)$.*

Proof (1) Clear.

(2) If g, b_i witness $D_G(p,\varphi,k) \geq n+1$, then clearly $g.h^{-1}, b_i$ witness $D_G(h.p,\varphi,k) \geq n+1$. For the converse, translate $h.p$ by h^{-1} to get p. The hence clause follows since $h.\mathrm{tp}(a/A) = \mathrm{tp}(h.a/A)$.

(3),(4) As in Lemma 2.3.4, those come easily by usual induction with compactness.

(5) Note that due to (2), if $D_G(p,\varphi,k) \geq n+1$, then there are $g \in G$ and b_i such that $D_G(g^{-1}.p \cup \{\varphi(x,b_i)\},\varphi,k) = D_G(p \cup g.\varphi(x,b_i),\varphi,k) \geq n$, and $\{g.\varphi(x,b_i) | i < \omega\}$ is k-inconsistent. Thus, if $D_G(G,\varphi,k) \geq n$ for all $n \in \omega$, then by induction $\psi(x;yz) = z.\varphi(x;y)$ produces the tree property (within $G(x)$); i.e., simplicity of T fails. This contradiction shows (5).

(6) We omit the proof as it will be similar to that of Proposition 2.3.9 or is a particular case of that of Lemma 7.2.4(5). ■

Proposition 7.1.6 *There is generic $p(x) \in S_G(A)$.*

Proof Due to 7.1.2, it suffices to show when $A = \emptyset$. Now let $\langle(\varphi_\alpha,k_\alpha)| \ \alpha \in |T|\rangle$ be an enumeration of $\{(\varphi(x,y),k)| \ \varphi \in \mathcal{L}, k \geq 2\}$. We inductively define an increasing sequence of partial types $p_\alpha(x)$ over $\emptyset$ and a sequence of natural numbers n_α as follows:

(i) $p_0(x) = G(x)$, and $n_0 = D_G(p_0, \varphi_0, k_0)$,

(ii) $p_{\alpha+1} = p_\alpha(x) \cup \{\neg\psi(x) \in \mathcal{L} \mid D_G(\psi(x), \varphi_\alpha, k_\alpha) < n_\alpha\}$ and $n_{\alpha+1} = D_G(p_{\alpha+1}, \varphi_{\alpha+1}, k_{\alpha+1})$,

(iii) For limit δ, $p_\delta = \bigcup\{p_\alpha \mid \alpha < \delta\}$ and $n_\delta = D_G(p_\delta, \varphi_\delta, k_\delta)$.

Note that each p_α is consistent (since if not then it violates 7.1.5(3)). Hence, $\bigcup\{p_\alpha \mid \alpha \in |T|\}$ is consistent, and there is a completion $p(x) \in S_G(\emptyset)$ of it.

We claim that $p(x)$ is a generic type of G: Note first that $D_G(p, \varphi_\alpha, k_\alpha) = n_\alpha$ since if not then again due to 7.1.5(3), it violates (ii). Let $g(\models p) \mathop{\smile\!\!\!\!\!|}\, a(\in G)$. Then from 7.1.5(6),(2), we have $D_G(g/a, \varphi_\alpha, k_\alpha) = n_\alpha = D_G(a.g/a, \varphi_\alpha, k_\alpha)$. Hence, $n_\alpha \leq D_G(\text{tp}(a.g), \varphi_\alpha, k_\alpha)$. Thus, it follows by induction, $a.g \models p_\alpha$ for each α in (i),(ii),(iii). Hence, $D_G(\text{tp}(a.g), \varphi_\alpha, k_\alpha) \leq D_G(p_\alpha, \varphi_\alpha, k_\alpha) = n_\alpha$. Therefore, by 7.1.5(6), we have $a.g \mathop{\smile\!\!\!\!\!|}\, a$; i.e., p is generic. ■

Proposition 7.1.7 *Let $p(x)$ be a partial type over A implying $G(x)$. The following are equivalent:*

(1) *The type p is generic.*

(2) *$D_G(p, \varphi, k) = D_G(G(x), \varphi, k)$, for each φ, k.*

(3) *For any $g \in G$, the type $g.p$ does not fork over $\emptyset$.*

(4) *For any $g \in G$, the type $g.p$ does not fork over A.*

(5) *For any $g \in G$, $g.p$ is generic.*

Proof (1)⇒(2) There is generic $g \models p$ over A. Given φ, k, there is $h \in G$ such that $D_G(G, \varphi, k) = D_G(h^{-1}/A, \varphi, k)$ and $g \mathop{\smile\!\!\!\!\!|}_A h$. Hence, $g.h \mathop{\smile\!\!\!\!\!|}_A h$, and

$$\begin{aligned} D_G(G, \varphi, k) &= D_G(h^{-1}/A, \varphi, k) = D_G(h^{-1}/A(g.h), \varphi, k) \\ &= D_G(g.h.h^{-1}/A(g.h), \varphi, k) \leq D_G(g/A, \varphi, k) \leq D_G(p, \varphi, k) \\ &\leq D_G(G, \varphi, k). \end{aligned}$$

(2)⇒(3) Assume (2). Now, if there is $g \in G$ such that $g.\psi(x, a)(\in g.p)$ k-divides over $\emptyset$, then it follows from the definition and 7.1.5(2), $D_G(G(x), z.\psi(x,y), k) \geq D_G(g.\psi(x,a), z.\psi(x,y), k) + 1 \geq D_G(g.p, z.\psi(x,y), k) + 1 = D_G(p, z.\psi(x,y), k) + 1 = D_G(G, z.\psi, k) + 1$, a contradiction.

(3)⇒(4) Clear.

(4)⇒(1) Choose $g \in G$ generic over A. By (4), $g.p$ does not fork over A. Hence, there is $d \models p$ such that $\text{tp}(g.d/Ag)$ does not fork over A. By symmetry, $\text{tp}(g/A(g.d))$ does not fork over A; thus due to 7.1.2(2), $\text{tp}(g/A(d^{-1}.g^{-1}))$ is generic as well. Again by 7.1.2, so are $\text{tp}(d^{-1}/A(d^{-1}.g^{-1}))$, $\text{tp}(d^{-1}/A)$, and $\text{tp}(d/A)$. Since $d \models p$, p is generic.

(1)⇔(5) Use (1)⇔(3). ■

Remark & Definition 7.1.8

(1) Hence, $g \in G$ is generic over A iff for any $a \in G$ with $a \mathop{\smile\hskip-0.9em\raise0.5ex\hbox{$|$}}_{Ag}$, we have $a.g \mathop{\smile\hskip-0.9em\raise0.5ex\hbox{$|$}}_A a$. Assume the right-hand side. Let $p = \mathrm{tp}(g/A)$. Given $a \in G$, we want to show $a.p$ does not fork over A. By extension, we can assume that $g \mathop{\smile\hskip-0.9em\raise0.5ex\hbox{$|$}}_A a$. Let $q = \mathrm{tp}(g/Aa)$. Now, we have $a.g \mathop{\smile\hskip-0.9em\raise0.5ex\hbox{$|$}}_A a$; i.e., $a.q = \mathrm{tp}(a.g/Aa)$ does not fork over A. Since $a.q \models a.p$, neither does $a.p$. Therefore, p is generic.

(2) Similarly to 5.1.7(2), one can show that for a type-definable subgroup H of G over A, p is generic for H iff $D_G(p, \varphi, k) = D_G(H, \varphi, k)$, for each φ, k.

(3) Any $h \in G$ is the product of two generic elements of G, each of which is independent from h. Choose generic $g(\mathop{\smile\hskip-0.9em\raise0.5ex\hbox{$|$}} h)$. Then $h.g(\mathop{\smile\hskip-0.9em\raise0.5ex\hbox{$|$}} h)$ is generic by 7.1.2, and $h = (h.g).g^{-1}$.

(4) We say a type-definable subgroup H of G has *bounded index* in G if $|G : H| < \bar{\kappa}$. Compactness shows (Lemma 7.1.9) that H over A has bounded index in G iff $|G : H| \leq 2^{|T|+|A|}$.

(5) For a set A, the *A-connected component* of G, denoted by G^0_A, is the intersection of all type-definable subgroups of G over A having bounded index. Hence, G^0_A is the smallest one among them. We say G is *connected* over A if $G^0_A = G$. As is well known if T is stable, then $G^0_\emptyset = G^0_A$ for any small set A. This no longer holds in simple context. It is not hard to see that in $(V, +, \cdot, 0; \mathbb{F}, \langle,\rangle)$, an infinite-dimensional vector space over a finite field $\mathbb{F}$ with some non-degenerate bilinear form $\langle,\rangle : V \times V \to \mathbb{F}$, which is simple as pointed out in Example 2.6.3, we can have $V^0_A = \{v \in V |\ \langle v, a\rangle = 0 \text{ for all } a \in A\}$ for any $A \subseteq V$ [120, 4.1.14].

However, always $G^0_A = G^0_{\mathrm{bdd}(A)}$ since G^0_A must be the (bounded) intersection of all the A-automorphic images of $G^0_{\mathrm{bdd}(A)}$.

(6) By 7.1.2, $g \in G$ is generic (over $\emptyset$) iff g is generic over some set. Now, given a type-definable subgroup H over A, the following are equivalent:

(a) H is generic in G.

(b) H has bounded index.

(c) $G^0_A \leq H$.

(d) $D_G(G(x), \varphi, k) = D_G(H, \varphi, k)$ for every φ, k.

(e) Any generic type in H is generic in G:

We only prove that (b)⇒(e) as the rest easily come from 7.1.7. Assume g is generic in H. If $p(x) = \mathrm{tp}(g/A)$ is not generic for G, again by 7.1.7 there is A-indiscernible $\langle a_i |\ i < \omega\rangle$ such that $\{a_i.p |\ i < \omega\}$ is k-inconsistent. Since the solution set of p is contained in H, that of $a_i.p$ is contained in a coset of H. Then since the index is bounded, all a_i must be in the same coset. Thus, there is $h \in G$ such that each $h_i := h^{-1}.a_i \in H$. In particular, $h_i.p$ forks over A; i.e., p is not generic in H, a contradiction.

(7) Given two type-definable subgroups $K \leq H$ of G, the arguments for (6) and 7.1.7(1)⇒(2) show the equivalence of the following:

(a) $|H : K|$ is bounded

(b) $D_G(H(x), \varphi, k) = D_G(K(x), \varphi, k)$ for every φ, k.

Lemma 7.1.9 *G_A^0 is a normal subgroup of G, and $|G : G_A^0| \leq 2^{|T|+|A|}$.*

Proof We claim that $G_A^0 = \mathrm{Ker}(\sigma_A)$, where $\sigma_A : G \to \mathrm{Sym}(G/G_A^0)$ is a homomorphism sending $g \in G$ to a permutation of the left-coset space G/G_A^0 mapping $h.G_A^0$ to $g.h.G_A^0$. Hence, G_A^0 is normal: Let $\{h_i\}_{i\in I}$ be a bounded set of some representatives of G/G_A^0. Then since

$$\mathrm{Ker}(\sigma_A) = \bigcap_{i\in I} \{g \in G | \exists g', g'' \in G_A^0 (g.h_i.g' = h_i.g'')\},$$

it is type-definable and A-invariant. Moreover, $|G : \mathrm{Ker}(\sigma_A)|$ must be bounded since, otherwise, G/G_A^0 is unbounded by the first isomorphism theorem. Hence, $G_A^0 \leq \mathrm{Ker}(\sigma_A)$. On the other hand, for $g \in \mathrm{Ker}(\sigma_A)$, $g.G_A^0 = G_A^0$. Thus, $\mathrm{Ker}(\sigma_A) \leq G_A^0$ as well.

Now that $|G : G_A^0| \leq 2^{|T|+|A|}$ comes from usual compactness. Namely, for each definable X_i with $G_A^0 := \bigcap_{i<|T|+|A|} X_i$, by compactness there do not exist infinitely many $g_j \in G$ $(j < \omega)$ such that $g_j^{-1}.g_{j'} \notin X_i$ $(j \neq j' < \omega)$. Thus, the conclusion follows from the Erdös–Rado theorem. ■

We now introduce the notion of the *stabilizers* of types in G, which deepens our theoretical development.

Definition 7.1.10 Let $p(x) \in S_G(A)$. We let

$$S(p) := \{g \in G | \, g.p(x) \cup p(x) \text{ does not fork over } A\},$$

and let $\mathrm{Stab}(p) := S(p).S(p)$, the *stabilizer* of p.

Remark 7.1.11 Note that $1_G \in S(p)$ and $S(p) \subseteq \mathrm{Stab}(p)$. Moreover,

$$\begin{aligned} g \in S(p) \quad & \text{iff there are } a, c \models p \text{ such that } a = g.c \text{ and } a \perp_A g \\ & \text{iff } g^{-1} \in S(p). \end{aligned}$$

For such c, $c \perp_A g$ as well, since

$$\begin{aligned} D_G(p, \varphi, k) &= D_G(g.c/A, \varphi, k) = D_G(g.c/Ag, \varphi, k) \\ &= D_G(c/Ag, \varphi, k) \text{ (by 6.1.5(2))} \leq D_G(c/A, \varphi, k) = D_G(p, \varphi, k). \end{aligned}$$

It also follows (from 4.2.13) $S(p)$ is type-definable over A, and so is $\mathrm{Stab}(p)$. Similarly one can define *right stabilizer* of p. We shall see that the two notions coincide when p is a generic Lascar strong type.

Lemma 7.1.12 *Let X be a non-empty type-definable subset over A of G containing 1_G. Let $H = X.X$. Suppose that for any $a, b \in X$ with $a \perp_A b$, $a^{-1}.b \in X$. Then H is a type-definable* *subgroup of G over A, and* $X(\subseteq H)$ *contains all generic elements for H. Moreover, for any* $h \in H$, $h = h_1.h_2$, *where* $h_i \in X$ *and* $h \mathop{\smile\!\!\!\!\!\!\,|}_A h_i$ $(i = 1, 2)$.

Proof

Claim 1 For $a \in X$, we have $a^{-1} \in X \subseteq H$. Hence, for any $a, b \in X$ with $a \mathop{\smile\!\!\!\!\!\!\,|}_A b$, $a.b \in X$: Claim 1 easily follows, since $1_G \in X$.

Claim 2 For any $a, b, c \in X$, $a.b.c \in H$: By the same construction (i), (ii), (iii) in the proof of 7.1.6, but putting $p_0(x) = X(x)$ and choosing $\neg\psi(x) \in \mathcal{L}(A)$, we obtain a type $p(x) \in S_G(A)$ extending $X(x)$, and numbers $n_\alpha = D_G(p, \varphi_\alpha, k_\alpha)$. (If A is a hyper-imaginary, work with real $\bar{A} \supseteq A$; $\neg\psi(x) \in \mathcal{L}(\bar{A})$; and obtain $\bar{p} \in S_G(\bar{A})$. Then put $p(x) = \bar{p}\lceil A$.) Now choose $g(\models p) \mathop{\smile\!\!\!\!\!\!\,|}_A abc$. We have $b.g \in X$ by Claim 1, and $b.g \mathop{\smile\!\!\!\!\!\!\,|}_{bA} ac$. Similar to the proof of 7.1.6,

$$n_\alpha = D_G(g/Ab, \varphi_\alpha, k_\alpha) = D_G(b.g/Ab, \varphi_\alpha, k_\alpha) \leq D_G(b.g/A, \varphi_\alpha, k_\alpha) \leq n_\alpha,$$

and $b.g \mathop{\smile\!\!\!\!\!\!\,|}_A b$. Hence, $b.g \mathop{\smile\!\!\!\!\!\!\,|}_A a$, and $a.b.g \in X$. Moreover, since $g \mathop{\smile\!\!\!\!\!\!\,|}_A c$, we have $g^{-1}.c \in X$ and $a.b.c \in H$ as desired.

Now, $1 = 1.1 \in H$ and by Claim 2, for $a = a_1.a_2$, $b = b_1.b_2 \in H$ with $a_i, b_i \in X$ we have $a.b = c.b_2 = c_1.c_2.b_2 \in H$, where $c_i \in X$; i.e., H is closed under the group operation. Moreover, for $c.d \in H$ with $c, d \in X$, by Claim 1, we have $d^{-1}.c^{-1} \in H$. Hence, H is a subgroup.

Claim 3 Any generic h for H is in X: There are $a, b \in X$ such that $h = a.b$. Now choose $f(\models p) \mathop{\smile\!\!\!\!\!\!\,|}_A ab$. Then $b.f \in X$, and moreover $b.f \mathop{\smile\!\!\!\!\!\!\,|}_A b$ since again $n_\alpha = D_G(b.f/Ab, \varphi_\alpha, k_\alpha) \leq D_G(b.f/A, \varphi_\alpha, k_\alpha) \leq n_\alpha$. Hence, $b.f \mathop{\smile\!\!\!\!\!\!\,|}_A a$, and $a.b.f = h.f \in X$. Since $h(= a.b \mathop{\smile\!\!\!\!\!\!\,|}_A f)$ is generic, $h.f \mathop{\smile\!\!\!\!\!\!\,|}_A f$. Therefore, $h = h.f.f^{-1} \in X$.

Claim 4 For $h \in H$, there is $h_i(\in X) \mathop{\smile\!\!\!\!\!\!\,|}_A h$ such that $h = h_1.h_2$: Let $h = g_1.g_2$ with $g_i \in X$. Choose $h_1(\in X) \mathop{\smile\!\!\!\!\!\!\,|}_A g_1g_2$. By Claim 3, h_1 is generic for H, so $h_1^{-1}.g_1 \mathop{\smile\!\!\!\!\!\!\,|}_A g_1g_2$, and $h_1^{-1}.g_1.g_2 \mathop{\smile\!\!\!\!\!\!\,|}_A g_1g_2$. Since $h_2 := h_1^{-1}.g_1.g_2$ is also generic for H, it is in X. Then, $h_i \mathop{\smile\!\!\!\!\!\!\,|}_A h$ and $h = h_1.h_2$. ■

Proposition 7.1.13 *Let* $p(x) \in S^L_G(A) = S_G(\mathrm{bdd}(A))$. *If* $g, h \in S(p)$ *with* $g \mathop{\smile\!\!\!\!\!\!\,|}_A h$, *then* $g^{-1}.h \in S(p)$. *Moreover,* Stab(p) *is a subgroup of G, and* $S(p)$ *contains all generic elements for* Stab(p).

Proof For $g, h \in S(p)$ with $g \mathop{\smile\!\!\!\!\!\!\,|}_A h$, since $g^{-1} \in S(p)$, it suffices to show that $g.h \in S(p)$. By Remark 7.1.11, there are $c, d \models p$ such that $c \mathop{\smile\!\!\!\!\!\!\,|}_A g$, $d \mathop{\smile\!\!\!\!\!\!\,|}_A h$, $g.c \mathop{\smile\!\!\!\!\!\!\,|}_A g$, $h.d \mathop{\smile\!\!\!\!\!\!\,|}_A h$, where $g.c, h.d \models p$. By the independence theorem, there is $e(\models \mathrm{tp}(g.c/Ag) \cup \mathrm{tp}(d/Ah)) \mathop{\smile\!\!\!\!\!\!\,|}_A gh$ (*). Hence, there is $c'(\models p) \mathop{\smile\!\!\!\!\!\!\,|}_A g$ such that $e = g.c'$ and $h.e(\models p) \mathop{\smile\!\!\!\!\!\!\,|}_A h$. Thus, from (*), we have $h.e \mathop{\smile\!\!\!\!\!\!\,|}_A gh$ where $h.e = h.g.c'$. Therefore, as in 7.1.11, $h.g \in S(p)$. Now, the moreover clause follows from 7.1.12. ■

Proposition 7.1.14 *Let $p(x) \in S^L_G(A)$. Then $\mathrm{Stab}(p) \leq G^0_{\mathrm{bdd}(A)}$, and they are equal iff p is generic for G.*

Proof The equivalence relation $x^{-1}.y \in G^0_{\mathrm{bdd}(A)}$ is type-definable over $\mathrm{bdd}(A)$ having boundedly many classes. Thus, each class (a coset of $G^0_{\mathrm{bdd}(A)}$) is type-definable over $\mathrm{bdd}(A)$. Hence, the solution set of $p(x)$ must be contained in a coset. Now, if $g \in S(p)$, then $g = c^{-1}.a$ for $a, c \models p$. Hence, $g \in G^0_{\mathrm{bdd}(A)}$. Thus, $\mathrm{Stab}(p) \leq G^0_{\mathrm{bdd}(A)}$.

Assume now $\mathrm{Stab}(p) = G^0_{\mathrm{bdd}(A)}$. Then, since $S(p)$ is generic in $\mathrm{Stab}(p)$, which is generic for G, by 7.1.8(6) there is $g \in S(p)$ generic in G over A. Thus, there are $a, c \models p$ such that $a = g.c$, $g \perp_A a$, and $g \perp_A c$, $a \perp_A c$ (7.1.11). Hence, each of $\mathrm{tp}(g/Ac)$, $\mathrm{tp}(g.c/Ac) = \mathrm{tp}(a/Ac)$ (7.1.2), and $\mathrm{tp}(a/\mathrm{bdd}(A)) = p$ is generic for G.

Conversely, assume p is generic. There are $a, c \models p$ such that $a \perp_A c$. Thus, $g = a.c^{-1}$ is generic over A, and $g \perp_A a$. Hence, $g \in S(p)$, and both $S(p)$ and $\mathrm{Stab}(p)$ are generic. Hence, $\mathrm{Stab}(p) \geq G^0_{\mathrm{bdd}(A)}$, and the equality holds. ■

We finish this section by stating a theorem true in any theory. The theorem will be used later. We recall a group theoretic notion of *commensurability*. Two groups H_1 and H_2 are said to be *commensurable* if $|H_i : H_1 \cap H_2| < \omega$ $(i = 1, 2)$. We tentatively call a family $\mathcal{F}$ of groups *uniformly commensurable* if there is $n < \omega$ such that for any $H_1, H_2 \in \mathcal{F}$, $|H_i : H_1 \cap H_2| \leq n$. Note that when H_1, H_2 are subgroups of G, $H_1.H_2 = \{h_i.h_2 | \, h_i \in H_i\}$ need not be a subgroup of G but $|H_1.H_2 : H_1|$ still makes sense and is equal to $|H_2 : H_1 \cap H_2|$. As usual for $g \in G$, and a set $A \subseteq G$, $A^g := \{a^g := g^{-1}.a.g | \, a \in A\}$.

Fact 7.1.15 *Let H, F be subgroups of a group K. Then*

$$(F \leq)F_H := \bigcap_{g \in F} (H.F)^g (\subseteq H.F)$$

is a subgroup of K.

Proof That $F \subseteq F_H \subseteq H.F$ is clear (take $g = 1_K$). Let $x, y \in F_H$ and let $g \in F$. To show $x.y \in F_H$, it suffices to see that $(x.y)^g = x^g.y^g \in H.F$. Now, since $x = (h.f)^{g^{-1}}$ for some $h \in H, f \in F$, we have $x^g = h.f$. Similarly, $y^{g.f^{-1}} = h'.f'$ for $h' \in H, f' \in F$. So $(x.y)^g = h.f.(h'.f')^f = (h.h').(f.f') \in H.F$. Similarly, since $(x^{-1})^{g.f^{-1}} = (f^{-1}.h^{-1})^{f^{-1}} = h^{-1}.f^{-1} \in H.F$, we have $x^{-1} \in F_H$. ■

Theorem 7.1.16 *Let $X \supseteq G$ be defined by σ_0. Assume a uniformly commensurable family $\mathcal{F} = \{H_i \leq K | \, i \in I\}$ is given such that each H_i is an $\emptyset$-definable subgroup of $K \subseteq X$ and $\mathcal{F}$ is G-invariant (i.e., for each $g \in G$, $\mathcal{F} = \mathcal{F}^g := \{H_i^g | \, i \in I\}$). Then there is a definable G-invariant group $(\bigcap_{i \in I} H_i \leq) N (\leq K)$ commensurable with any H_i.*

If instead $\mathcal{F}$ is G-multiplication invariant (i.e., $g.\mathcal{F} = \mathcal{F}$ for $g \in G$) we can choose so is N.

Proof Let n be a bound witnessing the uniform commensurability of $\mathcal{F}$. Fix F, some k-intersection of members H_i in $\mathcal{F}$. Then for each $H \in \mathcal{F}$, $|H.F : H| \leq |H.H_i : H| = |H_i : H \cap H_i| \leq n$. We call F *strong* if $m(F) := \max\{|H.F : H| : H \in \mathcal{F}\}$ attains the minimal possible value, say m_0. For strong F, we let

$$\mathcal{F}_F := \left\{H \in \mathcal{F} : |H.F : H| = m_0\right\}.$$

For the rest, we assume F is strong. The following claim follows from our definitions.

Claim Suppose that F' is another finite intersections of members in $\mathcal{F}$ such that $F' \leq F$. Then $m(F') \leq m(F)$ so F' is strong too, and $\mathcal{F}_{F'} \subseteq \mathcal{F}_F$. Thus, for any $H \in \mathcal{F}_{F'}$, $H.F = H.F'$ since $|H.F : H| = |H.F' : H| = m_0$.

Now we take the notation in Fact 7.1.15. We let

$$N(F) := \bigcap\{F_H | \, H \in \mathcal{F}_F\}.$$

Assume that $F' \leq F$ is as in the above claim. Then for any $H \in \mathcal{F}_{F'} \subseteq \mathcal{F}_F$, $F_H \leq F'_H$ by the claim, so $(F \leq)N(F) \leq N(F')$; moreover, for any such H, again by the above claim,

$$|N(F') : F| \leq |H.F' : F| = |H.F : F| = |H : H \cap F| \leq k.n.$$

Therefore, since given a strong F'', $F \cap F''$ is strong as well, there must exist the unique maximum group N of the form $(F \leq)N(F_0)$ for a strong $F_0(\geq F)$. Note now N is definable since F has finite index in N; and N is G-(multiplication) invariant, since so is $\mathcal{F}$. It remains to see that $N = N(F_0)$ is commensurable with any $H \in F$: $|H : N \cap H| \leq |H : F \cap H| \leq k.n$. On the other hand, $|N : N \cap H| = |N.H : H| \leq |H'.H : H| = |H' : H' \cap H| \leq n$, where $H' \in \mathcal{F}_{H \cap F_0}$ so that $N = N(F_0) = N(H \cap F_0) \subseteq H'.(H \cap F_0)$ and $N.H \subseteq H'.H$. ■

7.2 Hyperdefinable Groups

Now we study groups in more general context, namely hyperdefinable groups in simple theories. We fix $E = \{\epsilon(x', y') | \, \epsilon(x', y') \in E(x, y)\}$ (x, y possibly infinite), a type-definable equivalence relation over $\emptyset$, while taking the convention 4.1.2(3). Fix $\Gamma(x) = \Gamma(x_E)$, an E-type over $\emptyset$, and put $G := \Gamma/E$, where Γ is the set type-defined by $\Gamma(x)$. Moreover, assume $(G, ., 1_G)$ is a hyperdefinable group over $\emptyset$; i.e., the graph of the group operation is a hyperdefinable relation on G^3 (see 4.1.4(1)). We work with the group G throughout this section.

Fix some more notations. For a hyperimaginary set A, $S_G(A) = S_{G,E}(A) := \{p(x) = p(x_E) \in S_E(A) | \, p(x) \models \Gamma(x)\}$. Given a (real) type $\pi(x)$ over A, put an E-type $\pi_E(x) := \exists y(E(x, y) \wedge \pi(y))$ and another E-type $\pi_G(x) := \Gamma(x) \cup \pi_E(x)$. If $\pi(x)$ is already an E-type then $\pi(x) \equiv \pi_E(x)$. For $g \in G$, $g.\pi(x) = g.\pi(x_E)$ denotes $\exists y_E(x_E = g.y_E \wedge$

$\pi_G(y_E))$, an E-type over Ag implying Γ. Similarly, we define $\pi(x).g$. Note that $g.\pi(x) = g.\pi_E(x) = g.\pi_G(x)$ and $1_G.\pi(x) \equiv \pi(x).1_G \equiv \pi_G(x)$. Also for a formula $\varphi(x,b)$ (in this section we write x in a formula though it should be a finite subvariable), both $\varphi_G(x,b)$ $(= \{\varphi(x,b)\}_G)$ and $g.\varphi(x,b)$ are partial E-types (need not be formulas) implying $\Gamma(x_E)$. Put $\varphi(g.x,b) := g^{-1}.\varphi(x,b)$, and $\varphi(x.g,b) := \varphi(x,b).g^{-1}$.

As in the previous section we start with the introduction of *generic* elements.

Definition 7.2.1 A type $p(x) \in S_G(A)$ is said to be *left-generic* if for any $g \in G$ realizing p, and $a \in G$ with $g \perp_A a$, we have $a.g \perp Aa$. We say g is *left-generic* over A for G, if $\mathrm{tp}(g/A)$ is so. Right-generic types and elements are defined similarly.

The same proofs of 7.1.2 and 7.1.3 give the following.

Proposition 7.2.2

(1) *Suppose that $g, a \in G$ and $\mathrm{tp}(g/Aa)$ is left-generic. Then so is $\mathrm{tp}(a.g/Aa)$.*

(2) *Let $A \subseteq B$, and let $p(x)(\in S_G(A)) \subseteq q(x) \in S_G(B)$. Then $q(x)$ is left-generic iff $q(x)$ is a non-forking extension of left-generic $p(x)$.*

(3) *If $g \in G$ is left-generic over A, then so is g^{-1}.*

(4) *$p \in S_G(A)$ is left-generic iff it is right-generic.*

(5) *Suppose that $I = \langle a_i \mid i < \alpha \rangle$ is some A-independent sequence from G such that each a_i is left-generic over A. Then I is left-generic for the group G^α over A.*

So, from now on, we simply say *generic* types or elements of G. As in type-definable group case, we aim to prove the existence of a generic type. For this we introduce the $D_G(-,\varphi,\epsilon,k)$-rank, a variant of the $D(-,\varphi,k)$-rank. For $k \geq 1$, $\varphi(x,y) \in \mathcal{L}$, and $\epsilon(x,y) \in E(x,y)$, we have the formula

$$\varphi^k_\epsilon(y_1,\ldots,y_k) := \neg\exists x \bigwedge_{i=1}^{k} \exists z_i\, (\epsilon(x,z_i) \wedge \varphi(z_i,y_i)).$$

Given a real type $p(x)$, we let $D_G(p(x),\varphi(x,y),\epsilon,k) \geq 0$ if $p(x) \wedge \Gamma(x)$ is consistent, and let $D_G(p(x),\varphi(x,y),\epsilon,k) \geq n+1$ if there are $g \in G$ and real b_i $(i < \omega)$ such that $D_G(g.p \cup \{\varphi(x,b_i)\},\varphi,\epsilon,k) \geq n$ for all $i < \omega$, and $\varphi^k_\epsilon(b_{i_1},\ldots,b_{i_k})$ holds for each $i_1 < \ldots < i_k \in \omega$.

Remark 7.2.3 Note by compactness and induction that $D_G(p,\varphi,\epsilon,k) \geq n+1$ is a type-definable condition. Hence, in the definition, we can replace ω by any infinite κ. Moreover, for b_i $(i < \kappa)$, if $\varphi^k_\epsilon(b_{i_1},\ldots,b_{i_k})$ holds for each $i_1 < \ldots < i_k \in \kappa$, then the set of E-types $\{\varphi_G(x,b_i) \mid i \in \kappa\}$ is k-inconsistent.

Lemma 7.2.4

(1) *For a type $p(x)$, and $h \in G$, we have $D_G(p,\varphi,\epsilon,k) = D_G(p_G,\varphi,\epsilon,k) = D_G(h.p, \varphi,\epsilon,k)$. Hence, $D_G(a/A,\varphi,\epsilon,k) = D_G(a_E/A,\varphi,\epsilon,k) = D_G(h.a_E/A,\varphi,\epsilon,k)$, for $a \models \Gamma$ and $h \in \mathrm{dcl}(A) \cap G$.*

(2) *$D_G(p_1(x), \varphi, \epsilon, k_1) \leq D_G(p_2(x), \varphi, \epsilon, k_2)$ if $p_1 \vdash p_2$ and $k_1 \leq k_2$. For every $p(x)$, $\varphi(x, y)$, k, there is a finite subset $q(x)$ of p such that $D_G(p, \varphi, \epsilon, k) = D_G(q, \varphi, \epsilon, k)$; $D_G(p(x) \cup \bigvee_{i<n} q_i(x), \psi, \epsilon, k) = \max\{D_G(p(x) \cup q_i(x), \psi, \epsilon, k) | \ i < n\}$.*

(3) *Let $\varphi(x, y)$, k be given. For a type $p(x)$ over A, there is a completion $p'(x) \in S(A)$ of p such that $D_G(p, \psi(x, y), k) = D_G(p', \psi(x, y), k)$.*

(4) *$D_G(\Gamma, \varphi, \epsilon, k)$ is finite.*

(5) *Let $q(x) \in S_G(B)$. Then $q(x)$ is a non-forking extension of $p(x) (\in S_G(A),\ A \subseteq B)$ iff $D_G(p, \varphi, \epsilon, k) = D_G(q, \varphi, \epsilon, k)$ for every $\epsilon \in E$, φ, k.*

Proof (1) Since $g.p = g.p_G$ for $g \in G$, we clearly have the first equality. Now, if g, b_i witness $D_G(p, \varphi, \epsilon, k) \geq n + 1$, then clearly $g.h^{-1}$, b_i witness $D_G(h.p, \varphi, \epsilon, k) \geq n + 1$. The converse similarly holds. Lastly, the hence clause follows since $\text{tp}_E(a/A) = \text{tp}(a_E/A)$ and $h.\text{tp}(a/A) = \text{tp}(h.a_E/A)$.

(2),(3) As in 2.3.4, these come easily by usual induction with compactness.

(4) From Remark 7.2.3 and (2), we have that if $D_G(p, \varphi, \epsilon, k) \geq n + 1$, then there are $g \in G$ and b_i such that $D_G(g^{-1}.p \cup \{\varphi(x, b_i)\}, \varphi, \epsilon, k) = D_G(p_E \cup g.\varphi(x, b_i), \varphi, \epsilon, k) \geq n$, and $\{g.\varphi(x, b_i) | \ i < \kappa\}$ is k-inconsistent. Thus, if $D_G(\Gamma, \varphi, \epsilon, k) \geq n$ for all $n \in \omega$, then by compactness, for infinite κ, there is a collection $\{g_\alpha \in G, b_\alpha | \alpha \in \kappa^{<\kappa}\}$ such that

(i) For each $\alpha \in \kappa^{<\kappa}$, the set of E-types $\{g_\alpha.\varphi(x, b_{\alpha^\frown i}) | i \in \kappa\}$ is k-inconsistent, and

(ii) For every $\beta \in \kappa^\kappa$, $\Gamma \cup \{g_{\beta\lceil\alpha}.\varphi(x, b_{\beta\lceil\alpha}) | \alpha < \kappa\}$ is consistent.

Hence, there is a forking chain of length arbitrary κ; i.e., simplicity of T fails. This contradiction shows (4).

(5) If q is a forking extension of p, then for any infinite κ, there is k-inconsistent $\{q_i | \ i < \kappa\}$ such that each $q_i(x)$ is an A-conjugate of q. In particular, $\bigwedge_{\ell=1}^{k} \exists z(E(x, z) \wedge q_{i_\ell}(z))$ $(i_1 < \ldots < i_k)$ is inconsistent. Since κ is arbitrarily large, by Ramsey argument there are some infinite $\tau \subseteq \kappa$; $\epsilon(x, y) \in E$, and $\varphi(x, b_i) \in q_i$ such that $\varphi^k_\epsilon(b_{i_1}, \ldots, b_{i_k})$ holds for any $i_1 < \ldots < i_k \in \tau$. Hence, it follows that $D_G(q, \varphi, \epsilon, k) \geq n$ implies that $D_G(p, \varphi, \epsilon, k) \geq n + 1$. Thus, by (4), $D_G(p, \varphi, \epsilon, k) > D_G(q, \varphi, \epsilon, k)$.

Conversely, assume that q is a non-forking extension of p, and $D_G(p, \varphi, \epsilon, k) \geq n + 1$. We want to show that $D_G(q, \varphi, \epsilon, k) \geq n + 1$, with the induction hypothesis for n. Now, there are $c \models \Gamma$ and cA-indiscernible $I = \langle b_i | \ i < \omega\rangle$ such that $D_G(c_E.p \cup \{\varphi(x, b_0)\}, \varphi, \epsilon, k) \geq n$, and $\varphi^k_\epsilon(b_{i_1}, \ldots, b_{i_k})$ holds for each $i_1 < \ldots < i_k \in \omega$. By (1),(3), there is $p' \in S_G(Acb_0)$ implying that $c_E.p \cup \varphi_G(x, b_0)$ such that $D_G(p', \varphi, \epsilon, k) \geq n$. Thus, there must be $a_E \models p$ such that $p' = \text{tp}(c_E.a_E/Acb_0)$. By possibly taking an A-conjugate of B, we can assume that $q = \text{tp}(a_E/B)$ (so $a_E \mathop{\smile\!\!\!\!\!\vert}_A B$) and $cb_0 \mathop{\smile\!\!\!\!\!\vert}_{Aa_E} B$ (*). Hence, $c_E.a_E \mathop{\smile\!\!\!\!\!\vert}_{Acb_0} B$. Then by the induction hypothesis, $D_G(c_E.a_E/Bcb_0, \varphi, \epsilon, k) \geq n$, and $D_G(\text{tp}(c_E.a_E/Bc) \cup \varphi_G(x, b_0), \varphi, \epsilon, k) = D_G(\text{tp}(c_E.a_E/Bc) \cup \{\varphi(x, b_0)\}, \varphi, \epsilon, k) \geq n$. Moreover, by (*), we have $b_0 \mathop{\smile\!\!\!\!\!\vert}_{Ac} B$, and by 2.1.6, we can assume that I is Bc-indiscernible. Thus, it witnesses $D_G(c_E.a_E/Bc, \varphi, \epsilon, k) \geq n+1$. Hence, $D_G(a_E/B, \varphi, \epsilon, k) \geq D_G(a_E/Bc, \varphi, \epsilon, k) =$ (by (1)) $D_G(c_E.a_E/Bc, \varphi, \epsilon, k) \geq n+1$. ■

Now by the similar proof of Proposition 7.1.6, we have the existence of generic types.

Proposition 7.2.5 *There is generic $p_E(x) \in S_G(A)$.*

Proof Due to 7.2.2(2), it suffices to show when $A = \emptyset$. Let $\langle (\varphi_i(x_i, z_i), \epsilon_i(x_i, y_i), k_i) |$ finite $x_i \subseteq x$, $1 \leq k_i$, $\epsilon_i \in E$, $\varphi_i \in \mathcal{L}$, $i < |T| + |x| \rangle$ be an enumeration. Inductively, define partial types $p_i(x)$ over A and natural numbers n_i as follows:

(i) $p_0(x) = \Gamma(x)$, and $n_0 = D_G(p_0, \varphi_0, \epsilon_0, k_0)$,

(ii) $p_{i+1} = p_i(x) \cup \{\neg\psi(x) \in \mathcal{L} | \ D_G(\psi(x), \varphi_i, \epsilon_i, k_i) < n_i\}$ and $n_{i+1} = D_G(p_{i+1}, \varphi_{i+1}, \epsilon_i, k_{i+1})$, and

(iii) For limit i, $p_i = \bigcup\{p_j | \, j < i\}$ and $n_i = D_G(p_i, \varphi_i, \epsilon_i, k_i)$.

Note that each p_i is consistent since if not then it violates 7.2.4(2). Hence, $\bigcup\{p_i | \, i \in |T| + |x|\}$ is consistent, and there is a real completion $p(x) \in S(\emptyset)$.

We claim that g_E with $g \models p$ is generic for G: Let $a_E (\in G) \downarrow g_E$. Then, due to 7.2.4,

$$\begin{aligned} n_i &\leq D_G(g, \varphi_i, \epsilon_i, k_i) \ \text{(otherwise contradicting (ii))} \\ &= D_G(g_E, \varphi_i, \epsilon_i, k_i) = D_G(g_E/a_E, \varphi_i, \epsilon_i, k_i) = D_G(a_E.g_E/a_E, \varphi_i, \epsilon_i, k_i) \\ &\leq D_G(a_E.g_E, \varphi_i, \epsilon_i, k_i) = D_G(c, \varphi_i, \epsilon_i, k_i), \ \text{where real } c \models \mathrm{tp}(a_E.g_E) \\ &\leq n_i \ \text{since by induction } c \models p_i \text{ in (i),(ii),(iii).} \end{aligned}$$

Therefore again by 7.2.4(5), we have $a_E.g_E \downarrow a_E$; i.e., g_E is generic. ■

Definition 7.2.6 A *real* partial type $p(x)$ implying $\Gamma(x)$ is said to be *generic* if $q(x) \models p_E(x)$ for some generic $q \in S_G(A)$.

Now, the similar arguments in the proofs of 7.1.7 and 7.1.8(1) imply the following.

Proposition 7.2.7 *Let $p(x)$ be a partial type over A implying $\Gamma(x)$. The following are equivalent:*

(1) The type p is generic.

(2) $D_G(p, \varphi, \epsilon, k) = D_G(\Gamma, \varphi, \epsilon, k)$ for each φ, ϵ, k.

(3) For any $g \in G$, the type $g.p$ does not fork over $\emptyset$.

(4) For any $g \in G$, the type $g.p$ does not fork over A.

(5) For any $g \in G$, $g.p$ is generic.

Moreover, $g \in G$ is generic over A iff for any $a \in G$ with $a \downarrow_A g$, we have $a.g \downarrow_A a$.

Remark & Definition 7.2.8

(1) Any $h \in G$ is the product of two generic elements such that each of which is independent from h (see 7.1.8(3)).

(2) As in Section 7.1, we say a hyperdefinable subgroup H of G has *bounded index* in G if $|G : H| < \bar{\kappa}$. Again, compactness shows (Lemma 7.2.9) that H over A has bounded index in G iff $|G : H| \leq 2^{|T|+|A|+|x|}$.

(3) For a hyperimaginary A, the *A-connected component* of G, denoted by G^0_A, is the intersection of all hyperdefinable subgroups of G over A having bounded index. Hence, G^0_A is the smallest one among such. We say G is *connected* over A if $G^0_A = G$. As pointed out in 7.1.8(5), G can be connected over $\emptyset$ but need not be over any set. However, always $G^0_A = G^0_{\mathrm{bdd}(A)}$.

(4) The similar proof of 7.1.8(6) implies the equivalence of the following:

   (a) An hyperdefinable subgroup H over A is generic in G.

   (b) H has bounded index.

   (c) $G^0_A \leq H$.

   (d) $D_G(H, \varphi, \epsilon, k) = D_G(\Gamma, \varphi, \epsilon, k)$ for each φ, ϵ, k.

   (e) Any generic type in H is generic in G.

(5) Similarly to 7.1.8(7), for two hyperdefinable subgroups $K \leq H$ of G, it follows that $D_G(H, \varphi, \epsilon, k) = D_G(K, \varphi, \epsilon, k)$ for each φ, ϵ, k iff $|H : K|$ is bounded. Also a type $p \models H$ is generic for H iff $D_G(H, \varphi, \epsilon, k) = D_G(p, \varphi, \epsilon, k)$ for each φ, ϵ, k.

(6) We write $S^0_G(A)$ for $S_{G^0_A}(A)(\subseteq S_G(A))$. A generic type/element of G^0_A over A is said to be a *principal* generic type/element of G over A. Of course, there can exist a non-principal generic type in $S_G(A)$. We shall see that for any principal generic $p \in S^0_G(A)$, and $B \supseteq A$, there is a non-forking extension $q \in S^0_G(B)$ of p which is also principal generic (Proposition 7.2.16).

(7) It is easy to see that $SU(g/A) = SU(h.g/A)$ for $g, h \in G$ and $h \in \mathrm{dcl}(A)$. Now assume $p \in S_G(A)$ is generic and $SU(p) = \alpha$. Then $\alpha \geq SU(q)$ for any $q \in S_B(G)$. So any generic type has SU-rank α. Conversely, if a type has SU-rank α, then it is generic. Let $C = AB$ and choose $g \models p, h \models q$ such that $g \perp_A B, h \perp_B A$, and $g \perp_C h$. Then $h.g \perp_C h$ and $\alpha = SU(g/Ch) = SU(h.g/Ch) = SU(h.g/C) \geq SU(h.g/Cg) = SU(h/Cg) = SU(q)$. The converse follows similarly.

So, $SU(G)$ defined in 4.2.13(2) is equal to the SU-rank of any/some generic type of G. If H is a hyperdefinable subgroup of G, then $SU(H) = SU(g.H)$ for any $g \in G$ (see the remark after 2.5.19); if H' is another hyperdefinable subgroup commensurate with H, then $SU(H') = SU(H)$. Similarly, for a hyperdefinable $K \leq H$, we have $SU(K) = SU(H)$ iff K has bounded index in H.

We say now G is *minimal* if $SU(G) = 1$. If G is minimal, then every hyperdefinable subgroup H over $\mathrm{bdd}(\emptyset)$ of G having unbounded index is contained in $\mathrm{bdd}(\emptyset)$.

Now the similar proof of Lemma 7.1.9 shows the following.

Lemma 7.2.9 *G_A^0 is a normal subgroup of G and $|G : G_A^0| \leq 2^{|T|+|A|+|x|}$.*

As in the case of type-definable groups, we define the stabilizer of a type as follows.

Definition & Remark 7.2.10 Let $p(x) \in S_G(A)$. We let

$$S(p) := \{g \in G |\, g.p(x) \cup p(x) \text{ does not fork over } A\},$$

and let $\mathrm{Stab}(p) := S(p).S(p)$, the *stabilizer* of p. Note again that $1_G \in S(p) \subseteq \mathrm{Stab}(p)$, and

$$\begin{aligned} g \in S(p) \quad &\text{iff} \quad \text{there are } a, c \models p \text{ such that } a = g.c \text{ and } a \mathop{\smile\!\!\!\!\!\!\,|}_A g \\ &\text{iff} \quad g^{-1} \in S(p). \end{aligned}$$

For such c, we have $c \mathop{\smile\!\!\!\!\!\!\,|}_A g$, since again

$$\begin{aligned} D_G(p, \varphi, \epsilon, k) &= D_G(g.c/A, \varphi, \epsilon, k) = D_G(g.c/Ag, \varphi, \epsilon, k) \\ &= D_G(c/Ag, \varphi, \epsilon, k) \leq D_G(c/A, \varphi, \epsilon, k) = D_G(p, \varphi, \epsilon, k). \end{aligned}$$

Furthermore, from 4.2.13, $S(p)$ is hyperdefinable over A, and so is $\mathrm{Stab}(p)$.

We also have the following 7.2.11 and 7.2.12 by the same proofs in 7.1.12, 7.1.13 but working with an enumeration $\langle (\varphi_i(x_i, y_i), \epsilon_i(x_i, y_i), k_i) |$ finite $x_i \subseteq x,\ 1 \leq k_i,\ \epsilon_i \in E,\ \varphi_i \in \mathcal{L},\ i < |T| + |x|\, \rangle$, and ranks $D_G(-, \varphi_i, \epsilon_i, k_i)$.

Lemma 7.2.11 *Let X be a non-empty hyperdefinable subset over A of G containing 1_G. Let $H = X.X$. Suppose that for any $a, b \in X$ with $a \mathop{\smile\!\!\!\!\!\!\,|}_A b$, $a^{-1}.b \in X$. Then H is a hyperdefinable subgroup of G over A, and $X(\subseteq H)$ contains all generic elements for H. Moreover, for any $h \in H$, $h = h_1.h_2$, where $h_i \in X$ and $h \mathop{\smile\!\!\!\!\!\!\,|}_A h_i$ $(i = 1, 2)$.*

Recall that $S^L_{G,E}(A) = S^L_G(A)$ is the set of Lascar E-types over A implying $\Gamma(x)$. Hence, $S^L_G(A) = S_G(\mathrm{bdd}(A))$.

Proposition 7.2.12 *Let $p(x) \in S^L_G(A)$. If $g, h \in S(p)$ with $g \mathop{\smile\!\!\!\!\!\!\,|}_A h$, then $g^{-1}.h \in S(p)$. Hence, $\mathrm{Stab}(p)$ is a hyperdefinable subgroup of G over A, and $S(p)$ contains all generic elements for $\mathrm{Stab}(p)$. Moreover, $\mathrm{Stab}(p) \leq G^0_{\mathrm{bdd}(A)}$, and they are equal iff p is generic for G.*

Now as promised we aim to show that any principal generic type over a set has a principal generic non-forking extension over any larger set (Proposition 7.2.16).

Definition 7.2.13 Let H be a hyperdefinable subgroup of G over A, and $g \in G$. We say $p \in S_G(B)$ is *generic* for a coset $g.H$ over B if $p \models g.H$ and $g^{-1}.p$ is generic for H.

Proposition 7.2.14 *Let $p_0, p_1, p_2 \in S^0_G(A)$ be principal generic types of G over $A = \mathrm{bdd}(A)$. Then there are $h_i \models p_i$ $(i = 0, 1, 2)$ such that $h_0 \mathop{\smile\!\!\!\!\!\!\,|}_A h_1$ and $h_2 = h_0.h_1$.*

Proof Choose $g_1 \models p_1, g_2 \models p_2$ such that $g_2 \mathop{\smile\!\!\!\!\!\!\,|}_A g_1$. Then due to genericity, $a = g_2.g_1^{-1}$ is principal generic over A and $g_1 \mathop{\smile\!\!\!\!\!\!\,|}_A a$. Now choose $h_0(\models p_0) \mathop{\smile\!\!\!\!\!\!\,|}_A a$, then $b = a^{-1}.h_0$ is also principal generic over A, and $a \mathop{\smile\!\!\!\!\!\!\,|}_A b$. Moreover, due to 7.2.12, we have $b \in S(p_1)$. Hence, by the definition 7.2.10, there is $g_1'(\models p_1) \mathop{\smile\!\!\!\!\!\!\,|}_A b$ such that $b^{-1}.g_1' \models p_1$ and

$b^{-1}.g_1' \mathop{\smile\!\!\!\!\!\!\,|}_A b$ (*). By type amalgamation, we have $h_1' \models \text{tp}(g_1/Aa) \cup \text{tp}(g_1'/Ab)$ such that $\{h_1', a, b\}$ is A-independent. Thus, there is $h_2 \models p_2$ with $g_1g_2 \equiv_{Aa} h_1'h_2$ so that $a = h_2.(h_1')^{-1}$. Put $h_1 := b^{-1}.h_1' = h_0^{-1}.h_2 \models p_1$. Lastly, since $b \mathop{\smile\!\!\!\!\!\!\,|}_A h_1$ by (*), and $a \mathop{\smile\!\!\!\!\!\!\,|}_{Ab} h_1'$, it follows that $a \mathop{\smile\!\!\!\!\!\!\,|}_{Ab} h_1$; $ab \mathop{\smile\!\!\!\!\!\!\,|}_A h_1$; and $h_0 \mathop{\smile\!\!\!\!\!\!\,|}_A h_1$. ■

Corollary 7.2.15 *Let F be a hyperdefinable field in a simple structure, and let K be a hyperdefinable multiplicative subgroup of F having bounded index. Then* $F^\times = K_0 + K_1$ *for any cosets* K_i $(i = 0, 1)$ *of K.*

Proof Assume F, K are A-hyperdefinable. Let $c \in F^\times$. Since K has bounded index, there is a model M containing cA and all the representatives a_i $(i < \kappa)$ of cosets of K. So each coset $K_i := a_i.K$ (possibly $a_0 = a_1$) is over M. Now let $p' \in S_G(M)$ be principal *additive* generic for F over M. Thus $p' \models K_j$, for some j. Now for any $d \in F^\times$ and $B \supseteq Ad$, we have $(F^+)^0_B \leq d.(F^+)^0_B \leq (F^+)^0_B$ since the $D_G(-, \varphi, \epsilon, k)$ ranks are preserved (7.2.8(5), 7.2.4(1)) (*). Hence, for some $a_{j'} (\subseteq M)$ in $K_{j'}$ containing a_j^{-1}, we have that $p := a_{j'}.p'$ is again a principal additive generic complete type for F over M; and $p \models K$ (**). Again by (*), each of $c.p$, $a_i.p \in S_G(M)$ are principal additive generic for F as well. Then by 7.2.14, there are x, y, z realizing p such that $a_0x + a_1y = cz$. Therefore, $c = a_0(x/z) + a_1(y/z)$ and by (**), $c \in K_0 + K_1$. ■

Proposition 7.2.16 *Given* $p \in S^0_G(A)$ *and* $B \supseteq A$, *there is* $q \in S^0_G(B)$, *a principal generic non-forking extension of p over B.*

Proof We can assume $A = \text{bdd}(A)$ (7.2.8(3)), and 7.2.14 says that there are $g', h \models p$ such that $g' \mathop{\smile\!\!\!\!\!\!\,|}_A h$ and $k' = g'.h \models p$. Moreover, we can assume $h \mathop{\smile\!\!\!\!\!\!\,|}_A B$, and by genericity $k' \mathop{\smile\!\!\!\!\!\!\,|}_A h$. Then by type amalgamation there is $k \models \text{tp}(k'/Ah) \cup \text{tp}(h/B)$ such that $k \mathop{\smile\!\!\!\!\!\!\,|}_A Bh$ and $kh \mathop{\smile\!\!\!\!\!\!\,|}_A B$. There now is g with $kg \equiv_{Ah} k'g'$ so that $k = g.h$; both $k, g \models p$; and $g \mathop{\smile\!\!\!\!\!\!\,|}_A B$ (*). Let $q' = \text{tp}(k/B) = \text{tp}(h/B)$, a non-forking extension of p, so generic over B. Moreover, $k \mathop{\smile\!\!\!\!\!\!\,|}_B h$ implies that $k \mathop{\smile\!\!\!\!\!\!\,|}_B g$. Hence, from 7.2.10 and 7.2.12, we have $g \in S(q') \subseteq G^0_B$, so by (*) $q := \text{tp}(g/B)$ is a desired type. ■

7.3 Commensurativity and Local Connectedness

We keep working with the hyperdefinable group $G = \Gamma/E$. Let H be a subgroup of G. By a *G-conjugate* of H or $h \in G$ we mean $H^g = g^{-1}.H.g$ or $h^g = g^{-1}.h.g$ for some $g \in G$. Let $\mathcal{H}$ be the family of all hyperdefinable subgroups of G. We say two subgroups $H_1, H_2 \in \mathcal{H}$ are *commensurate* if $H_1 \cap H_2$ has bounded index in H_i $(i = 1, 2)$. Then since *commensurativity* is transitive (an easy exercise), it gives an automorphism invariant equivalence relation on $\mathcal{H}$. Note that this commensurativity notion is a generalization of the commensurability notion in group theory aforementioned at the end of Section 7.1.

In each commensurativity (equivalence) class, we aim to find the *canonical* representative. Before finding such we need to clarify what we mean by "canonical." Note that the two typical actions on $\mathcal{H}$ preserving the $D_G(-,\varphi,\epsilon,k)$ ranks are the structural automorphism action and the G-conjugation. (As declared in the introduction of this chapter, *automorphisms* refer to those of the structure $\mathcal{M}$.)

Definition 7.3.1 A hyperdefinable subgroup H of G is said to be *locally connected* if for any automorphic image or G-conjugate H' of H, whenever H, H' are commensurate, then $H = H'$.

Remark 7.3.2

(1) In general when we intersect hyperdefinable objects, the number of objects needs to be (reduced) bounded (i.e., $< \bar{\kappa}$) in order for the intersection to make sense independent from the choice of our saturated model $\mathcal{M}$. Thus, in the future proofs where some intersection arguments are introduced, we occasionally check whether the intersection number is bounded if not obvious.

Note now that if $\{H_i | i < \kappa\}$ is some bounded set of hyperdefinable subgroups of G in the same commensurativity class; i.e., for $i, j < \kappa$, H_i is commensurate with H_j, then $\bigcap_{i<\kappa} H_i$ is an hyperdefinable subgroup of G commensurate with each H_i.

(2) Note that the local connectedness of a subgroup H (of G) depends on the ambient group G. Obviously, any hyperdefinable group is a locally connected subgroup of itself. In simple theories, a connected subgroup need not be locally connected, while that is true for type-definable groups in stable structures. The example V_A^0 in 7.1.8(5) for non-empty A is *not* locally connected subgroup of V.

We shall show that each commensurativity class of hyperdefinable subgroups of G contains the smallest locally connected subgroup (Theorem 7.3.6).

(3) If $H = H_a$ is a hyperdefinable subgroup over a, then on $\text{tp}(a)$, the equivalence relation R such that $R(b,c)$ holds for $b, c \models p$ iff H_b, H_c are commensurate, is $\emptyset$-type-definable since $R(b,c)$ holds iff $D_G(H_b \cap H_c, \varphi, \epsilon, k) \geq D_G(H_a, \varphi, \epsilon, k)$ for each $\varphi, \epsilon \in E, k$.

Lemma 7.3.3 *Assume that H is a hyperdefinable locally connected subgroup of G. Then there is a canonical parameter u for H; i.e., for any automorphism σ, σ fixes H setwise iff it fixes the hyperimaginary u. Any coset $g.H$ (or $H.g$) with $g \in G$ also has a canonical parameter u_g. Moreover, if $p, q \in S_G^L(A)$ are generic for H and $g.H$, respectively, then $u \subseteq \text{Cb}(p) \subseteq \text{bdd}(u)$ and $u_g \subseteq \text{Cb}(q) \subseteq \text{bdd}(u_g)$.*

Proof Let $H = H_a$ be a-hyperdefinable. As in Remark 7.3.2(3), the commensurativity equivalence relation R on $\text{tp}(a)$ is type-definable. Then due to the local connectedness of H, $u := a_R$ is the canonical parameter for H. Similarly, the equivalence relation R' on $G \times \text{tp}(a)$ such that $(f,b)R'(h,c)$ holds iff $f.H_b = h.H_c$, is $\emptyset$-type-definable since iff $D_G(f.H_b \cap h.H_c, \varphi, \epsilon, k) \geq D_G(H_a, \varphi, \epsilon, k)$ for each $\varphi, \epsilon \in E, k$. (Note that $H_b \cap f^{-1}.h.H_c = H_b \cap h'.H_c = h'.(H_b \cap H_c)$ for any $h' \in H_b \cap f^{-1}.h.H_c$, and $(f,b)R'(h,c)$ implies bRc.) Hence, $u_g := (g,a)_{R'}$ is the canonical parameter for $g.H$.

Now, since p is generic for H, p does not fork over u. Hence, $\mathrm{Cb}(p) \subseteq \mathrm{bdd}(u)$. Moreover, any automorphism fixes $\mathrm{Cb}(p)$ iff it fixes the parallelism class $\mathcal{P}_p$. In particular, for an automorphism σ, if $\sigma(p) = p' \in S_G^L(A')$ is parallel with p, then for $H' := \sigma(H)$, we have $D_G(H \cup H', \varphi, \epsilon, k) = D_G(p \cup p', \varphi, \epsilon, k) = D_G(H, F, \epsilon, k)$ for each $\varphi, \epsilon \in E, k$. Due to the local connectedness of H, it follows $H = H'$, so $\sigma(u) = u$. This shows that $u \in \mathrm{dcl}(\mathrm{Cb}(p))$.

Similarly, q cannot fork over u_g by the similar reason described in 7.1.8(6), so $\mathrm{Cb}(q) \subseteq \mathrm{bdd}(u_g)$. Also due to the arguments we used above, it similarly follows that $u_g \subseteq \mathrm{Cb}(q)$. ■

Remark & Definition 7.3.4

(1) If H is a locally connected subgroup of G with the canonical parameter a, then a is the canonical parameter for H_a^0, too. But we do not know whether H_a^0 is locally connected in G as well.

Now if g is generic for G over $(a \in)A$, then $\mathrm{tp}(g/Av)$ is generic for $g.H$, where v is the canonical parameter for $g.H$. Choose generic h for H over Ag, so $h \perp_{A} g$; $g.h \perp Ah$, and $g.h$ is generic for $g.H$ over $Ag(\supseteq Av)$. Now since $g.h.H = g.H$, we have $v \subseteq A(g.h)$, too. Hence, $g.h \perp_{Av} h$, and $g.h$ is generic for $g.H$ over Avh. Then since $D_G(g.H, \varphi, \epsilon, k) = D_G(g.h/Avh, \varphi, \epsilon, k) \leq D_G(g/Av, \varphi, \epsilon, k) \leq D_G(g.H, \varphi, \epsilon, k)$, $\mathrm{tp}(g/Av)$ is generic for $g.H$.

(2) Let H, K be hyperdefinable subgroups of G over A. According to the similar comparison of the $D_G(-, \varphi, \epsilon, k)$ ranks as in 7.3.2(3), the *approximate normalizer* of H in K

$$\tilde{N}_K(H) := \{g \in K : H \text{ and } H^g \text{ are commensurate}\}$$

is a hyperdefinable subgroup of K over A. In particular, if H is locally connected in G over a, then $N_G(H) = \{h \in G : H^h = H\}$, the normalizer of H in G, is equal to $\tilde{N}_G(H)$ so a-hyperdefinable.

(3) Let $\mathcal{F}$ be a family of subgroups of G. For a subgroup $F \leq G$, we say $\mathcal{F}$ is *F-invariant* if $\mathcal{F} = \mathcal{F}^g := \{H^g : H \in \mathcal{F}\}$ for any $g \in F$ (while, of course, for a hyperimaginary A, "A-invariance" refers to "being invariant under A-automorphisms").

We call $\mathcal{F}$ *uniformly hyperdefinable over A* if there are partial types $\pi(y)$ over A and $H(x; y)$ over $\emptyset$ such that for each $a \models \pi$, $H(x, a)$ is an E-type, and $\mathcal{F} = \{H_a \leq G : a \models \pi(x)\}$, where each H_a is hyperdefined by $H(x, a)$. For example, given a subgroup K_a hyperdefined by a type $K(x, a)$, the class $\mathcal{K}$ of G-conjugates of automorphic images of K_a commensurate with it is uniformly hyperdefinable over a, since $K_b^g \in \mathcal{K}$ for $g \in G, b \models \mathrm{tp}(a)$ iff $D_G(K_b^g \cap K_a, \varphi, \epsilon, k) \geq D_G(K_a, \varphi, \epsilon, k)$ for each $\varphi, \epsilon \in E, k$. Indeed, $\mathcal{K}$ is hyperdefinable over a_R where R is the type-definable commensurativity relation on $\mathrm{tp}(a)$ (7.3.2(3)), since $\mathcal{K}$ is a_R-invariant.

Lemma 7.3.5 *Let $\mathcal{F}$ be a uniformly A-hyperdefinable family of subgroups of G such that any two groups in $\mathcal{F}$ are commensurate.*

(1) *There is a hyperdefinable subgroup N of G over A commensurate with every $H \in \mathcal{F}$.*

(2) *Suppose that F is a hyperdefinable subgroup of G over A such that $\mathcal{F}$ is F-invariant. Then we can choose the above N, which is additionally F-invariant.*

Proof (1) There is a partial type $\pi(y)$ over A so that $\mathcal{F} = \{H(x,a) | a \models \pi(y)\}$. Write H_a for the group hyperdefined by $H(x,a)$. Let $p(y)$ be a completion of $\pi(y)$ over bdd(A), and let

$$K := K_p = \left\{g \in G | \exists b \models p \left(b \underset{A}{\perp} g \wedge g \in H_b\right)\right\}.$$

Then by 4.2.13, K is hyperdefinable over bdd(A), and $1_G \in K$. Moreover, if $g_1, g_2 \in K$ and $g_1 \perp_A g_2$, then there are $b_i \models p$ such that $b_i \perp_A g_i$ and $g_i \in H_{b_i}$ $(i = 1, 2)$. Thus by the independence theorem, we can assume that $b_1 = b_2 = b \perp_A g_1^{-1}.g_2$, and thus $g_1^{-1}.g_2 \in K$. Hence, by 7.2.11, $N_p = K.K$ is a hyperdefinable subgroup of G over bdd(A). Now let

$$N := \bigcap \{N_p | p \text{ is a completion of } \pi(y) \text{ over bdd}(A)\}.$$

Then N is hyperdefinable over A as it is invariant under A-automorphisms. By transitivity, it remains to see whether N is commensurate with some $H \in \mathcal{F}$. Indeed, by Remark 7.3.2(1), it suffices to see whether each N_p is so. Now if h is generic for N_p, then according to 7.2.11, $h \in K$ and there is some $b \models p$ such that $h \perp_A b$ and $h \in H_b$. Hence, tp$(h/\text{bdd}(A)b)$ is generic in N_p and implies that $H(x,b)$. Thus, $|N_p : N_p \cap H_b|$ is bounded. Conversely, choose $a \models p$ with $a \perp_A b$ and generic g of $H_a \cap H_b$ over abA. Then $g \perp_{ab} A$. Since H_a, H_b are commensurate, g is generic for both H_a and H_b, and $g \perp_{aA} b$. Hence, $g \perp_A b$ and $g \in K \subseteq N_p$ and $|H_b : H_b \cap N_p|$ is bounded, too.

(2) Let $\mathcal{F}' = \{N^h | h \in F\}$. Note that H^h and N^h are commensurate for $h \in F$ and $H \in \mathcal{F}$. So are N^h and N, since $H^h \in \mathcal{F}$. Hence, we can apply the proof of (1) to $\mathcal{F}'$. In particular, fix $q(x) \in S_G(\text{bdd}(A))$ generic for F, and let

$$K' := K'_q = \left\{g \in G | \exists h \models q \left(h \underset{A}{\perp} g \wedge g \in N^h\right)\right\}.$$

Then $M_q = K'.K'$ is a hyperdefinable subgroup of G over bdd(A) commensurate with each group in $\mathcal{F}'$ and $\mathcal{F}$. Hence, again for $h \in F$, M_q^h is commensurate with a group in $\mathcal{F}$. Now consider $N_q = \bigcap\{M_q^h | h \in F\}$. It clearly is normalized by F, and hyperdefinable if the intersection number is bounded. It is so by the following claims.

Claim 1 For $g \in K'$ and generic $h \in F^0_{\text{bdd}(A)}$, if $g \perp_A h$, then $g^h \in K'$: There is $h_0 \models q$ such that $h_0 \perp_A g$ and $g \in N^{h_0}$. Now from 7.2.12, $h \in S(q(x^{-1}))$, and thus there is $h_1(\models q(x)) \perp_A h$ and $h_1.h(\models q(x)) \perp_A h$. By the independence theorem, there is

$h' \models tp(h_0/bdd(A)g) \cup tp(h_1/bdd(A)h)$ such that $\{h', g, h\}$ is A-independent. Then $g^h \in N^{h'.h}$, and $g^h \perp_{Ah} h'$; $g^h \perp_A h'.h (\models q)$. Therefore, $g^h \in K'$.

Claim 2 M_q is normalized by $F^0_{bdd(A)}$: Since any element in $F^0_{bdd(A)}$ is a product of two generic elements (7.2.8(1)), it suffices to show $g^h \in M_q$ for $g \in M_q$ and generic $h \in F^0_{bdd(A)}$. By 7.2.11, there is $g_i (\in K') \perp_A g$ such that $g = g_1.g_2$. By extension we can further assume that $g_1 g_2 \perp_{gA} h$, and so $g_i \perp_A h$. Thus, by Claim 1, $g^h \in M_q$.

Hence, $N_q = \bigcap \{M_q^h \mid h \in F'\}$ is hyperdefinable over $bdd(A)$ and commensurate with $\mathcal{F}$, where F' is a *bounded* set of representatives for $F/F^0_{bdd(A)}$. Now let

$$N' = \bigcap \{N_q \mid q \in S_G(bdd(A)) \text{ is generic for } F\}.$$

Then N' is the desired hyperdefinable subgroup over A. ■

Theorem 7.3.6 *In each commensurativity class containing a hyperdefinable subgroup H of G, there is the unique smallest locally connected subgroup H^c of G. H^c is hyperdefined over the same parameter for H.*

Proof Let $\mathcal{F}$ be the class of G-conjugates of automorphic images of $H = H_a = H(x, a)$ commensurate with H. As in 7.3.4(3), $\mathcal{F}$ is uniformly hyperdefinable over a_R, where is R is the commensurativity equivalence relation on $tp(a)$. Now let

$$F := \{g \in G \mid H \text{ and } H^g \text{ are commensurate}\},$$

a hyperdefinable subgroup. Clearly, $\mathcal{F}$ is F-invariant. Now $g \in F$ iff $\exists a' \equiv a(a'Ra \wedge H^g_{a'} \in \mathcal{F})$, and thus F is hyperdefinable over a_R too. Hence, by 7.3.5, there is an F-invariant hyperdefinable subgroup H_0 over a_R commensurate with H.

Claim H_0 is locally connected: For f an automorphism, suppose H_0 and $f(H_0)$ are commensurate. Now, $f(H)$ and $f(H_0)$ (so H) are commensurate. Thus, f fixes a_R and $H_0 = f(H_0)$. Now, for $g \in G$, if H_0 and H_0^g are commensurate then $g \in F$, since, otherwise, neither H, H^g nor H_0, H_0^g are commensurate. Hence again $H_0 = H_0^g$.

Now let $\mathcal{H}$ be the class of all hyperdefinable locally connected subgroups of G commensurate with H. Then $H_0 \in \mathcal{H} \neq \emptyset$, and we let $H^c = \bigcap \mathcal{H}$. Now, if $H_1 \in \mathcal{H}$, then H_1 is over a_R, since by H_1 being locally connected, $f(H_1) = H_1$ for an automorphism f fixing a_R. Hence, H^c is obtained by a bounded intersection (i.e., an intersection of $< \bar{\kappa}$-many objects) and so hyperdefinable over a_E. ■

Remark & Definition 7.3.7 We call the above H^c the *locally connected companion* of H (H^c need not be a subgroup of H). It follows from Theorem 7.3.6 that for f an automorphism, f setwise fixes the commensurativity class of H iff it fixes H^c setwise iff it fixes the canonical parameter of H^c. In particular, H and H^c are commensurate, and if H_1, H_2 are commensurate, then $H_1^c = H_2^c$.

For the rest of this section we point out some frequently used elementary facts in group theory, and introduce hyperdefinably expressible group theoretic notions.

Definition & Fact 7.3.8

(1) For $h \in G$, its *centralizer* in G,

$$C_G(h) := \{g \in G \mid h.g = g.h\} = \{g \in G \mid h^g = h\},$$

is a hyperdefinable subgroup of G over h. But for hyperdefinable $H, K \leq G$, $C_H(K) := \{h \in H \mid k.h = h.k \text{ for all } k \in K\}$, the *centralizer* of K in H need not be hyperdefinable unless K is bounded. Similarly, $Z(G)$, the *center* of G, need not be hyperdefinable. Recall that $H.K := \{h.k \mid h \in H,\ k \in K\}$.

(2) For $g, h \in G$, $[g, h] := g^{-1}.h^{-1}.g.h$ is the *commutator* of g and h. For subsets A, B of G, $[A, h] := \{[g, h] \mid g \in A\}$ a *subset* of G, while $[A, B]$ denotes the *subgroup* generated by $\{[h, g] \mid h \in A,\ g \in B\}$. In particular, $G^{(1)} := [G, G]$ is the *commutator* or *derived* subgroup of G.[1] $G^{(1)}$ is normal in G. Indeed, any subgroup N of G containing $G^{(1)}$ is normal.

Let $F \lhd G$. Then G/F abelian iff $G^{(1)} \leq F$. Moreover, $F^{(1)} \lhd G$ as well, since for $n, m \in F$, $g \in G$ we have $g^{-1}.n^{-1}.m^{-1}.n.m.g = n_0^{-1}.m_0^{-1}.n_0.m_0$, where $n_0 = n^g, m_0 = m^g$. Similarly, both $[F, G] \lhd G$ and $[F, G] \lhd F$ hold.

(3) If $N \lhd G$ and $K \leq G$, then $N \cap K \lhd K$, $N.K = K.N \lhd G$, and $K/N \cap K \cong N.K/N$.

(4) For $g \in G$, we have $|[G, h]| = |G : C_G(h)|$ via the map $g^{-1}.h^{-1}.g.h \mapsto g.C_G(h)$. Similarly, we have that

$$|\{[g, h] \mid g, h \in G\}| \leq |G/Z(G)|^2$$

since $g.Z = k.Z$ iff for all $h \in G$, $[g, h] = [k, h]$ and $[h, g] = [h, k]$.

Definition 7.3.9 Let H, K be hyperdefinable subgroups of G.

(1) The *approximate centralizer* of K in H is

$$\tilde{C}_H(K) := \{h \in H \mid |K : C_K(h)| \text{ is bounded}\}.$$

(2) The *approximate center* $\tilde{Z}(G)$ of G is $\tilde{C}_G(G)$.

Recall that for certain conditions P and Q, a group is said to be *P-by-Q* if the group has a normal subgroup having the property P such that the quotient group satisfies Q.

Lemma 7.3.10 *Let H, K be hyperdefinable subgroups of G over A.*

(1) $\tilde{C}_H(K)$ is a hyperdefinable subgroup over A of H, and $\tilde{C}_K(\tilde{C}_H(K))$ is generic in K.

(2) $\tilde{Z}(G)$ is a bounded-by-abelian hyperdefinable normal subgroup of G over $\emptyset$. Indeed, $[\tilde{Z}(G), \tilde{Z}(G) \cup G^0_\emptyset]$ is bounded, and centralized by $G^0_\emptyset$.

1. We avoid using G' to denote the derived subgroup, and save it to mean another group.

Proof (1) By the similar $D_G(-,\varphi,\epsilon,k)$-rank comparison in 7.3.2(3), it follows that $\tilde{C}_H(K)$ is hyperdefinable over A. Moreover, $\tilde{C}_H(K)$ forms a group because for $g,h\in H$, $C_K(g)\cap C_K(h)\leq C_K(g.h)$. Now choose a generic $h\in\tilde{C}_H(K)$ (over A). Then $C_K(h)$ is generic in K (over A). Thus, there is $k\in C_K(h)$ which is generic for K over Ah so that $k\perp_A h$ and $k.h=h.k$. Hence, $\mathrm{tp}(h/Ak)\models C_H(k)$ and $|\tilde{C}_H(K):\tilde{C}_H(K)\cap C_H(k)|$ is bounded. Note that $\tilde{C}_H(K)\cap C_H(k)=C_{\tilde{C}_H(K)}(k)$. Thus, $(k\in)\tilde{C}_K(\tilde{C}_H(K))$ is generic in K.

(2) By (1), $\tilde{Z}(G)$ is hyperdefinable over $\emptyset$. It is normal since if $|G:C_G(g)|$ is bounded, then for any $h\in G$, $C_G(g^h)=(C_G(g))^h$ has bounded index, too.

Note now for $h\in\tilde{Z}(G)$, each of $[h,G]$ and $[G,h]$ is bounded (so $\subseteq\mathrm{bdd}(h)$) since there is a bijection between $[h,G]$ and the coset space $G/C_G(h)$. From (1), we have $G^0_\emptyset\leq\tilde{C}_G(\tilde{C}_G(G))=\tilde{C}_G(\tilde{Z}(G))$; i.e., for $g\in G^0_\emptyset$, the coset space $\tilde{Z}(G)/C_{\tilde{Z}(G)}(g)$ is bounded. Hence, again due to a bijection, the set $[\tilde{Z}(G),g]$ is bounded (*). Therefore, for $h\in\tilde{Z}(G)$ and $g\in G^0_\emptyset\cup\tilde{Z}(G)$ with $h\perp g$, we have $[h,g]\in\mathrm{bdd}(h)\cap\mathrm{bdd}(g)=\mathrm{bdd}(\emptyset)$ (**).

Claim $[\tilde{Z}(G),\tilde{Z}(G)\cup G^0_\emptyset]$ is bounded: Given $h\in\tilde{Z}(G)$ and $g\in G^0_\emptyset\cup\tilde{Z}(G)$, we will show $[h,g]\in\mathrm{bdd}(\emptyset)$. Now, $h=h_1.h_2$ with generic $h_i(\perp h)$ in $\tilde{Z}(G)$ according to 7.2.8(1). Hence, we can assume $h_i\perp g$ $(i=1,2)$. Note that $[h,g]=[h_1.h_2,g]=[h_1,g]^{h_2}.[h_2,g]=[h_1,g].[[h_1,g],h_2].[h_2,g]$. From (**), we have $[h_i,g]\in\mathrm{bdd}(\emptyset)$. It suffices to see that $[[h_1,g],h_2]\in\mathrm{bdd}(\emptyset)$. Note that

$$|\tilde{Z}(G):C_{\tilde{Z}(G)}(h_1)|(\leq|G:C_G(h_1)|)$$

is bounded and thus $C_{\tilde{Z}(G)}([h_1,g])(\geq C_{\tilde{Z}(G)}(h_1)\cap C_{\tilde{Z}(G)}(g))$ is generic in $\tilde{Z}(G)$. Hence, by the same reason in (*), $[[h_1,g],\tilde{Z}(G)]$ is bounded. Therefore, $[[h_1,g],h_2]\in\mathrm{bdd}(\emptyset)$, and the claim follows.

By the claim, $[\tilde{Z}(G),\tilde{Z}(G)]$ is bounded, and $\tilde{Z}(G)$ is bounded-by-abelian. It remains to show that $G^0_\emptyset\leq C_G(H)$ for H, the subgroup of G generated by $[\tilde{Z}(G),G^0_\emptyset\cup\tilde{Z}(G)]$. Now, $C_G(H)=\bigcap\{C_G(h)\mid h\in H\}$, is hyperdefinable over $\emptyset$ as it is $\emptyset$-automorphism invariant and by the claim the intersection number is bounded. Since each of $\tilde{Z}(G)$ and $G^0_\emptyset$ is normal in G, so is H in G. Hence, for $h\in H$, we have $[G,h]\subseteq H$ and the index $|G:C_G(h)|(=|[G,h]|)$ is bounded. Therefore, $G^0_\emptyset\leq C_G(H)$. ■

7.4 1-Based Groups

We continue to work with the $\emptyset$-hyperdefinable group G. In this section we study the group when it is 1-based (see Section 6.1); i.e., for any (finite) tuple a from G, and any hyperimaginary A, we have $\mathrm{Cb}(a/A)\subseteq\mathrm{bdd}(a)$. So until Proposition 7.4.5, we assume G is 1-based. We shall see that if 1-based G is minimal and connected, then it is canonically endowed with an infinite vector space structure where linear independence corresponds to non-forking independence.

Proposition 7.4.1 *If a hyperdefinable subgroup H of G is locally connected, then H is over* $\mathrm{bdd}(\emptyset)$.

Proof Let u be a canonical parameter for H. There are generic $h \in H$ over u, and generic $g \in G$ over hu. Hence, $h.g$ is generic for $H.g$ over gu, and generic for G over hu. Now from 7.3.3, $v \in \mathrm{Cb}(h.g/gu)$, where v is the canonical parameter for $H.g$. Hence, by 1-basedness, $v \in \mathrm{bdd}(h.g)$. Note that $(H.g)^{-1} = (g^{-1}).H$ is also over v, so is $H = (H.g).(H.g)^{-1}$, and thus $u \in \mathrm{dcl}(v)$. Now since $h.g \perp u$, we have $u \in \mathrm{bdd}(\emptyset)$. ■

Proposition 7.4.2 $(G^0_\emptyset)^{(1)}$ *is bounded and contained in* $Z(G^0_\emptyset)$. *Hence,* $G^0_\emptyset$ *is bounded-by-abelian.*

Proof For $g \in G$, let $H_g = \{(h, h^g) \mid h \in G\}$ be a hyperdefinable subgroup of $G \times G$. Note that $G \times G$ is 1-based, too. Thus by Proposition 7.4.1, the locally connected companion H^c_g is over $\mathrm{bdd}(\emptyset)$ (*). For $g, g', h \in G$, we have $h \in C_G(g'.g^{-1})$ iff $(h, h^g) = (h, h^{g'}) \in H_g \cap H_{g'}$. Hence, it easily follows that the groups H_g and $H_{g'}$ are commensurate iff $C_G(g'.g^{-1})$ has bounded index iff $g'.g^{-1} \in \tilde{Z}(G)$. Thus, there is a bijection between the cosets of $\tilde{Z}(G)$ in G and the commensurativity classes partitioning $\{H_g \mid g \in G\}$. Then by (*), $|G : \tilde{Z}(G)|$ is bounded, and $\tilde{Z}(G) \geq G^0_\emptyset$. Therefore, from 7.3.10(2), the commutator subgroup $(G^0_\emptyset)^{(1)}$ is bounded, and $\leq Z(G^0_\emptyset)$. ■

Note that $(G^0_\emptyset)^{(1)}$ need not be hyperdefinable. Now since $G^0_\emptyset$ is normal (7.2.9), we have the following.

Proposition 7.4.3 *G is (bounded-by-abelian)-by-bounded. If G is finitary type-definable, then G is (finite-by-abelian)-by-finite.*

Example 7.4.4 Let $\mathbb{F}$ be a finite field with odd prime $|\mathbb{F}|$ (so that for any non-zero $n \in \mathbb{F}$, $n \neq -n$). Let V be an infinite-dimensional vector space over $\mathbb{F}$. Take a non-degenerate alternating bilinear form $\langle , \rangle$ on V over F. Note that $\mathbb{V} = (V, +, \cdot, 0; \mathbb{F}, \langle , \rangle)$ is ω-categorical and 1-based having SU-rank 1 (see Chapter 8). Now we define a multiplication . on $\mathbb{F} \times V$ as

$$(n, v).(m, w) = (n + m + \langle v, w\rangle, v + w).$$

Then, $G_1 := (\mathbb{F} \times V, .)$ forms a (because of the choice of $\mathbb{F}$) non-abelian group. Also it is not hard to check $Z(G_1) = \mathbb{F} \times \{0\} = [G_1, G_1]$. Hence, G_1 is abelian-by-finite, and ω-categorical 1-based (since it is interpretable in $\mathbb{V}$). Now, for $m \in \mathbb{F}$, $(n, v) \mapsto (mn, m \cdot v)$ induces a homomorphism from $\mathbb{F}$ to $\mathrm{Aut}(G_1)$. We put $G := G_1 \rtimes \mathbb{F}$ with respect to the homomorphism. Hence, now G is (finite-by-abelian)-by-finite and again 1-based ω-categorical with $SU(G) = 1$. This example shows that the above proposition is a maximal possible result.

Proposition 7.4.5 *Assume that G is 1-based, and $p \in S_G(A)$ with boundedly closed A. Then there is a locally connected subgroup H over $\mathrm{bdd}(\emptyset)$ of G such that p is generic for some coset over A of H.*

Proof Let $a \models p$. Choose g_1 generic for G over A, and $a \perp_A g_1$. Now $h_1 (= g_1.a) \perp_e Ag_1$, where $e = \mathrm{Cb}(h_1/Ag_1)$. Hence, $h_1 \perp_C g_1$ where $C = \mathrm{bdd}(eA) (\subseteq \mathrm{bdd}(Ag_1))$. Choose $h_2g_2 (\equiv_C h_1g_1)$ with $g_2 \perp_C g_1$. Then by type amalgamation, there is $h \models \mathrm{tp}(h_i/Cg_i)$ $(i = 1, 2)$ with $h \perp_C g_1g_2$. Hence, we have $a_i \models p$ such that $h = g_i.a_i$ and $a_i \perp_A g_i$. Then, it follows that $\{a_i, g_1, g_2\}$ is C-independent for each $i = 1, 2$. So we have $a_1 \perp_{Cg_1} g_2^{-1}.g_1$; $a_1 \perp_A Cg_1$; and $a_1 \perp_A g_2^{-1}.g_1$. Then since $g_2^{-1}.g_1.a_1 = a_2 \models p$, we have $g_2^{-1}.g_1 \in S(p)$ and $g_1.\mathrm{Stab}(p) = g_2.\mathrm{Stab}(p)$ (*).

Now, let H be the locally connected companion of $\mathrm{Stab}(p)$. By 7.4.1, H is over $\mathrm{bdd}(\emptyset)$. Moreover, from (*) and that H and $\mathrm{Stab}(p)$ are commensurate, it follows the canonical parameter u for $g_1.H$ is in $\mathrm{bdd}(Ag_2)$. Hence, $u \in \mathrm{bdd}(g_1) \cap \mathrm{bdd}(Ag_2) \subseteq \mathrm{bdd}(Ag_1) \cap \mathrm{bdd}(Ag_2) \subseteq C$, and due to 1-basedness, $u \in C = \mathrm{bdd}(eA) \subseteq \mathrm{bdd}((g_1.a_1)A)$.

Let v be the canonical parameter for $g_1.H.a_1$. Since $g_1.a_1 \perp_A a_1$, $g_1.a_1$ is generic for $g_1.a_1(a_1^{-1}.H.a_1) = g_1.H.a_1$ over Av, by the argument in 7.3.4.(1). Thus, $v \in \mathrm{Cb}(g_1.a_1/Av)$ from 7.3.3, and due to 1-basedness, $v \in \mathrm{bdd}(g_1.a_1)$.

Now, let w be the canonical parameter for $H.a_1$. Then since $H.a_1 = (g_1.H)^{-1}.(g_1.H.a_1)$, $w \in \mathrm{dcl}(u, v) \subseteq \mathrm{bdd}(A, g_1.a_1)$. Thus, from $g_1.a_1 \perp_A a_1$ and $w \in \mathrm{dcl}(A, a_1)$, we have $w \perp_A a_1$; $w \perp_A w$; and $w \in \mathrm{bdd}(A) = A$. Then since $a_1 \models p \models H.a_1$, $D_G(p, \varphi, \epsilon, k) \leq D_G(H.a_1, \varphi, \epsilon, k)$.

Claim $D_G(H.a_1, \varphi, \epsilon, k) \leq D_G(p, \varphi, \epsilon, k)$ for all φ, ϵ, k, and hence p is generic for $H.a_1$: There is generic g for $\mathrm{Stab}(p)$ over A such that $g \perp_A a_1$ and $g.a_1 \models p$. Then $g.a_1$ is generic for $\mathrm{Stab}(p).a_1$ over Aa_1. Now, $D_G(H, \varphi, \epsilon, k) = D_G(\mathrm{Stab}(p), \varphi, \epsilon, k) = D_G(\mathrm{Stab}(p).a_1, \varphi, \epsilon, k) = D_G(g.a_1/Aa_1, \varphi, \epsilon, k) \leq D_G(g.a_1/A, \varphi, \epsilon, k) = D_G(p, \varphi, \epsilon, k)$.

■

Proposition 7.4.6 *The hyperdefinable group G over $\emptyset$ is 1-based iff for each positive n, any $p \in S_{G^n}(A)$ with boundedly closed A is generic for some coset of a hyperdefinable subgroup of G^n.*

Proof $(\Rightarrow)$ This is clear from Proposition 7.4.5, since G^n is a hyperdefinable 1-based group as well.

$(\Leftarrow)$ Assume the right-hand side, and fix $n < \omega$.

Claim Any locally connected hyperdefinable subgroup K of G^n is over $\mathrm{bdd}(\emptyset)$.

Proof of Claim Let v be the canonical parameter for K, and let q be a generic type for K over $\mathrm{bdd}(v)$. Now, consider a Morley sequence $\langle a_i | i < \omega.4\rangle$ of q over $\mathrm{bdd}(v)$. As in 7.2.2(5), both $\langle a_i |\ i \in \omega.2\rangle$ and $\langle a_{\omega.2+i} |\ i \in \omega.2\rangle \models q^{\omega.2}$ are generic for $K^{\omega.2}$ over $\mathrm{bdd}(v)$. Then, $I = \langle c_i | i \in \omega.2\rangle$ with $c_i = a_i.a_{\omega.2+i}^{-1}$ is Morley over $\mathrm{bdd}(v)$ (*). Let

$I_1 = \langle c_i | \ i < \omega \rangle$ and $I_2 = \langle c_i | \ \omega \leq i < \omega.2 \rangle$, so that $I = I_1 I_2$. Since $I \in S(q^{\omega.2})$, I is principal generic over $\mathrm{bdd}(v)$ for $K^{\omega.2}$, by 7.2.12. Note that from (*) and 4.3.9(5), we have $v \in \mathrm{Cb}(q) \subseteq \mathrm{dcl}(I_i)$ $(i = 1, 2)$ (**). By applying 7.2.14 to I, we can find $J(\equiv^L_v I) \downarrow_v I$ such that $J' = I.J \equiv^L_v I$ and $J' \downarrow_v I$. Now, by compactness applied to the assumption with indiscernibility, it can be seen that $\mathrm{Ltp}(I/\emptyset)$ also is generic for some coset $k.N$ (in $(G^n)^{\omega.2}$) hyperdefinable over A. By the extension axiom, we can assume that $I \downarrow_\emptyset A$ and so $I_1 I_2 \downarrow_v A$. By type amalgamation we can further assume that $\{I_1, I_2, A\}$ is v-independent and $I_1 \equiv^L_{vA} I_2$ (***). Note that $J, J', I \in k.N$ and $J = I^{-1}.J' \in N$. Hence, $k.N = N$, and $r = \mathrm{Ltp}(I/A)$ is generic for N. There is $I'_1 I'_2 v' \equiv^L_A I_1 I_2 v$ such that $I'(= I'_1 I'_2) \downarrow_A I$ (+). Hence, v' is the canonical parameter for a locally connected group K'. From (**), we have $v' \in \mathrm{dcl}(I'_i)$ $(i = 1, 2)$, as well.

Sub Claim $K = K'$: Let $I^*_i = I'_i . I_i^{-1}$ (in $G^{n.\omega}$) for $i = 1, 2$, and let $I^* = I^*_1 I^*_2$ so that $I^* = I'.I^{-1}$. Now from (***),(+), we have $I'_1 \downarrow_{v'} A I_2 I'_2$. Hence, $q' = \mathrm{Ltp}(I'_1 / A I_2 I'_2) = \mathrm{Ltp}(I'_1 / A I'_2 I^*_2)$ is generic for $(K')^\omega$. Moreover, since $(I, I' \models) r$ is generic for N, from (+), we have $I^* \downarrow_A I$ and $I^* \downarrow_A I'$ (‡). Hence, $I_1 \downarrow_{A I_2 I^*_2} I^*_1$, $I_1 \downarrow_{A I_2 I'_2} I^*_1$, and $I_1 \downarrow_v A I_2 I^*$ hold. Therefore, $I^*_1 . \mathrm{Ltp}(I_1 / A I_2 I^*) = \mathrm{Ltp}(I'_1 / A I_2 I^*) = \mathrm{Ltp}(I'_1 / A I'_2 I^*)$ is generic for a coset $I^*_1 . K^\omega$ (in $G^{n.\omega}$), and due to (‡), so is q'. It follows that K^ω, $(K')^\omega$ are commensurate and so are K, K'. As they are locally connected, $K = K'$. We have proved the subclaim.

From (+), $v \downarrow_A v'$ and $v \downarrow A$. Hence $v \downarrow v'$ and by the subclaim, we have $v \in \mathrm{bdd}(\emptyset)$. We have shown the claim.

Now, let $p \in S_{G^n}(A)$ with boundedly closed A. By the assumption, there are a hyperdefinable subgroup H over B of G^n, and $g \in G^n$ such that p is generic for $g.H$. We take a model M containing ABg, and p' a non-forking extension of p over M with $a \models p'$. By the claim, the locally connected companion H^c is over $\mathrm{bdd}(\emptyset)$ (*). Now, the coset $a.H^c$ has the canonical parameter say u. Since H, H^c are commensurate, we have $u \in \mathrm{bdd}(gB) \subseteq \mathrm{bdd}(M) = M$. Hence, $p' = \mathrm{tp}(a/M)$ is generic for $a.H^c$ and $a \downarrow_u M$. According to (*), we have $u \in \mathrm{bdd}(a)$. Then, for $e = \mathrm{Cb}(p) = \mathrm{Cb}(p')$, we have $e \subseteq \mathrm{bdd}(u) \subseteq \mathrm{bdd}(a)$. Therefore, G is 1-based. ■

Assume now the group G over $\emptyset$ is connected, 1-based. Hence, $G = G^0_{\mathrm{bdd}(\emptyset)}$ by 7.2.8(3). Now, let G_0 be $G \cap \mathrm{bdd}(\emptyset)$, an $\emptyset$-invariant subgroup of G. Then according to 7.3.8(2) and 7.4.2, G_0 is normal and G/G_0 is abelian. We now call f an endomorphism of G/G_0, an *endogeny* of G/G_0 if the graph of f is *hyperdefinably induced* over $\mathrm{bdd}(\emptyset)$; i.e., there is a hyperdefinable subset S_f over $\mathrm{bdd}(\emptyset)$ of $G \times G$ such that

$$\{(a + G_0, b + G_0) | \ (a, b) \in S_f\}$$

forms the graph of f.

Now let $D(G) = \{f | \ f \text{ is an endogeny of } G/G_0\}$. With addition and composition, it is easy to check $D(G)$ is a ring. Assume additionally that G is *minimal* (7.2.8(7)). Then $D(G)$ forms a division ring. If $f \in D(G)$ is non-zero, then there is unbounded $p(x) \in S_G(\mathrm{bdd}(\emptyset))$ such that $p(x) \models \exists y \ S_f(y, x)$. Hence, $SU(p) = 1$ and p is generic for G. From 7.2.12,

$G = \mathrm{Stab}(p)$. Then since $\{a + G_0 | \, a \models p\}$ is contained in ran$f$, so are $\{a + G_0 | \, a \in S(p)\}$ and $\{a + G_0 | \, a \in \mathrm{Stab}(p)\}$, and thus f is surjective. Similarly, Ker$f = 0$. Hence, f indeed is an automorphism.

Hence, G/G_0 is a $D(G)$-vector space. Moreover, the following says non-trivial forking independence is linear independence there.

Proposition 7.4.7 *Assume that 1-based hyperdefinable group G over $\emptyset$ is connected and minimal. Then for $a_0, \ldots, a_n \in G \setminus G_0$, the following are equivalent:*

(1) $\{a_0, \ldots, a_n\}$ is dependent.

(2) There are $f_0, \ldots, f_n \in D(G)$ not all zero such that $f_0(a_0 + G_0) + \cdots + f_n(a_n + G_0) = 0$ in G/G_0.

(3) $a_i \in \mathrm{bdd}(\{a_0, \ldots, a_n\} \setminus \{a_i\})$ for some $i \leq n$.

Proof Clearly (2) $\Rightarrow$ (3) $\Rightarrow$ (1). We shall show (1) implies (2). By induction we can suppose that any proper subtuple of $\bar{a} = (a_0, \ldots, a_n)$ is independent. Now, by 7.4.5, $\mathrm{tp}(\bar{a}/\mathrm{bdd}(\emptyset))$ is generic for some coset $H + \bar{h}$ over $\mathrm{bdd}(\emptyset)$ of a locally connected subgroup H over $\mathrm{bdd}(\emptyset)$ of G^{n+1}. Now, for $i < n$, define subgroups over $\mathrm{bdd}(\emptyset)$

$$F_i = \{(x_i, x_n) \in G \times G | \, (0, \ldots, 0, x_i, 0, \ldots, 0, x_n) \in H\}$$

and

$$F = \{x \in G | \, (0, \ldots, 0, x) \in H\}.$$

Claim 1 $\mathrm{Pr}_1(F_i) = \mathrm{Pr}_2(F_i) = G$, and $F < G_0$: Note that since each a_i is unbounded, it is generic for minimal G by 7.2.8(7). Now let say $i = 0$. There is $a_0' a_n' (\equiv^L_{a_1 \ldots a_{n-1}} a_0 a_n)$ such that $a_0' a_n' \perp_{a_1 \ldots a_{n-1}} a_0 a_n$. Then $(a_0', a_1, \ldots, a_{n-1}, a_n') \in H + \bar{h}$, and $(a_0' - a_0, 0, \ldots, 0, a_n' - a_n) \in H$. Now since each of $\{a_0, \ldots, a_{n-1}\}$, $\{a_1, \ldots, a_n\}$ is independent, we have $a_0 \perp a_0'$ and $a_n \perp a_n'$. Hence, each of $a_0' - a_0$, $a_n' - a_n$ is generic for G. Thus, the subgroups $\mathrm{Pr}_1(F_0), \mathrm{Pr}_2(F_0)$ are generic in connected G and so the first part of Claim 1 is proved.

Note now that F is a subgroup over $\mathrm{bdd}(\emptyset)$ of connected minimal G. Hence, either $F = G$ or a subgroup of G_0. If $F = G$, then there is $a' \in F$ generic for G over $\bar{a}$, so that $(0, \ldots, 0, a') \in H$, $(a_0, \ldots, a_{n-1}, a' + a_n) \in H + \bar{h}$, and $\{a_0, \ldots, a_{n-1}, a' + a_n\}$ is independent. Hence, $SU(\bar{a}) < SU(a_0, \ldots, a_{n-1}, a' + a_n) = n + 1$ and $\bar{a}$ cannot be generic for $H + \bar{h}$ by 6.2.8(7), a contradiction. Thus, Claim 1 follows.

Claim 2 $f_i = \{(a + G_0, b + G_0) | \, (a, b) \in F_i\}$ is the graph of a non-zero endogeny of G/G_0: Based on the first assertion of Claim 1 and that F_i is a subgroup of $G \times G$, it suffices to show that $(a, b) \in F_i$ with $a \in G_0$ implies that $b \in G_0$ (*). Now, if $(a, b) \in F_i$ with $a \in G_0$ but $b \notin G_0$ (so b generic in G), then there is generic $b' (\equiv^L_a b)$ with $b' \perp_a b$, so $(a, b') \in F_i$ and $b' \perp b$. Hence, $(0, b' - b) \in F_i$ with generic $b' - b$. Therefore, again it suffices to check (*) for $a = 0$. But this follows from the second part of Claim 1.

Claim 3 $f_0(a_0 + G_0) + \ldots + f_{n-1}(a_{n-1} + G_0) + f_n(a_n + G_0) = 0$ where f_n is an endogeny such that $f_n(x + G_0) = -x + G_0$: Let $\bar{h} = (h_0, \ldots, h_n)$. By Claim 1, there are g_i, g_i' $(i < n)$ such that where $F_i(a_i, g_i), F_i(h_i, g_i')$ hold (so $f_i(a_i + G_0) = g_i + G_0, f_i(h_i + G_0) = g_i' + G_0$). Then both $(a_0, \ldots, a_{n-1}, \sum_{i<n} g_i)$ and $(h_0, \ldots, h_{n-1}, \sum_{i<n} g_i') \in H$, and

$$\left(a_0 - h_0, \ldots, a_{n-1} - h_{n-1}, \sum_{i<n} g_i - \sum_{i<n} g_i'\right) \in H.$$

Note also that $(a_0 - h_0, \ldots, a_n - h_n) \in H$. Thus, by Claim 1, $k = \sum_{i<n} g_i - \sum_{i<n} g_i' - (a_n - h_n) \in G_0$. Therefore, in G/G_0 with $h = f_0(h_0 + G_0) + \cdots + f_n(h_n + G_0)$,

$$\left(\sum_{i\leq n} f_i(a_i + G_0)\right) - h = \left(\sum_{i<n} f_i((a_i - h_i) + G_0)\right) - ((a_n - h_n) + G_0) = k + G_0 = 0.$$

As we can assume $\bar{a} \perp \bar{h}$, we have, as desired, $\sum_{i\leq n} f_i(a_i + G_0) = 0$. ■

7.5 Supersimple Groups

In this section we shall show two main results. One is that a type-definable group in a supersimple theory is an intersection of definable groups (Theorem 7.5.4). The other is that any definable group in an ω-categorical supersimple theory is (finite-by-abelian)-by-finite and having finite SU-rank (Corollary 7.5.13). Let us briefly say why we are interested in the two results.

As is well known, any type-definable group in a stable theory is a conjunction of definable groups [119]. So the first result says this can be generalized in the supersimple context. It is still open whether the same holds in every simple theory.

As mentioned in Chapter 6, Zilber [126] and Cherlin et al. [29] showed that any ω-categorical superstable theory is 1-based and having finite SU-rank. So naturally one can ask whether the same holds in supersimple context. As pointed out in Chapter 6, this is no longer true. Namely, there is a non-1-based supersimple ω-categorical structure. But still one may ask whether any supersimple ω-categorical theory has finite SU-rank. This question is still open, but at least the second result says that this is true when a the whole structure of the theory is equipped with a definable group.

There is another important result by Pillay, Scanlon, and Wagner [104] that any type-definable division ring D in a simple theory with $SU(D) < \infty$ is commutative. The proof uses a fact they show ahead that if such D is a field then its Brauer group is trivial. We omit the detailed proof, however. This generalizes Cherlin's result that any superstable division ring is a field [28].

We now fix a hyperdefinable group G over $\emptyset$ in **simple** T (except Fact 7.5.6) throughout this section. We say G is *supersimple* if T is supersimple. (But the term "simple groups" refers to the group theoretic notion as stated in Chapter 1.)

For the first result, we need a couple of preparatory results. Recall that $SU(G)$ is equal to the SU-rank of any/some generic type of G (7.2.8(7)).

Remark 7.5.1 Let H, K be hyperdefinable subgroups of G over A. In general, G/H, a left coset space, and $K.H$ are not groups but still hyperdefinable, so $SU(G/H)$ and $SU(K.H)$ are defined as in 4.2.13(2). Now, the following inequalities hold:

(1) $SU(H) + SU(K.H/H) \leq SU(K.H) \leq SU(H) \oplus SU(K.H/H)$: This is a good exercise for the reader to check using the argument in 7.2.8(7) and the Lascar inequalities. We sketch the ideas. Choose generic elements $h \in H$ and $g \in K$ over A for H, K, respectively, such that $h \perp_A g$. Then $SU(K.H) = SU(f/A)$ where $f = g.h$. Let F be the equivalence relation $x^{-1}.y \in H$. Then $SU(K.H/H) = SU(f_F/A)$. Moreover, f is generic for $g.H$ over Af_F. Hence, $SU(H) = SH(g.H) = SU(f/Af_F)$. Also $SU(f/A) = SU(ff_F/A)$. Hence, the equality comes from 2.5.19(1).

(2) $SU(H) + SU(G/H) \leq SU(G) \leq SU(H) \oplus SU(G/H)$. This is a corollary of (1).

Proposition 7.5.2 *Assume that* $SU(G) = \sum_{i\leq k} \omega^{\alpha_i}.n_i$ *in normal form. Then for each* $j \leq k$, *there is a hyperdefinable normal subgroup* $N_j(\leq G)$ *over* $\emptyset$, *unique up to commensurativity, such that* $SU(N_j) = \sum_{i\leq j} \omega^{\alpha_i}.n_i$.

Proof Let $\beta_j := \sum_{i\leq j} \omega^{\alpha_i}.n_i$. As in the remark before 2.5.19, we can choose a model M and $p \in S_G(M)$ such that $SU(p) = \beta_j$.

Claim 1 $SU(\text{Stab}(p)) = SU(p) = \beta_j$: Fix $a, b \models p$ with $a \perp_M b$, and let $c = a.b^{-1}$. Note first that

$$SU(c/M) \geq SU(c/Mb) = SU(a/Mb) \text{ (by 7.2.8(7))} = SU(a/M) = SU(p). \quad (*)$$

It suffices to show that $SU(c/M) = SU(p)$, since then $c \perp_M b$ holds, so $c \in S(p)$ and $SU(p) = SU(c/M) = SU(\text{Stab}(p))$. Now from (*) and 2.4.2, we have an indiscernible sequence I in $\text{tp}(c/Mb)$ such that $SU(c/MI) = SU(p)$ with $I \perp_{Mc} ab$ (**). Then again it suffices to show that $c \perp_M J$ for given a finite subsequence J of I. Due to (**), $SU(J/Mabc) = SU(J/Mab) = SU(J/Mc)$. Note now that $SU(c/M) \leq SU(G) < \beta_j + \omega^{\alpha_j} \leq SU(c/MJ) + \omega^{\alpha_j}$. Hence, 2.5.19(4) implies both that $SU(J/M) < SU(J/Mc) + \omega^{\alpha_j} = SU(J/Mab) + \omega^{\alpha_j}$ and that

$$SU(ab/M) = \beta_j \oplus \beta_j < SU(ab/MJ) + \omega^{\alpha_j}.$$

Then clearly due to the property of ordinal sum, we must have $SU(ab/M) = SU(ab/MJ)$. Therefore, as desired, $c \perp_M J$ and Claim 1 is proved.

Now let $H = \text{Stab}(p)$, and let

$$\mathcal{H} := \{H' \leq G \mid H' \text{ is an } \emptyset\text{-automorphic image of } H^g \text{ for some } g \in G\}.$$

Claim 2 Any $H' \in \mathcal{H}$ is commensurate with H: Due to 7.2.8(7), it easily follows that any $H' \in \mathcal{H}$ has SU-rank β_j. Now to lead a contradiction, assume that $H \cap H'$ has unbounded index in, say, H. So $SU(H \cap H') < SU(H)$. Then since $SU(H) \leq SU(H \cap H') \oplus SU(H/H \cap H')$ as in 7.5.1, due to the property of ordinal sum, we must have $\omega^{\alpha_j} \leq SU(H/H \cap H')$. Note that $H/H \cap H'$ is isomorphic to HH'/H'. Then a contradiction is induced as follows. $SU(G) < \beta_j + \omega^{\alpha_j} \leq SU(H') + SU(H/H \cap H') \leq SU(H') + SU(HH'/H') \leq SU(HH') \leq SU(G)$. We have shown Claim 2.

Hence, $\mathcal{H}$ is uniformly hyperdefinable over $\emptyset$ (Claim 2 says the R-relation in 7.3.4(3) is trivial). As $\mathcal{H}$ is G-invariant, by applying 7.3.5 to Claim 2, we have a hyperdefinable normal subgroup $N_j(\leq G)$ over $\emptyset$ commensurate with H. Hence, $SU(N_j) = \beta_j$. Uniqueness (up to commensurativity) comes by the similar rank calculation in the proof of Claim 2. ■

In the rest of this section, when G is type-definable, its arity is finite real.

Proposition 7.5.3 *If G is type-definable with $SU(G) = \omega^\alpha.n$, suppose that $D(G) < \infty$, then G is an intersection of definable groups over $\emptyset$.*

Proof As pointed out in the introduction of Section 7.1, we can assume that $G = \bigcap_{i<\kappa} X_i$, where each X_i is an $\emptyset$-definable set such that $1_G \in X_i$, associativity holds inside X_i, and every $x \in X_i$ has the unique inverse in X_i (*). Due to compactness we more suppose that $X_2^2(= X_2.X_2) \subseteq X_1, X_1^2 \subseteq X_0$, and $D(G) = D(X_0)$ (**).

Claim 1 $SU(X_i) = SU(G)$ for any $X_i \subseteq X_1$: Suppose not; say there is $p(x)(\in S(\emptyset)) \models X_1(x)$ with $SU(p) > SU(G)$. Let $\langle a_i \rangle$ be a Morley sequence in p. Then from (**), we have $a_i.G \subseteq X_0$. Now, for $j \neq i$, $a_j.G \cap a_i.G = \emptyset$ since, otherwise, $a_j \in a_i.G$ and $SU(G) < SU(p) = SU(a_j/a_i) \leq SU(a_i.G) = SU(G)$, a contradiction. Hence, there is a definable set X' with $G \subseteq X' \subseteq X_1$ such that $\langle a_i.X' \rangle (\subseteq X_0)$ is 2-inconsistent. It then leads to $D(G) \leq D(X') < D(X_0) = D(G)$, a contradiction. Therefore, Claim 1 is proved.

Claim 2 Suppose that $\varphi(x, y)(\in \mathcal{L}) \models X_2(x)$. Then for some $\theta(y) \in \mathcal{L}$,

(i) $\models \theta(b)$ iff (ii) there is some $q(x)(\in S(b)) \models \varphi(x, b)$ with $SU(q(x)) = SU(G)$
iff for a generic $p \in S_G(\emptyset)$,
(iii) $\exists g \models p$ ($g \perp b$ and $\varphi(g.x, b)$ does not fork over b).

Firstly, show the equivalence of (ii) and (iii). From (ii), there are $x_0 \models q$, $g \models p$ with $g \perp bx_0$, so $g^{-1}.x_0 \models \varphi(g.x, b)$, and

$$SU(G) = SU(q) = SU(x_0/bg) = SU(g^{-1}.x_0/bg) \leq SU(g^{-1}.x_0)$$
$$\leq SU(X_1) = SU(G) \text{ (by (**) and Claim 1).}$$

Thus, $g^{-1}.x_0 \perp_b g$, and (iii) follows. Conversely, if (iii) holds then we find $x_0 \models \varphi(g.x, b)$ with $x_0 \perp_b g$ yielding $SU(G) = SU(g/bx_0) = SU(g.x_0/bx_0) \leq SU(g.x_0) \leq$

$SU(X_2) = SU(G)$. By taking $q = \mathrm{tp}(g.x_0/b)$, (ii) is obtained. Secondly, from 2.2.9, it easily follows that the condition (iii) on b is $\emptyset$-type-definable. Hence, to find $\theta(x)$ in (i), it suffices to verify the negation of (iii) is also type-definable over $\emptyset$. Indeed, we subclaim that $\neg$(iii) is equivalent to

$$\exists g \models p\ (g \perp b \text{ and } \varphi(g.x, b)\ (n+1)\text{-divides over } b), \quad (\dagger)$$

which clearly is type-definable: If $(\dagger)$ holds, then it follows from the similar rank calculation in the proof of (ii)$\Rightarrow$(iii), that *any* $q(\in S(b)) \models \varphi(x, b)$ has SU-rank $< SU(G)$. Hence, (iii) must not hold. Conversely, if we deny (iii), then there is $g_0(\models p) \perp b$ such that $\varphi(g_0.x, b)$ divides over b. It remains to check whether, in fact, it $(n+1)$-divides. Now given $\langle g_i | i < \omega\rangle$, a Morley sequence of $\mathrm{tp}(g_0/b)$, $\{\varphi(g_i.x, b) | i \in \omega\}$ is inconsistent. To lead a contradiction suppose some $x_0 \models \{\varphi(g_i.x, b) | i \leq n\}$. Thus, $x_0 \not\perp_{bg_0 \ldots g_{i-1}} g_i$ for $i \leq n$. Hence, $SU(g_i/x_0bg_0 \ldots g_{i-1}) < SU(g_i/bg_0 \ldots g_{i-1}) = SU(g_i) = \omega^\alpha.n$, and $SU(g_i/x_0bg_0 \ldots g_{i-1}) + \omega^\alpha \leq SU(g_i/bg_0 \ldots g_{i-1})$. Then, from 2.5.19(4), we have $SU(x_0/bg_0 \ldots g_i) + \omega^\alpha \leq SU(x_0/bg_0 \ldots g_{i-1})$, and so $\omega^\alpha.(n+1) \leq SU(x_0/b)$. Since $x_0 \in X_2^2 \subseteq X_1$, it contradicts Claim 1. We have verified Claim 2.

Choose, say, $X = X_{i_0}$ such that $X^3 \subseteq X_2$. By Claim 2, we have $\theta(y, z)$ such that $\theta(a, b)$ holds iff $a \in X$ and $b \in X^3$ and the definable set $a.X \cap b.X$ contains a solution set of some $q \in S(ab)$ with $SU(q) = SU(G)$.

Claim 3 The formula $\theta(y, z)$ is stable: Let $\langle a_i b_i\rangle$ be an indiscernible sequence, and let $\theta(a_0, b_0)$ hold. Thus, there is

$$q(x, a_0 b_0) \models x \in a_0.X \cap b_0.X(\subseteq X_2)$$

with $SU(q) = SU(G)$. According to Claim 1, clearly q does not fork over $\emptyset$. Hence, from 3.2.4, $\bigcup_i q(x, a_i b_i)(\models x \in \bigcap_i a_i.X \cap b_i.X)$ does not fork over $\emptyset$, and so has the same SU-rank as q. It then follows that $\theta(a_i, b_j)$ holds for every $i, j \in \omega$. We have proved Claim 3.

Now there are finitely many θ-definitions $\psi_j(y)$ for generic and complete θ-types over $\mathrm{acl}^{\mathrm{eq}}(\emptyset)$. Let Y be a set defined by $\bigwedge_j \psi_j(y)$. Then as Y is $\emptyset$-automorphism invariant, Y is $\emptyset$-definable. Note that $b \in Y$ iff for any generic h for G over b, $\theta(h, b)$ holds.

Claim 4 $G \subseteq Y \subseteq X^3 \subseteq X_2$ and for any $g \in G, b \in Y$, we have $g.b \in Y$: We know that $Y \subseteq X^3 \subseteq X_2$. For any $g, g' \in G$, we have $G \subseteq g.X \cap g'.X$, so by 7.2.8(7), $\theta(g, g')$ holds and $G \subseteq Y$. Now for $b \in Y$, to show $g.b \in Y$, it suffices to verify that for any generic h for G over gb, $\theta(h, g.b)$ holds. Indeed, since $g^{-1}.h$ is generic for G over gb as well, $\theta(g^{-1}.h, b)$ holds. Namely, $g^{-1}.h \in X$, $b \in X^3$, and $g^{-1}.h.X \cap b.X$ contains a solution set of some type of SU-rank $\omega^\alpha n$. By g-translation, it follows that $h.X \cap g.b.X$ $(\subseteq X_2)$ contains a type of SU-rank $\omega^\alpha.n$. Moreover, some $x', x \in X$ with $h.x' = g.b.x \in h.X \cap g.b.X \neq \emptyset$ witness $g.b = h, x'.x^{-1} \in X^3$ (see (*)). Therefore, $\models \theta(h, g.b)$ holds, and Claim 4 is shown.

According to Claim 4 and (*), if we let $Y_1 := \{y \in Y | \ \forall y' \in Y, y.y' \in Y\}$ ($\subseteq Y \subseteq X_2$) and $Y_2 := \{y \in Y_1 | \ y^{-1} \in Y_1\}$, then both Y_1 and Y_2 are definable supersets of G closed under the group operation. In particular, $Y_2(\subseteq X_2)$ is a definable group over $\emptyset$. Finally, as the whole arguments work for an arbitrary X_i, G must be the intersection of $\emptyset$-definable groups. ■

Theorem 7.5.4 *If G is type-definable and supersimple, then it is an intersection of $\emptyset$-definable groups. More precisely, $G = \bigcap_{i<\kappa} G_i$ where each $G_i(\geq G)$ is an $\emptyset$-definable group and for $i_0 < \ldots < i_n$, $\bigcap_{k\leq n} G_{i_k}$ is a subgroup of each G_{i_k}.*

Proof We prove using induction on $SU(G) = \omega^\alpha.n + \beta$, where $\beta < \omega^\alpha$. Again we take the convention described in the introduction of Section 7.1. So let $X(\supseteq G)$ be the set defined by the formula σ_0 in $G(x)$. According to 7.5.2 and 7.5.3, there now is a type-definable normal subgroup N of G over $\emptyset$ such that $SU(N) = \omega^\alpha.n$, and there are definable groups $N_i(\geq N)$ over $\emptyset$ such that $N = \bigcap_{i<\kappa} N_i$. Due to compactness we can, of course, assume each N_i is a subset of X.

Claim There is an $\emptyset$-definable G-invariant group $\bar{N}$ commensurable with N (so that $SU(\bar{N}) = SU(N)$) such that the type-definable *group* $G.\bar{N} = \{g.n | \ g \in G, n \in \bar{N}\}$ $\subseteq X$: By compactness we can clearly assume the type-definable *set* $G.N_0 \subseteq X$. Again by compactness, we more assume that for all $g \in G$ and $i > 0$, we have $(N \leq) N_i^g \leq N_0$. Now, since T is supersimple, as in the proof of Claim 1 in the proof of 7.5.3, we can suppose that $SU(N_0) = SU(N)$ for all i, so $SU(N_i^g) = SU(N)$. Then for a fixed $i > 0$, clearly $\{N_i^g | \ g \in G\}$ is G-invariant and uniformly commensurable. Hence, due to Theorem 7.1.16, there is a G-invariant $\emptyset$-definable group $(N \leq)\bar{N}(\leq N_0)$ commensurable with N_i and N. Then $G.\bar{N} = \{g.n | \ g \in G, n \in \bar{N}\}$ is a type-definable group contained in the set $G.N_0 \subseteq X$. Hence, the claim is proved.

In the claim, we have $\bar{N} \lhd G.\bar{N} \subseteq X_{\bar{N}} \subseteq X$, where $X_{\bar{N}}$ is a normalizer of $\bar{N}$ in X (i.e., $= \{x \in X | \ \bar{N}^x = \bar{N}\}$). Then $G\bar{N}/\bar{N}$ is a $\emptyset$-type-definable group in the set $X_{\bar{N}}/\bar{N}$ (of imaginaries). Moreover, since $G.\bar{N}/\bar{N} \cong G/G \cap \bar{N}$ and $SU(\bar{N}) \geq SU(G \cap \bar{N}) \geq SU(N \cap \bar{N}) = SU(\bar{N})$, due to 7.5.1, we have

$$SU(G.\bar{N}/\bar{N}) = SU(G/G \cap \bar{N}) = SU(G/N) = \beta < SU(G).$$

Hence, by the induction hypothesis, $G.\bar{N}/\bar{N} = \bigcap_j \bar{G}_j$ for some $\emptyset$-definable groups $\bar{G}_j \subseteq X_{\bar{N}}/\bar{N}$. Then clearly the pre-image $G_0(\subseteq X)$ of $\bar{G}_0$ is an $\emptyset$-definable group containing $G\bar{N}$ (so G); i.e., $G \leq G_0 \subseteq X$. Since the choice of $X(x) \in G(x)$ can be arbitrary, G must be an intersection of definable groups over $\emptyset$. Moreover, since the group multiplication for G_0 coincides with that for X in the introduction of Section 7.1, the second sentence follows. ■

As a corollary we have the definability of the type-definable supersimple division ring. We use the fact mentioned at the beginning of this section that any supersimple division ring is a field.

Corollary 7.5.5 *Let F be a type-definable division ring in a supersimple theory. Then F is a definable field.*

Proof We shall show F is definable. Clearly, we can assume F is infinite. By 7.5.4, the additive group $F^+ = \bigcap \mathcal{F}$, where $\mathcal{F} = \{F_i^+ | i < \kappa\}$ such that each $F_i^+(\geq F)$ is definable and $\mathcal{F}$ is closed under finite conjunctions. By the same argument in the proof of Claim 1 of Proposition 7.5.3, we can assume that $SU(F) = SU(F_i^+)$ and $F_i^+ \leq F_0^+$ for all $i < \kappa$ (*). Moreover, as in the introduction of Section 7.1 (basically, by compactness), we can further assume multiplication is well defined inside F_0^+, which has no zero-divisors, and the multiplicative associativity and distributive laws hold inside F_0^+, though the multiplication need not be closed there. Now by compactness we can find, say, F_1^+ such that for any $g, h \in F_1^+$ and $g.h \in F_0^+$. In particular, for any $g \in F$, $(F \subseteq)g.F_1^+ \subseteq F_0^+$. Thus, for $g, g' \in F$, $g.F_1^+$ and $g'.F_1^+$ are commensurable. Hence, due to Theorem 7.1.16, there must exist a subgroup $(F^+ \leq)H$ of F_0^+ such that $H = g.H$ for all $g \in F$. But H need not yet be closed under the multiplication. So we reduce further by putting

$$H_0 := \{h \in H : h.H \leq H\}.$$

Then for $g, h \in H_0$, we have $g.h \in g.H \leq H$ and $g.h.H \leq H$, so $g.h \in H_0$. Hence, H_0 is a definable ring containing F. Now since the theory does not have the strict order property, for any non-zero $h \in H_0$, $\langle h^n.H_0 : n < \omega\rangle$ cannot descend infinitely; i.e. for some $m \geq 0$ and all $k \geq 0$, we have $h^m.H_0 = h^{m+k}.H_0$. In particular, $h^m.(1 - h^k.g) = 0$ for some $g \in H_0$. Hence, $h^k.g = 1$ and h is invertible. Thus, H_0 indeed is a division ring. By doing the same starting with F_i^+ instead of F_0^+, we conclude F is an intersection of definable division rings, say H_i. But if $H_i \lneq H_0$ for some i, then for $a \in H_0 - H_i$, and distinct $h_1, h_2 \in H_i$, it follows that $h_1.a + H_i$ and $h_2.a + H_i$ are distinct additive cosets in H_0. Hence, $|H_0^+ : H_i^+|$ is infinite and $SU(H_i) < SU(H_0)$, contradicting (*). Therefore, $F = H_0$. ■

We move on to our second result: Any definable ω-categorical supersimple group is (finite-by-abelian)-by-finite having finite SU-rank. Note that if $SU(G) < \omega$, then by the Lascar inequalities, $SU(G^n) < \omega$ for $n < \omega$.

In the rest of this section the group G is **ω-categorical, definable** over $\emptyset$, and acl and dcl are computed in $\mathcal{M}^{eq}$. We frequently use that for a finite real tuple b, any b-invariant subset of G is b-definable. We state first a well-known fact (without a proof) by D. McPherson.

Fact 7.5.6 *[89] If the ω-categorical group G does not have the strict order property, then it has an $\emptyset$-definable nilpotent normal subgroup having finite index; i.e., G is nilpotent-by-finite.*

Of course, G may have additional definable structures more than a pure group. But clearly the same conclusion holds if we consider G as a definitional reduct, a pure group that does not have the strict order property.

Now we introduce some of preparatory notions for the second result.

Definition & Remark 7.5.7

(1) Call a subgroup H of G *minimum* if it is infinite, definable, but having no infinite definable subgroup of infinite index. If H_1 and H_2 are subgroups of minimum H having finite index in H, then so is $H_1 \cap H_2$, which must be infinite.

(2) Fix definable abelian minimum subgroups H and K of G. By a *(virtual) isogeny* from H to K, we mean a definable isomorphism $f : H_1/C \to K_1/D$, where each $|H : H_1|, |K : K_1|$ is finite, and both C and D are finite subgroups of H_1 and K_1, respectively. If $H = K$, the isogeny is called a *(virtual) endogeny* of H.

(3) We say two virtual isogenies $f_i : H_i/C_i \to K_i/D_i$ $(i = 1, 2)$ from H to K are *equivalent* if the induced definable maps

$$H_1 \cap H_2 \hookrightarrow H_i/C_i \xrightarrow{f_i} K_i/D_i \hookrightarrow (K_1 + K_2)/(D_1 + D_2)$$

agree on a definable subgroup of H having finite index. One can check that it forms an equivalence relation on the set of virtual isogenies from H to K.

(4) By naming parameters, assume H and K above are defined over $\emptyset$ and $f = f_a$ be a virtual isogeny defined over a between them. The equivalence relation among isogenies induces a *definable* (due to ω-categoricity) equivalence relation on $\mathrm{tp}(a)$, and its equivalence class $[f]$ is in G^{eq}.

(5) Let $\mathcal{R}_H$ be the set of all such equivalence classes of virtual *endogenies* of H over $\emptyset$, so $\mathcal{R}_H \subseteq G^{\mathrm{eq}}$. Addition and composition endow a canonical ring structure to $\mathcal{R}_H^0 := \mathcal{R}_H \cup \{0\}$. More precisely, given $[f_i] \in \mathcal{R}_H$ with $f_i : H_i/C_i \to K_i/D_i$ $(i = 1, 2)$, firstly, define

$$\begin{aligned} f_1 + f_2 : H_1 \cap H_2 &\to (K_1 + K_2)/(D_1 + D_2) \\ x &\mapsto f_1(x + C_1) + f_2(x + C_2) + (D_1 + D_2). \end{aligned}$$

Then the induced isomorphism

$$\overline{f_1 + f_2} : (H_1 \cap H_2)/\mathrm{Ker}(f_1 + f_2) \to \mathrm{Im}(f_1 + f_2)$$

is another virtual endogeny if $\mathrm{Ker}(f_1 + f_2)$ is finite. It is easy to see that $[f_2] = [-f_1]$ if $\mathrm{Ker}(f_1 + f_2)$ is infinite. Define $[f_1] + [f_2] := [\overline{f_1 + f_2}]$. One can check then it is well defined and so $\mathcal{R} \cup \{0\}$ forms an abelian group. Secondly, one can easily find an endogeny f_2' such that $[f_2'] = [f_2]$ and

$$\mathrm{dom}(f_2') = (H_2 + D_1)/(C_2 + D_1).$$

Now let

$$\begin{aligned} f_2.f_1 : f^{-1}(K_1 \cap (H_2 + D_1)) &\to \mathrm{Im} f_2' \\ x &\mapsto f_2' \circ f_1(x + C_1) + (C_2 + D_1). \end{aligned}$$

It is a well-defined map inducing an endogeny $\overline{f_1.f_2}$. We let $[f_1].[f_2] := [\overline{f_1.f_2}]$. Again, the multiplication is well defined and distributes over addition. Moreover, $[f_i^{-1}]$ is an inverse of $[f_i]$. Therefore, $\mathcal{R}_H^0$ forms a division ring.

(6) Let $A = \{[f_{a_1}], \ldots, [f_{a_n}]\}$ be a finite subset of $\mathcal{R}_H$. Observe from (5), to define an element in R_H generated by A, it suffices to consider definable subgroups of G over the fixed parameter $a_1, \ldots, a_n$. According to ω-categoricity, there are only finitely many such subgroups of G. Hence, A generates a finite subring that must be a finite field. Therefore, $\mathcal{R}_H^0$ is a field of positive characteristic and every element should be algebraic. Since any $\emptyset$-automorphism is a field automorphism of $\mathcal{R}_H^0$, we have $\mathcal{R}_H \subseteq \mathrm{acl}(\emptyset)$.

More generally, if $f = f_a$ is a virtual isogeny from H to K, then $[f] \in \mathrm{acl}(\emptyset)$ too. For $a' \models \mathrm{tp}(a), f' = f_{a'}$ is a virtual isogeny between H and K, too. We can similarly define $[f^{-1}].[f']$, which is in $\mathcal{R}_H$. Since $[f] \in \mathrm{dcl}([f'], [f^{-1}].[f'])$, we have $[f] \in \mathrm{acl}([f'])$. Hence, $\mathrm{tp}([f])$ must have boundedly many so finitely many realizations, and, thus, $[f] \in \mathrm{acl}(\emptyset)$.

Proposition 7.5.8 *Suppose that G is an abelian ω-categorical group with $SU(G) < \omega$, and H is a minimum subgroup of G. Then there are only finitely many commensurability classes among all the automorphic images of H.*

Proof Consider the sequence $\langle H = H_0, H_1, \ldots \rangle$ such that each H_n $(n > 0)$ is an automorphic image of H, and $H_n \cap (H_0 + \cdots + H_{n-1})$ is finite. Note that $SU(H^n) = SU(H).(n+1)$, where $H^n := H_0 + \cdots + H_n$. According to the finiteness of the rank, the sequence must stop at a finite stage, say at some maximal m. Then $\{H_i | \ i \leq m\}$ is almost linearly independent; i.e., $H_i \cap \sum_{j(\neq i) \leq m} H_j$ is finite, otherwise $SU(H^m) = SU(\sum_{j(\neq i)\leq m} H_j) < SU(H).(m+1)$, a contradiction.

Now let K be an automorphic image of H. Let

$$I_K := \left\{ i \leq m | \ H_i \cap \left(K + \sum\nolimits_{j(\neq i) \leq m} H_j \right) \text{ is infinite} \right\}.$$

Note that $I_K \neq \emptyset$ since, otherwise, $SU(H^m + K) > SU(H^m)$, contradicting the choice of m.

Claim 1 $SU(\sum_{i\in I_K} H_i) = SU(K + \sum_{i\in I_K} H_i) = SU(K + \sum_{j(\neq k)\in I_K} H_j)$ for $k \in I_K$. Hence, any proper subset of $\{K\} \cup \{H_i | \ i \in I_K\}$ is almost linearly independent. If $SU(K + \sum_{i\in I_K} H_i) > SU(\sum_{i\in I_K} H_i)$, then

$$SU\left(\left(K + \sum_{i\in I_K} H_i\right) + \sum_{j\notin I_K} H_j\right) > SU(H^m),$$

a contradiction. Moreover, if $SU(K + \sum_{i\in I_K} H_i) > SU(K + \sum_{j(\neq k)\in I_K} H_j)$, then $K \cap \sum_{j(\neq k)\in I_K} H_j$ has finite index in K, so it follows that $SU(H^m) = SU(K + \sum_{j(\neq k)\leq m} H_j)$ (since $k \in I$) $= SU(H).m$, a contradiction. Thus, Claim 1 follows.

Now for $i \neq j \in I_K$ and $x \in H_i \cap (H_j + K + D)$, where $D = D_{ij} := \sum_{l(\neq i,j)\in I_K} H_l$, define

$$f_K^{ij}(x) = H_j \cap (x + K + D) \in H_j/(H_j \cap (K + D)).$$

Claim 2 f_K^{ij} is a virtual isogeny from H_i to H_j: According to Claim 1, $H_i \cap (H_j + K + D)$ has finite index in H_i. Note that $\mathrm{Ker} f_K^{ij} = H_i \cap (K + D)$, which is finite by, again, Claim 1. Therefore, f_K^{ij} induces a virtual isogeny.

Claim 3 Let F be another automorphic image of H. Assume that $I_K = I_F \overset{\text{let}}{=} I$, and for a fixed $i \in I$ and each $j(\neq i) \in I$, f_K^{ij} and f_F^{ij} are equivalent. Then K and F are commensurable. Due to the assumption, there exist definable subgroups B of H_i of finite index, and finite subgroups $C_j(\geq f_K^{ij}(0) + f_F^{ij}(0))$ of H_j for $j(\neq i) \in I$ such that for $x \in B$, $f_K^{ij}(x) + C_j = f_F^{ij}(x) + C_j$. Let

$$K_i := \left\{ x - \sum_{j(\neq i)\in I} y_j \,\middle|\, x \in B, \text{ choose one } y_j \in f_K^{ij}(x) \text{ for } j(\neq i) \in I \right\}.$$

Similarly define F_i where $f_K^{ij}(x)$ is replaced by $f_F^{ij}(x)$. Note that $f_K^{ij}(x) = \{y \in H_j | x - y \in K + D\}$. Hence, it follows K_i, F_i are definable subgroups of G. Moreover, given typical $g = x - \sum_{l(\neq i)\in I} y_l \in K_i$, we have $x - y_j = g + \sum_{l(\neq i,j)\in I} y_l = k' + d_j \in K + D_{ij}$, where $k' \in K$, $d_j \in D_{ij}$. Hence, $K_i \leq \bigcap_{j(\neq i)\in I} K + D_{ij} \leq K + \sum_{j(\neq i,l)\in I}(H_j \cap (K + \sum_{k(\neq i,j)\in I} H_k))$ for any $l \neq i \in I$. Since the summand is finite by Claim 1, and

$$\left\{k \in K |\, k + \sum\nolimits_{j(\neq i)\in I} y_j \in B \text{ for some } y_j \in H_j\right\} (\leq K_i)$$

has finite index in K, we have that K and K_i are commensurable, so are F and F_i. Lastly, F_i and K_i are equal modulo the finite group $\sum_{j(\neq i)\in I} C_j$. Hence, they are commensurable, and Claim 3 is proved.

Applying 7.5.7(6) to Claim 2, we see that there are only finitely many isogenies of the form $[f_K^{ij}]$ from H_i to H_j. Hence, Claim 3 yields that there are only finitely many commensurability classes among all the automorphic images of H. ■

The following lemma says if G is (finite-by-abelian)-by-finite then it is witnessed in an $\emptyset$-definable manner.

Lemma 7.5.9 (G ω-categorical over $\emptyset$) *If G is (finite-by-abelian)-by-finite, then it is witnessed by $\emptyset$-definable $N \lhd G$. Namely, there is such N of finite index with finite $N^{(1)} \subseteq \mathrm{acl}(\emptyset)$.*

Proof Choose $N \lhd G$ of the least finite index having finite $N^{(1)}$. Note that $N^{(1)} \lhd G$ too by 7.3.8(2). Now for any automorphic image K of N, we have $K.N = N.K \lhd G$ and finite $N^{(1)}.K^{(1)} = K^{(1)}.N^{(1)} \lhd G$. Then it follows that $(K \cap N).N^{(1)}.K^{(1)}/N^{(1)}.K^{(1)}$ is central in $N.K/N^{(1)}.K^{(1)}$. Note that since $K/K \cap N \cong K.N/N$, clearly $|K.N : K \cap N|$ is finite. Hence, as pointed out in 7.3.8(4), $(N.K/N^{(1)}.K^{(1)})^{(1)}$ is finite and so is $(N.K)^{(1)}$; i.e., $N.K$ is another normal subgroup of G having finite index with finitely many commutators. By the choice of N, N and K must be equal; thus, due to ω-categoricity, N is $\emptyset$-definable and the finite $N^{(1)} \subseteq \mathrm{acl}(\emptyset)$. ∎

Corollary 7.5.10 *Suppose that G is a (finite-by-abelian)-by-finite ω-categorical group with $SU(G) < \omega$. Then every definable subgroup H of G is commensurable with some definable subgroup over* $\mathrm{acl}(\emptyset)$. *(Hence, there are only finitely many commensurability classes among all the automorphic images of H.)*

Proof The last sentence in the parentheses clearly follows from the second sentence. Let us first prove it when G is abelian. Use induction on the rank of G. If $SU(G) = 0$, so G is finite; nothing to prove. Thus, we can assume that $SU(G) > 0$, and H is infinite. Choose K some infinite definable subgroup of H having the least positive SU-rank. Clearly, K is a minimum subgroup of G. According to 7.5.8 (G abelian) and 7.3.6, its locally connected companion K^c must be definable over $\mathrm{acl}(\emptyset)$. Then $H.K^c/K^c \leq G/K^c$ (both are groups since G abelian) and $SU(G/K^c) < SU(G)$. By the induction hypothesis, $H.K^c/K^c$ is commensurable with some group F/K^c, where $(K^c \leq)F(\leq G)$ is $\mathrm{acl}(\emptyset)$-definable. Since K, K^c are commensurable, so are H, HK^c, F.

Now, for (finite-by-abelian)-by-finite G, by 7.5.9, there is $\emptyset$-definable normal N of finite index with finite $N^{(1)} \subseteq \mathrm{acl}(\emptyset)$. If definable subgroups H_i $(i = 1, 2)$ of G are not commensurable, then since $|G : N|$ is finite, neither are $(H_i \cap N).N^{(1)}/N^{(1)}$ in abelian $N/N^{(1)}$. Therefore, our corollary must follow. ∎

Recall that the *descending central series* of G is

$$\ldots \lhd \gamma_1(G) \lhd \gamma_0(G) = G,$$

where $\gamma_{i+1}(G) = [\gamma_i(G), G]$, in particular $\gamma_1(G) = G^{(1)}$. Note that each $\gamma_n(G) \lhd G$ by 7.3.8(2); and $\gamma_{n-1}(G)/\gamma_n(G) \leq Z(G/\gamma_n(G))$.

Theorem 7.5.11 *If G is an ω-categorical group with $SU(G) < \omega$, then G is (finite-by-abelian)-by-finite.*

Proof Assume that there is a counterexample G. We shall lead a contradiction. Note that $G^0_{\emptyset}$, an $\emptyset$-definable subgroup of G, is also a counterexample. By replacing G by this, we can assume G does not have a proper $\emptyset$-definable subgroup of finite index (*). Recall from 7.5.6, G is nilpotent, and each subgroup $G_n := \gamma_n(G)(\lhd G)$ in the descending central series is $\emptyset$-definable. Hence, both $G^{(1)}(\lneq G)$ and $G/G^{(1)}$ must be infinite, and there is the least $k \geq 2$ such that $G^{(1)}/G_k$ is infinite (**).

Claim 1 $H := G/G_k$ is another counterexample: Clearly, H (in G^{eq}) is ω-categorical having finite SU-rank by 7.5.1. Suppose not Claim 1. Then from 7.5.9, there is $\emptyset$-definable $N(\lhd G)$ containing G_k such that $N/G_k(\lhd H)$ has a finite index in H and $(N/G_k)^{(1)} = N^{(1)}.G_k/G_k(\lhd H)$ is finite. Hence, N has a finite index in G too.

According to (*), $N = G$ and $N^{(1)}.G_k/G_k = G/G_k$ is finite. It contradicts (**). Hence, Claim 1 is shown.

Claim 2 $H^{(1)}/H^{(1)} \cap Z(H)$ is finite. It follows that both $K := H/Z(H)$ and $F := K \times Z(H)$ are finite-by-abelian: From (**), $G^{(1)}/G_{k-1}$ is finite, and $G_{k-1}/G_k \subseteq Z(G/G_k) = Z(H)$. Hence,

$$|H^{(1)}/H^{(1)} \cap Z(H)| \left(\leq |(G^{(1)}/G_k)/(G_{k-1}/G_k)| = |G^{(1)}/G_{k-1}|\right)$$

is finite. Now, $|K^{(1)}| = |H^{(1)}.Z(H)/Z(H)| = |H^{(1)}/H^{(1)} \cap Z(H)|$, and since $Z(H)$ is abelian, $|F^{(1)}| = |K^{(1)}| < \omega$. Therefore, Claim 2 is proved.

For $h \in H$, write $h_Z := h.Z(H) \in K$. Let $K_h := C_K(h_Z)(\leq K)$. Note that for $g, g', h' \in H$ with $g_Z = g'_Z$, $h_Z = h'_Z$, clearly we have that $[g,h] = [g',h']$, and that $g_Z \in K_h$ iff $[g,h] = [g',h'] \in Z(H)$. Hence, it easily follows that $F_h := \{(g_Z, [g,h])| \, g_Z \in K_h\}$, the graph of the assignment, is a well-defined subgroup of F. Thus, from 7.5.10 with Claim 2, there are only finitely many commensurability classes in $\{F_h| \, h \in H\}$. Then by the extension axiom, we can clearly find generics $h_0, h_1 \in H$ with $h_0 \mathop{\smile\!\!\!\!\!\mid} h_1$ such that F_{h_0} and F_{h_1} are commensurable. Then $\mathrm{Pr}_1(F_{h_0} \cap F_{h_1})(\subseteq \mathrm{Pr}_1(F_{h_0}) \cap \mathrm{Pr}_1(F_{h_1}) = K_{h_0} \cap K_{h_1})$ must be generic in $K_{h_0} \cap K_{h_1}$. Note now by Claim 2 with 7.3.8(4), for $h \in H$, $|K : K_h| = |[K, h_Z]| \leq |K^{(1)}|$ is finite. It says then $\mathrm{Pr}_1(F_{h_0} \cap F_{h_1})$ is generic in K, too. Thus, there is generic $g \in H$ over h_0h_1 such that $g_Z \in \mathrm{Pr}_1(F_{h_0} \cap F_{h_1})$, and so $[g, h_0] = [g, h_1]$, $[g, h_0.h_1^{-1}] = 1$. Hence, if $h' \equiv_g h_0.h_1^{-1}$, then $[g, h'] = 1$ and $h' \in C_H(g)$. Since $h_0.h_1^{-1}$ is generic of H over g, it means $C_H(g)$ must be generic in H, yielding $g \in \tilde{Z}(H)$. This time g is generic, so $|H : \tilde{Z}(H)|$ is finite. Recall that $\tilde{Z}(H)$ is finite-by-abelian by 7.3.10(2). Hence, H is (finite-by-abelian)-by-finite, and it violates Claim 1. We have completed the proof. ■

Theorem 7.5.12 *Let G be an ω-categorical supersimple group. Then $SU(G)$ is finite.*

Proof Assume not. From the remark before 2.5.19 and supersimplicity, there is $p \in S_G(A)$ for some finite real A such that $SU(p) = \omega$. By naming A (while preserving ω-categoricity), we assume $A = \emptyset$. Since $S_G(\emptyset)$ is finite, there is

$$n := \max\left\{SU(g)| \, g \in G \text{ and } SU(g) \text{ is finite}\right\}.$$

Now, there is a forking extension q of p such that $SU(q)$ is finite $> n$. For a Morley sequence $\langle g_i| \, i < \omega\rangle$ of q, due to 2.4.2, there must be $k \geq 1$ such that $\bar{g}_0 = \{g_i| \, i < k\}$ is independent, whereas $\bar{g}_0 \not\mathop{\smile\!\!\!\!\!\mid} g_k$. Note that

$$\omega = SU(p) = SU(g_k) > SU(g_k/\bar{g}_0) \geq SU(q) > n. \ (*)$$

Claim There is an indiscernible sequence $\langle \bar{g}_i|i < \omega\rangle$ in $\mathrm{tp}(\bar{g}_0)$ such that $\bar{g}_0 \mathop{\smile\!\!\!\!\!\mid} \bar{g}_1$ and for some m, $SU(\bar{g}_m/\bar{g}_0 \ldots \bar{g}_{m-1}) < \omega$. Due to ω-categoricity, there is

$$n' := \max\left\{SU(g/h_1 \ldots h_{2k-1})| \, g \models p, \text{each } h_i \in G, \text{ and } SU(g/h_1 \ldots h_{2k-1}) < \omega\right\}.$$

Now since each $g_i \models p$, there is a parameter A_i such that $n' < SU(g_i/A_i) < \omega$. Moreover, since $\bar{g}_0 = \{g_0, \ldots, g_{k-1}\}$ is independent, due to extension and other basic non-forking axioms, we can additionally assume that $\{g_iA_i | \ i < k\}$ is independent (**). By the Lascar inequality, $SU(\bar{g}_0/A)$ is finite, where $A = \bigcup\{A_i | \ i < k\}$. Hence, for a Morley sequence $\langle \bar{g}_i | i < \omega \rangle$ of $\mathrm{tp}(\bar{g}_0/A)$, again from 2.4.2 and supersimplicity, there is some m such that $SU(\bar{g}_m/\bar{g}_0 \ldots \bar{g}_{m-1}) = SU(\bar{g}_0/A) < \omega$. It remains to show $\bar{g}_0 \downarrow \bar{g}_1$. From the choice of n', it suffices to prove for $j < k$, $SU(g_j/\bar{g}_0^j\bar{g}_1) > n'$, where $\bar{g}_0^j = \bar{g}_0 - \{g_j\}$ (so that $\omega = SU(g_j/\bar{g}_0^j\bar{g}_1) = SU(g_j) = SU(p)$ and $g_j \downarrow \bar{g}_0^j\bar{g}_1$). In fact, since $\bar{g}_0 \downarrow_A \bar{g}_1$ and $g_j \downarrow_{A_j} A\bar{g}_0^j$ from (**), we have that $SU(g_j/\bar{g}_0^j\bar{g}_1) \geq SU(g_j/A\bar{g}_0^j\bar{g}_1) = SU(g_j/A\bar{g}_0^j) = SU(g_j/A_j) > n'$. The claim is verified.

We now let

$$F := \{h \in G | \ SU(h/\bar{g}_0 \ldots \bar{g}_{m-1}) < \omega\}, \text{ and } H := \{h \in G | \ SU(h/\bar{g}_m) < \omega\}.$$

Then it follows that $H \subseteq F$ since for $h \in H$, due to the claim and the Lascar inequality, we have $SU(h\bar{g}_m/\bar{g}_0 \ldots \bar{g}_{m-1}) = SU(h/\bar{g}_0 \ldots \bar{g}_m) + SU(\bar{g}_m/\bar{g}_0 \ldots \bar{g}_{m-1}) = SU(\bar{g}_m/h\bar{g}_0 \ldots \bar{g}_{m-1}) + SU(h/\bar{g}_0 \ldots \bar{g}_{m-1}) < \omega$. Indeed, again by the Lascar inequalities with ω-categoricity, both $H \leq F$ are definable subgroups of G having finite SU ranks. Hence, from 7.5.11, both are (finite-by-abelian)-by-finite. Then from 7.5.10 and the claim, there are $m_1 > m_2 \geq m$ such that $H_1 = f_{m_1}(H), H_2 = f_{m_2}(H)$ are commensurable, where f_{m_i} is an automorphism fixing $\bar{g}_0 \ldots \bar{g}_{m-1}$ (so fixing the group F setwise) and sending $\bar{g}_m$ to $\bar{g}_{m_i}$. Hence, H_1 and H_2 have the same locally connected companion, say $H^c(\leq G)$. Moreover, as each $H_i(\leq G)$ is definable over $\bar{g}_{m_i}$, H^c has the canonical parameter v in $\mathrm{dcl}(\bar{g}_{m_1}) \cap \mathrm{dcl}(\bar{g}_{m_2})$. Again due to the claim, $\bar{g}_{m_1} \downarrow \bar{g}_{m_2}$ and so $v \downarrow v$ and $v \in \mathrm{acl}(\emptyset)$. Therefore, for a generic element g' of H^c over v, we have from (*)

$$\omega > SU(g') = SU(g'/v) = SU(H^c) = SU(H) > n.$$

It contradicts the choice of n. ■

Corollary 7.5.13 *Any ω-categorical supersimple group is (finite-by-abelian)-by-finite, and has finite SU-rank.*

Example 7.4.4 says Corollary 7.5.13 is a maximal result.

7.6 Generically Given Partial Groups and Homogeneous Spaces

This section shows techniques how to obtain a full hyperdefinable group or a homogeneous space from data supplying a *partial* hyperdefinable multiplication or an action only between *independent* pairs of elements in given domains. Such data are called a *generically given (hyperdefinable partial)* group or homogeneous space. The main point is how to extend the partial map to all non-independent pairs of elements. These techniques will be crucially used in the group configuration theorem in Chapter 9.

Throughout this section by omitting or suppressing parameters we assume all sets and relations are (hyper)defined over $\emptyset$, and we use the same notation for a hyperdefinable type and its set of realizations. The proofs will only use the symmetry, transitivity, and extension of non-forking independence; the results can possibly be extended to non-simple context.

We first state a result for a generically given group case.

Theorem 7.6.1 *Suppose that there is a hyperdefinable function* $. : \Gamma'(x,y) \to H(z)$, *where* Γ' *and* H *are hyperdefinable and* Γ' *contains the set*

$$\Gamma = \left\{(x,y) : H(x) \wedge H(y) \wedge x \perp y\right\},$$

such that

(1) (Generic Independence) *For* $(\alpha,\beta) \in \Gamma$, $\alpha.\beta$ *is independent from* α, *and* β, *respectively;*

(2) (Generic Associativity) *For independent* $\{\alpha,\beta,\gamma\} \subset H, \alpha.(\beta.\gamma) = (\alpha.\beta).\gamma$; *and*

(3) (Generic Surjectivity) *For* $(\alpha,\beta) \in \Gamma$, *there are* $\gamma, \gamma' \in H$ *each independent from* α *such that* $\gamma.\alpha = \beta$ *and* $\alpha.\gamma' = \beta$.

Now let $\sim$ *be a binary symmetric relation on* H^2 *such that*

$$(\alpha_1,\beta_1) \sim (\alpha_2,\beta_2) \text{ if there are } \gamma_1,\gamma_2 \in H \text{ such that}$$
$$\gamma_i \perp \alpha_1\alpha_2\beta_1\beta_2, \text{ and } \gamma_1.\alpha_1 = \gamma_2.\alpha_2,\ \gamma_1.\beta_1 = \gamma_2.\beta_2.$$

Then $\sim$ *is a hyperdefinable equivalence relation on* H^2. *Let* $G = H^2/\sim$ *and write* $[\alpha,\beta] = (\alpha,\beta)/\sim \in G$.

(i) For $\gamma(\in H) \perp \alpha\beta$, *we have* $[\alpha,\beta] = [\gamma.\alpha,\gamma.\beta]$, *and there is* $\delta_i(\perp \alpha,\beta) \in H$ *such that* $[\alpha,\beta] = [\gamma,\delta_1] = [\delta_2,\gamma]$.

(ii) Hence, it makes sense to define an operation $\circ$ *on* G *such that* $[\alpha,\beta] \circ [\beta,\gamma] = [\alpha,\gamma]$. *It is well defined, and* $(G,\circ)$ *forms a hyperdefinable group with the identity* $[\alpha,\alpha]$ *for any* $\alpha \in H$.

(iii) An hyperdefinable map $\iota : \alpha(\in H) \mapsto [\beta,\beta.\alpha](\in G)$ *for any* $\beta \perp \alpha$ *is a well-defined embedding such that for* $(\alpha,\beta) \in \Gamma$, $\iota(\alpha.\beta) = \iota(\alpha) \circ \iota(\beta)$ *holds, and* $H' := \iota(H)$ *is the set of all generic elements in* G. *For* $g \in G$, *there is* $\alpha_i(\perp g) \in H$ *such that* $g = \iota(\alpha_1) \circ \iota(\alpha_2)$.

(iv) Suppose there are a group $(G',\circ')$ *and an embedding* $\iota' : H \to G'$ *both hyperdefinable such that for* $(\alpha,\beta) \in \Gamma$, $\iota'(\alpha.\beta) = \iota'(\alpha) \circ' \iota'(\beta)$, *and* $\iota'(H)$ *is the set of all generic elements in* G'. *Then there is a hyperdefinable isomorphism from* G *to* G' *preserving the set of all generic elements.*

Proof Fix a complete type $p(x) \models H(x)$. Then as in 4.2.13(1), we have that $p(x) \wedge x \perp y$ is hyperdefinable (†).

Claim 1 Let $\Psi(y) := \exists\gamma \in H(\gamma \downarrow y_0 \wedge \Phi(y,\gamma))$, where $y_0 \subseteq y$ and $\Phi(y,x)$ is a partial type. Then $\Psi(y)$ is hyperdefined by $\exists\gamma' \in H, \exists\alpha \models p\ (\alpha \downarrow \gamma' y_0 \wedge \Phi(y,\gamma'.\alpha))$: Assume for $\gamma \in H$ and y, we have $\gamma \downarrow y_0 \wedge \Phi(y,\gamma)$. Now choose $\alpha(\models p) \downarrow \gamma y$. Then by (3), there is $\gamma'(\downarrow \alpha; \downarrow \gamma; \downarrow_{\alpha,\gamma} y_0) \in H$ such that $\gamma = \gamma'.\alpha$. Then we have $\gamma'\alpha \downarrow_\gamma y_0$, $\gamma'\alpha\gamma \downarrow y_0$, and $\alpha \downarrow \gamma' y_0$. Hence, $\Psi(y)$ is hyperdefined as suggested. To see the converse take $\gamma = \gamma'.\alpha$ and show that $\alpha \downarrow \gamma' y_0$ implies that $\gamma \downarrow y_0$. But it follows from (1).

Claim 2 $\sim$ is hyperdefinable and forms an equivalence relation: Iterative applications of Claim 1 imply that $\sim$ is hyperdefinable. Now suppose that $(\alpha_1,\beta_1) \sim (\alpha_2,\beta_2)$ witnessed $\gamma_i \in H$ $(i = 1,2)$, and that $(\alpha_2,\beta_2) \sim (\alpha_3,\beta_3)$ by $\delta_j \in H$. We want to see that $(\alpha_1,\beta_1) \sim (\alpha_3,\beta_3)$. Let A be the set of all mentioned elements. Take $\beta \in H$ independent from A. Then $\beta \downarrow_{\gamma_2} A$, $\beta.\gamma_2 \downarrow_{\gamma_2} A$ and, by generic independence, $\beta.\gamma_2 \downarrow A$ hold. Thus, by generic surjectivity, there is β' such that $\beta'.\delta_2 = \beta.\gamma_2$, $\beta' \downarrow_{\beta.\gamma_2,\delta_2} A$, and $\beta' \downarrow \delta_2$ hold. Hence, $\beta' \downarrow A$ easily follows. Now by generic associativity,

$$(\beta.\gamma_1).\beta_1 = \beta.(\gamma_1.\beta_1) = \beta.(\gamma_2.\beta_2) = (\beta.\gamma_2).\beta_2 = (\beta'.\delta_2).\beta_2 = \beta'.(\delta_3.\alpha_3) = (\beta'.\delta_3).\beta_3.$$

Similarly, $(\beta.\gamma_1).\alpha_1 = (\beta'.\delta_3).\alpha_3$. Moreover, again by generic independence, each of $\beta.\gamma_1, \beta'.\delta_3$ is independent from $\alpha_1\beta_1\alpha_3\beta_3$. Hence, $(\alpha_1,\beta_1) \sim (\alpha_3,\beta_3)$, and Claim 2 is proved.

Now we shall prove (i) to (iv).

Proof of (i): We use the same symbols in (i). Choose $\delta \downarrow \gamma\alpha\beta$. Then by (1) and (2), we have $\delta.\gamma \downarrow \gamma\alpha\beta$ and $\delta.(\gamma.\alpha) = (\delta.\gamma).\alpha$, $\delta.(\gamma.\beta) = (\delta.\gamma).\beta$. Therefore, $[\gamma.\alpha,\gamma.\beta] = [\alpha,\beta]$ (*). Now by generic surjectivity, there is δ' such that $\delta'.\alpha = \gamma$, $\delta' \downarrow \alpha$ and $\delta' \downarrow_{\gamma\alpha} \beta$. Hence, $\delta'\gamma \downarrow_\alpha \beta$ and $\delta' \downarrow \alpha\beta$ hold. Thus, by (*), $[\alpha,\beta] = [\delta'.\alpha,\delta'.\beta] = [\gamma,\delta_1]$ with $\delta_1 = \delta'.\beta(\downarrow \alpha\beta$ from (1)). Similarly, δ_2 exits, and (i) is proved.

We need the following before proving (ii).

Claim 3 If $\alpha_1.\beta = \alpha_2.\beta$ for some $\alpha_i \downarrow \beta \in H$, then $\alpha_1.\gamma = \alpha_2.\gamma$ for any $\gamma(\downarrow \alpha_i) \in H$: Note that it suffices to show when $\gamma \downarrow \alpha_i\beta$ $(i = 1,2)$, since then Claim 3 holds for $\gamma'(\downarrow \alpha_1\alpha_2\beta\gamma)$ and $\gamma \downarrow \alpha_i\gamma'$. Hence, assume $\gamma \downarrow \alpha_i\beta$. By (3), there is $\sigma(\downarrow \gamma; \downarrow \beta; \downarrow_{\gamma\beta} \alpha_1\alpha_2) \in H$ such that $\beta.\sigma = \gamma$. Then $\{\alpha_i,\sigma,\beta\}$ independent, and by (2) $\alpha_1.\gamma = \alpha_1.(\beta.\sigma) = (\alpha_1.\beta).\sigma = \alpha_2.(\beta.\sigma) = \alpha_2.\gamma$. Claim 3 is proved.

Proof of (ii): By (i), the operation is defined for any two elements of G. To check well-definedness, suppose that $[\alpha,\beta] = [\alpha',\beta']$ witnessed by γ,γ', and $[\beta,\epsilon] = [\beta',\epsilon']$ by δ,δ'. We shall show that $[\alpha,\epsilon] = [\alpha',\epsilon']$. In A, the set of all mentioned elements, we can assume that $\delta\delta' \downarrow_{\beta\beta'\epsilon\epsilon'} \alpha\alpha'\gamma\gamma'$. Hence, $\delta \downarrow A - \{\delta,\delta'\}$, and $\delta' \downarrow A - \{\delta,\delta'\}$. Due to generic surjectivity, there is $\sigma(\downarrow \gamma; \downarrow \delta; \downarrow_{\delta\gamma} A) \in H$ such that $\sigma.\gamma = \delta$. It then follows that each of $\{\sigma,\gamma,\beta\}, \{\sigma,\gamma',\beta'\}$, and $\{\sigma,\gamma,\alpha\}, \{\sigma,\gamma',\alpha'\}$ is independent. Thus, by (2),

$$\delta'.\beta' = \delta.\beta = (\sigma.\gamma).\beta = \sigma.(\gamma.\beta) = \sigma.(\gamma'.\beta') = (\sigma.\gamma').\beta'.$$

Then since $\delta' \perp \alpha'$ and $\sigma\gamma' \perp \alpha'$, by Claim 3 and (2),

$$\delta'.\alpha' = (\sigma.\gamma').\alpha' = \sigma.(\gamma'.\alpha') = \sigma.(\gamma.\alpha) = \delta.\alpha.$$

Hence, from (i), $[\alpha', \epsilon'] = [\delta'.\alpha', \delta'.\epsilon'] = [\delta.\alpha, \delta.\epsilon] = [\alpha, \epsilon]$.

Now, for any $\alpha, \beta \in H$, find $\gamma(\perp \alpha\beta)$ and $\sigma(\perp \alpha), \sigma'(\perp \beta) \in H$ so that $\alpha.\sigma = \gamma$, $\beta.\sigma' = \gamma$, and $[\alpha, \alpha] = [\alpha.\sigma, \alpha.\sigma] = [\gamma, \gamma] = [\beta.\sigma', \beta.\sigma'] = [\beta, \beta]$. Hence, any $[\alpha, \alpha]$ serves the identity, and the inverse of $[\alpha, \beta]$ is $[\beta, \alpha]$.

It remains to verify the associativity of the operation. Given $[\alpha, \alpha'], [\beta, \beta'], [\gamma, \gamma'] \in G$, we can replace $[\beta, \beta']$ by $[\sigma.\beta, \sigma.\beta']$ for some (any) $\sigma \perp \alpha\alpha'\beta\beta'\gamma\gamma'$. Thus, we can assume each of β, β' is independent from $\alpha\alpha'\gamma\gamma'$. Then again by (i), there are δ, δ' such that $[\alpha, \alpha'] = [\delta, \beta]$ and $[\gamma, \gamma'] = [\beta', \delta']$. Therefore,

$$\begin{aligned}([\alpha, \alpha'] \circ [\beta, \beta']) \circ [\gamma, \gamma'] &= [\delta, \beta'] \circ [\beta', \delta'] = [\delta, \delta'] = [\delta, \beta] \circ [\beta, \delta'] \\ &= [\alpha, \alpha'] \circ ([\beta, \beta'] \circ [\beta', \delta']) = [\alpha, \alpha'] \circ ([\beta, \beta'] \circ [\gamma, \gamma']).\end{aligned}$$

We have shown (ii).

Claim 4 For any $g \in G$, there is $\alpha_i \perp g$ such that $g = [\alpha_1, \alpha_2]$. For any $\alpha \perp \beta$, we have $[\alpha, \beta] \in H'$: If $g = [\alpha'_1, \alpha'_2]$ then select $\gamma' \perp g\alpha'_1\alpha'_2$ so that by (i), $g = [\gamma'.\alpha'_1, \gamma.\alpha'_2]$ and by (1), $\gamma'.\alpha'_i \perp g$. Secondly, $[\alpha, \beta] = [\alpha, \alpha.\gamma] = \iota(\gamma)$ for some $\gamma \perp \alpha$ and $\alpha.\gamma = \beta$.

Proof of (iii): We first show that for any $\gamma, \beta \perp \alpha$, $[\beta, \beta.\alpha] = [\gamma, \gamma.\alpha]$. (Then ι is hyperdefinable due to (+) above.) Take $\delta \perp \alpha, \beta, \gamma$. Now there are β', γ' such that $\gamma'.\gamma = \delta = \beta'.\beta$, and each of $\{\gamma', \gamma, \alpha\}$, $\{\beta', \beta, \alpha\}$ is independent. Thus, by (i) and (2), $[\gamma, \gamma.\alpha] = [\gamma'.\gamma, \gamma'.\gamma.\alpha] = [\beta'.\beta, \beta'.\beta.\alpha] = [\beta, \beta.\alpha]$.

Secondly, to show injectivity of ι, assume that $\iota(\alpha) = [\gamma, \gamma.\alpha] = [\gamma, \gamma.\alpha'] = \iota(\alpha')$ for some $\gamma \perp \alpha\alpha'$. By definition there is $\delta_i \perp \gamma\alpha\alpha'$ such that $\delta_1.\gamma = \delta_2.\gamma$ and $\delta_1.\gamma.\alpha = \delta_2.\gamma.\alpha'$; i.e., $\beta.\alpha = \beta.\alpha'$ for $\beta = \delta_1.\gamma \perp \gamma\alpha\alpha'$. Now take $\sigma \perp \beta\gamma\alpha\alpha'$. By generic surjectivity, there are ϵ, ϵ' such that $\sigma.\epsilon = \alpha$ and $\sigma.\epsilon' = \alpha'$, and each of $\{\beta, \sigma, \epsilon\}, \{\beta, \sigma, \epsilon'\}$ is independent. Let $\delta = \beta.\sigma$. Then $\delta.\epsilon = \beta.\alpha = \beta.\alpha' = \delta.\epsilon'$, and each $\{\delta, \sigma, \epsilon\}, \{\delta, \sigma, \epsilon'\}$ is independent. Again by surjectivity there is σ' such that $\sigma'.\delta = \sigma$ and $\sigma' \perp_{\delta\sigma} \epsilon\epsilon'$. Hence, each of $\{\sigma', \delta, \epsilon\}, \{\sigma', \delta, \epsilon'\}$ is independent, and $\alpha = \sigma.\epsilon = \sigma'.\delta.\epsilon = \sigma'.\delta.\epsilon' = \sigma.\epsilon' = \alpha'$.

Thirdly, for any $\alpha \perp \beta$ in H, choose $\gamma \perp \alpha\beta$, so that we have

$$\iota(\alpha.\beta) = [\gamma, \gamma.\alpha.\beta] = [\gamma, \gamma.\alpha] \circ [\gamma.\alpha, \gamma.\alpha.\beta] = \iota(\alpha) \circ \iota(\beta).$$

Fourthly, given arbitrary $h(= \iota(\alpha) \in H') \perp g = [\beta, \gamma] \in G$, we shall show that $g \circ h \in H'$, $g \circ h \perp g$ to conclude h is generic in G (**). By extension, we can assume

$\alpha \perp \beta\gamma$. Hence, $h = [\gamma, \gamma.\alpha]$, and $\gamma.\alpha \perp \beta\gamma$. Then by (3), there is $\delta(\perp \beta; \perp_{\gamma.\alpha,\beta}\gamma)$ such that $\beta.\delta = \gamma.\alpha$. Thus, $g \circ h = [\beta, \gamma.\alpha] = [\beta, \beta.\delta] = \iota(\delta) \in H'$, and $\delta \perp_\beta \gamma$; $\delta \perp [\beta, \gamma]$ hold. Therefore, $g \circ h \perp g$. Conversely, suppose that generic $h' = [\alpha', \beta'] \in G$ with $h' \perp \alpha'$ (Claim 4) is given (to be seen $h' \in H'$). Select $\gamma' \perp \alpha'\beta'h'$. Then $[\alpha', \gamma'] \perp h'$, and we have $[\alpha', \gamma'] \perp [\gamma', \beta'] = [\gamma', \alpha'] \circ h'$ since h' is generic. By Claim 4, $[\gamma', \beta'] \in H'$. Hence, from (**), $h' = [\alpha', \gamma'] \circ [\gamma', \beta'] \in H'$. We have proved that H' is the set of all generic elements of G. Then the last assertion of (iii) comes directly. Namely, for some generic $h(\in H')$ independent from $g(\in G)$, $h^{-1} \circ g(\perp g)$ is generic, so in H'. The proof of (iii) is finished.

Proof of (iv): We sketch the proof. First, take $\tau_{H'}$ as the composition of ι' and ι^{-1}. We want to extend its domain to G. For $g \in G$, as noted in (iii), there is generic $h_i(\perp g)$ such that $g = h_1 \circ h_2$. Then let $\tau_G(g) = \tau_{H'}(h_1) \circ' \tau_{H'}(h_2)$. It is not hard to show that τ_G is the desired, well-defined hyperdefinable isomorphism extending $\tau_{H'}$. For example, to check well-definedness, assume there is another generic $h_i'(\perp g)$ such that $g = h_1' \circ h_2'$. Due to extension, it suffices to verify when $h_1h_2 \perp_g h_1'h_2'$ (so that $h_1h_2 \perp h_1'$). Then $h = h_1^{-1} \circ h_1'(\perp h_1)$ is generic and $h_2 = h \circ h_2'$. Moreover, $h_2 \perp_{h_1} h_1'$, and $h_2 \perp h$. Since h_2 is generic, we have $h_2'(= h^{-1} \circ h_2) \perp h$. Thus, $\tau_G(h_1 \circ h_2) = \tau_{H'}(h_1) \circ' \tau_{H'}(h \circ h_2') = \tau_{H'}(h_1) \circ' \tau_{H'}(h) \circ' \tau_{H'}(h_2') = \tau_{H'}(h_1') \circ' \tau_{H'}(h_2') = \tau_G(h_1' \circ h_2')$. The similar arguments show the rest, and we leave the detail to the reader. ■

Now before stating the case of a generically given partial homogeneous space, we will clarify some of the terminologies.

Let $((G, \circ), X, \bullet)$ be (an $\emptyset$-)hyperdefinable homogeneous space (i.e., the hyperdefinable action $\bullet$ of the hyperdefinable group G on hyperdefinable X is transitive). Recall that it is said to be *principal* if the action is regular (i.e., for $x, y \in X$ there is unique $g \in G$ with $g \bullet x = y$). We say $a \in X$ is *generic*, if for $g \in G$ with $g \perp a$, $g \bullet a \perp g$ holds. Similarly to the group case, if $x(\in X)$ is independent with generic $f \in G$, then $y = f \bullet x$ is generic. Let $g(\in G) \perp y$. By extension we can assume that $g \perp_y fx$, so $\{g, f, x\}$ is independent. Since f is generic, $g \circ f \perp g$. Thus, $\{g \circ f, g, x\}$ is independent and $g \bullet y(= (g \circ f) \bullet x) \perp g$.

Hence, a generic element of X exists. Moreover, generic $f(\in G)$ is generic with respect to X as well. Namely, for $y(\in X) \perp f$, we have $y \perp f \bullet y$: Let $y' = f \bullet y$. Take $z(\in X) \perp y$. Then by the transitivity of the action, there is $g \in G$ such that $g \bullet z = y$. By extension, we can assume $f \perp ygz$. Hence, $f \circ g \perp_g y$, and since f is generic, $f \circ g \perp g, y$ and $f \circ g \perp z(= g^{-1} \bullet y), y$ hold. Thus, $f \circ g, z \perp y$ and since $y' = (f \circ g) \bullet z$, it follows that $y \perp y'$.

We are ready to state an extension of 7.6.1 to the homogeneous space case. In the following theorem and its proof, we take the same notation as in 7.6.1. We shall use the same symbol . for the partial multiplication and the action. This will not cause confusion since, unless stated otherwise, lower-case Greek letters and alphabets will represent members in H and X, respectively.

Theorem 7.6.2 *Suppose that there is a hyperdefinable function* $. : \Gamma' \to H$ *as in Theorem 7.6.1 with the same* Γ, H *satisfying (1),(2),(3). Also there are a hyperdefinable set* X *and* *a hyperdefinable map* $. : \Sigma' \to X$ *such that* Σ' *contains a set* $\Sigma = \{(x,y) : H(x) \wedge X(y) \wedge x \perp y\}$, *and*

- (Generic Independence) *For* $\alpha a \in \Sigma$, $\alpha.a \in X$ *and is independent from* a, *and* α *respectively;*
- (Generic Associativity) *For independent* $\alpha, \beta \in H$ *and* $a(\perp \alpha\beta) \in X$, $\alpha.(\beta.a) = (\alpha.\beta).a$;
- (Generic Surjectivity) *For* $\alpha b \in \Sigma$, *there is* $c(\perp b, \perp \alpha) \in X$ *such that* $\alpha.c = b$; *and*
- (Generic Transitivity) *For* $x \perp y \in X$ *there is* $\alpha \in H$ *such that* $\alpha \perp x$ *and* $\alpha.x = y$.

Then there are a set W, *a transitive group action* $*$ *of* G *on* W, *and an embedding* $\jmath : X \to W$, *all hyperdefinable over* $\emptyset$, *such that* $\jmath(X)$ *is the set of all generic elements in* W, *and for* $\alpha x \in \Sigma$, $\jmath(\alpha.x) = \imath(\alpha) * \jmath(x)$ *holds. In this sense,* $(G, W, *)$ *is unique up to hyperdefinable isomorphisms preserving the set of all generic elements (of the group and set).*

Moreover, if the generic action . *additionally satisfies*

- (Generic Regularity) *For* $x \perp y \in X$ *there is unique* $\alpha \in H$ *such that* $\alpha \perp x$ *and* $\alpha.x = y$;

then the transitive group action $*$ *is regular (so principal).*

Proof As the proof will be similar to that of 7.6.1, we will only sketch the proof. Let $\approx$ be a binary symmetric hyperdefinable relation on $H \times X$ such that:

$$(\alpha_1, a_1) \approx (\alpha_2, a_2) \text{ if there are } \gamma_1, \gamma_2 \in H \text{ such that}$$
$$\gamma_i \perp \alpha_1\alpha_2 a_1 a_2, \text{ and } \gamma_1.\alpha_1 = \gamma_2.\alpha_2,\ \gamma_1.a_1 = \gamma_2.a_2.$$

By reasons similar to those in 7.6.1 for $\sim$, it follows that $\approx$ is hyperdefinable and forms an equivalence relation. Now we write $[\alpha, a] := (\alpha, a)/\approx \in W := (H \times X)/\approx$. (This will not be confused with a member $[\alpha, \beta] \in G$, due to the notational agreement declared before this theorem.) Again due to generic independence and associativity, we have the following.

Claim 1 If $\gamma \perp \alpha a$, then $[\gamma.\alpha, \gamma.a] = [\alpha, a]$ and there is $b(\perp \alpha a)$ such that $[\alpha, a] = [\gamma, b]$.

Due to Claim 1 and 7.6.1(i), we can define an action $*$ of G on W by $[\alpha, \beta] * [\beta, a] = [\alpha, a]$. In particular, for the identity $e \in G$, $e * [\beta, a] = [\beta, a]$. It is well defined and associative action. Suppose that $[\alpha, \beta] = [\alpha', \beta']$ witnessed by γ, γ' and $[\beta, a] = [\beta', a']$ by δ, δ'. We want to show $[\alpha, a] = [\alpha', a']$. Now, we can assume that $\delta\delta' \perp_{\beta\beta' aa'} \alpha\alpha'\gamma\gamma'$. Hence, $\gamma \perp \delta$ and due to generic surjectivity, there is $\sigma \in H$ such that $\sigma.\gamma = \delta$. Now, since $\delta'.\beta' = \delta.\beta = \sigma.\gamma.\beta = \sigma.\gamma'.\beta'$, we have $\delta' = \sigma.\gamma'$. Hence, $\delta'.\alpha' = \sigma.\gamma'.\alpha' = \sigma.\gamma.\alpha = \delta.\alpha$. Note that $\delta' \perp \alpha' a'$ and $\delta \perp \alpha a$. Thus, from Claim 1, $[\alpha', a'] = [\delta'.\alpha', \delta'.a'] = [\delta.\alpha, \delta.a] = [\alpha, a]$. Similarly, one can show the action is associative.

Claim 2 Generic transitivity implies that

(i) For any $b(\perp \alpha a)$, there is $\beta(\perp \alpha a)$ such that $[\alpha, a] = [\beta, b]$; and

(ii) The group action $*$ is transitive.

(i) By generic transitivity there is γ such that $\gamma.a = b$ and $\gamma \perp a$. As we can assume that $\gamma \perp_{ab} \alpha$, it follows that $\gamma \perp \alpha a$. Then from Claim 1, $[\alpha, a] = [\gamma.\alpha, b]$ with $\gamma.\alpha(\perp \alpha a)$.

(ii) Given that $[\alpha, x], [\beta, y] \in W$, take $z \perp \alpha\beta xy$. Then by (i), there are $\alpha'\beta'$ such that $[\alpha, x] = [\alpha', z]$ and $[\beta, y] = [\beta', z]$ Thus, $[\beta', \alpha'] * [\alpha', z] = [\beta', z]$. Claim 2 is shown.

Claim 3 For $x \in X$ and any $\gamma, \alpha(\in H) \perp x$, $[\gamma, \gamma.x] = [\alpha, \alpha.x](\in W)$. Take $\beta \perp \alpha, \gamma, x$. Now, there are γ', α' such that $\gamma'.\gamma = \beta = \alpha'.\alpha$, and each of $\{\gamma', \gamma, x\}$, $\{\alpha', \alpha, x\}$ is independent. Thus, by Claim 1 and generic associativity, $[\gamma, \gamma.x] = [\gamma'.\gamma, \gamma'.\gamma.x] = [\beta, \beta.x] = [\alpha'.\alpha, \alpha'.\alpha.x] = [\alpha, \alpha.x]$. Claim 3 is proved.

From Claim 3, the map $\jmath : X \to W$ sending x to $[\gamma, \gamma.x]$, for any $\gamma(\in H) \perp x$, is well defined. It is hyperdefinable since, for some completion p of H, $x \perp \gamma \models p$ is hyperdefinable.

Claim 4 The map $\jmath$ is injective: Suppose that $\jmath(x) = \jmath(x')$. Hence, for some $\gamma \perp xx'$, $[\gamma, \gamma.x] = [\gamma, \gamma.x']$. By definition, there is $\alpha_i \perp \gamma xx'$ such that $\alpha_1.\gamma = \alpha_2.\gamma$ and $\alpha_1.\gamma.x = \alpha_2.\gamma.x'$. Then $\alpha_1 = \alpha_2$ and $\beta.x = \beta.x'$, where $\beta = \alpha_1.\gamma \perp \gamma xx'$. Now, take $\sigma \perp \beta\gamma xx'$. Then by surjectivity, there are z, z' such that $\sigma.z = x$ and $\sigma.z' = x'$. Note that each of $\{\sigma, \beta, z\}$ and $\{\sigma, \beta, z'\}$ is independent. Let $\delta = \beta.\sigma$. Then $\delta.z = \beta.x = \beta.x' = \delta.z'$, and each of $\{\delta, \sigma, z\}$ and $\{\delta, \sigma, z'\}$ is independent. Now by generic surjectivity there is σ' such that $\sigma'.\delta = \sigma$ and each of $\{\sigma', \delta, z\}$, $\{\sigma', \delta, z'\}$ is independent. Then $x = \sigma.z = \sigma'.\delta.z = \sigma'.\delta.z' = \sigma.z' = x'$. Hence, Claim 4 is proved.

Let $X' = \jmath(X)$, the embedded image of X in W. The following claim says that $W = H'(= \iota(H)) * X'$, X' is the set of all generic elements in X, and the group action $*$ of G on X is a generic extension of that . of H on X.

Claim 5

(i) For $\alpha \perp x$, we have that $[\alpha, x] \in X'$.

(ii) For $w \in W$, there are $\alpha', \alpha'' \in H$, $x' \in X$ such that $\alpha'\alpha'' \perp w$, $x' \perp w$, and $\iota(\alpha') * \jmath(x') = w$, $\iota(\alpha'') * w = \jmath(x')$.

(iii) For $\alpha \in H, x \in X$ with $\alpha \perp x$, $\jmath(\alpha.x) = \iota(\alpha) * \jmath(x)$.

(iv) If $w \in X'$, then for $g \in G$ with $g \perp w$, $g * w \in X'$ and $g * w \perp g$.

(v) $w \in X'$ iff $w(\in W)$ is generic for the action $*$.

(vi) Suppose there are ι', G' as in 7.6.1(iv), a group action $*'$ of G' on W', and an embedding $\jmath' : X \to W'$ all hyperdefinable over $\emptyset$ such that for $\alpha x \in \Sigma$, $\jmath'(\alpha.x) = \iota'(\alpha) *' \jmath'(x)$, and $\jmath'(H)$ is the set of all generic elements in W'. Then there are an isomorphisms $\tau_G : G \to G'$ and a bijection $\tau_W : W \to W'$, both hyperdefinable and preserving the set of all generic elements, such that for $g \in G$, $w \in W$, $\tau_W(g * w) = \tau_G(g) *' \tau_W(w)$ holds:

(i) comes easily by the same argument in the proof for Claim 4 of 7.6.1.

(ii) Given $w = [\beta, z] \in W$, due to Claim 1, we can assume that $\beta \mathop{\smile\!\!\!\!\!\!|}\, w$ and $z \mathop{\smile\!\!\!\!\!\!|}\, w$. Now from (i), for some $\delta \mathop{\smile\!\!\!\!\!\!|}\, \beta z$, $w = [\beta, \delta](\in H') * [\delta, z](\in X')$. Then $\delta \mathop{\smile\!\!\!\!\!\!|}_{\beta} w$, $\delta \mathop{\smile\!\!\!\!\!\!|}_{z} w$, and hence $[\delta, \beta] \mathop{\smile\!\!\!\!\!\!|}\, w$, $[\delta, z] \mathop{\smile\!\!\!\!\!\!|}\, w$ hold. Therefore, there are $\alpha', \alpha'' \in H$, $x' \in X$ with $\imath(\alpha') = [\beta, \delta], \imath(\alpha'') = [\delta, \beta], \jmath(x') = [\delta, z]$ satisfying (ii).

(iii) and (iv) come easily by the similar arguments showing (iii) of 7.6.1.

(v) ($\Rightarrow$) (iv) says it.

($\Leftarrow$) Suppose that generic $w = [\alpha, c] \in W$ is given. By Claim 1, we can assume that $w \mathop{\smile\!\!\!\!\!\!|}\, \alpha$. Now from (i) above, for some $\beta \mathop{\smile\!\!\!\!\!\!|}\, \alpha c w$, we have $[\alpha, \beta] \in H'$, $[\beta, c] \in X'$, and $[\beta, \alpha] \mathop{\smile\!\!\!\!\!\!|}\, w$. Since w is generic, $[\alpha, \beta] \mathop{\smile\!\!\!\!\!\!|}\, [\beta, c] = [\beta, \alpha] * w$. Hence, from (iv), $w = [\alpha, \beta] * [\beta, c] \in X'$.

(vi) The existence of τ_G is pointed out in 7.6.1. We similarly take $\tau_{X'}$ as the composition of $\jmath'$ and $\jmath^{-1}$. To extend its domain to W, for $w \in W$, let $\tau_W(w) = \tau_G(g) *' \tau_{X'}(w')$ for some generic $g(\mathop{\smile\!\!\!\!\!\!|}\, w)$ and generic $w' \in X'$ with $w = g * w'$ (*). (ii) says such g, w' exist. It is routine to see that τ_W is the desired, well-defined hyperdefinable bijection extending $\tau_{X'}$.

Claim 6 If the partial action . is generically regular, then the group action $*$ is regular: Suppose that $[\alpha, \beta] * [\gamma, a] = [\alpha', \beta'] * [\gamma, a]$. We want to show that $[\alpha, \beta] = [\alpha', \beta']$. From 7.6.1 (i) and Claim 1, by making δ independent from other elements we can find σ, σ', b such that $[\alpha, \beta] = [\sigma, \delta]$, $[\alpha', \beta'] = [\sigma', \delta]$ and $[\gamma, a] = [\delta, b]$. Thus, $[\sigma, b] = [\sigma', b]$. By the definition of $\approx$, there is $\sigma_i(\mathop{\smile\!\!\!\!\!\!|}\, \sigma\sigma' b)$ such that $\sigma_1.\sigma = \sigma_2.\sigma'$ and $\sigma_1.b = \sigma_2.b$. Then, by generic independence and regularity, $\sigma_1 = \sigma_2$ and $\sigma = \sigma'$. Therefore, $[\alpha, \beta] = [\alpha', \beta']$. We have proved Claim 6. ■

7.7 Bibliographical Remarks

F. O. Wagner has been a driving force in developing the theory of groups in simple theories. Most results in Section 7.1 are due to Pillay [98]. The group theoretic version of Theorem 7.1.16 is due to Schlichting [108]. Sections 7.2–7.4 are from [122],[121]. Theorem 7.5.4 is in [122], and Theorem 7.5.12 and Corollary 7.5.13 are in [39]. Theorem 7.6.1 is from [121], and Theorem 7.6.2 is in [70]. We do not cover the topics related to binding groups [110],[9],[46], and Zilber's indecomposability theorem with its applications to simple groups [120, 5.4.5–10].

8

A Geometry of Forking

For the explanation of theoretical and historical background of this chapter, we first state some more definitions related to pregeometries introduced in Section 4.4.

Definition 8.0.1 Let (S, cl) be a pregeometry.

(1) (S, cl) is *trivial* if $\mathrm{cl}(A) = \bigcup\{\mathrm{cl}(\{a\}) : a \in A\}$ for all $A \subseteq S$.

(2) (S, cl) is *modular* if X is independent from Y over $X \cap Y$ for all closed X, Y, or equivalently, if $\dim(X) + \dim(Y) = \dim(X \cup Y) + \dim(X \cap Y)$ for finite-dimensional closed X and Y.

(3) (S, cl) is *locally modular* if it is modular over some point in S.

(4) (S, cl) is *locally finite* if the closure of a finite set is finite.

(5) A 3-tuple $(ca_1a_2) \in S$ is said to be a *non-modular triple* if $\dim(ca_1a_2) = 3$, $\mathrm{cl}(a_1a_2) = \mathrm{cl}(a_1) \cup \mathrm{cl}(a_2)$ and there are $u_i \in \mathrm{cl}(ca_i) \smallsetminus (\mathrm{cl}(c) \cup \mathrm{cl}(a_i))$ for $i = 1, 2$.

Let S' be the geometry of a pregeometry S. S is trivial (modular, or locally modular, resp.) if and only if S' is (resp.). Moreover, each of the properties is preserved under localization. Thus, modularity implies local modularity, and obviously triviality implies modularity.

There are prototype examples of pregeometries in stability theoretic context. Assume T is *strongly minimal* (i.e., any definable subset (with parameters) of $\mathcal{M}^1$ is either finite or co-finite). Then it is easy to see that $(\mathcal{M}^1, \mathrm{acl})$ forms a geometry (where for a set $A \subseteq \mathcal{M}^1$, the closure is $\mathrm{acl}(A)(\cap \mathcal{M}^1)$). There, non-forking independence coincides with the dimensional independence. The following are the typical examples of such strongly minimal (pre)geometries.

Example 8.0.2

(1) An infinite set (having no relation other than equality): a trivial geometry.

(2) (a) A vector space $(V, +, 0, r)_{r \in F}$ as mentioned in Example 1.2.8: a modular geometry. The corresponding projective space has the equivalent geometry.

(b) An affine space $(V, \lambda_r, G)_{r \in F}$ over a field F, where $\lambda_r(u, v) = ru + (1 - r)v$, $G(u, v, w) = u - v + w$: a locally modular geometry. But this geometry is not modular as two parallel lines (each having dimension[1] 2) in a plain (having dimension 3) do not satisfy the equation in 8.0.1(2).

(3) The complex field $\mathbb{C} = (\mathbb{C}, +, -, \times, 0, 1)$: a non-modular geometry.

As mentioned in the introduction of Chapter 6, Zilber's principle (early 1980s) roughly says that the above are, more or less, all the examples of strongly minimal structures. Namely, any strongly minimal structure is trivial, locally modular, or interpreting an infinite field (which must be algebraically closed field by MacIntyre's result [87]).

Hrushovski (early 1990s) constructed the counterexamples of Zilber's Principle using his ingenious construction method, which are not locally modular where no infinite group is interpretable (= definable in $\mathcal{M}^{\mathrm{eq}}$) [52]. But later on, he and Zilber suggested the so-called "Zariski condition," and *proved* Zilber's Principle under the constraint [59]. It is well known that Hrushovski (mid-1990s) solved the function field version Mordell–Lang conjecture in number theory, by the striking application of the of the mentioned results [53].

Zilber's Principle also makes sense in the o-minimal theory context, and Peterzil and Starchenko solved it fully and positively [96].

Now, if $\mathcal{M}^1$ is strongly minimal, then there is a unique complete 1-type p over $\emptyset$ having SU-rank 1. p, indeed, is a Lascar strong type. If D is the solution set of the type,[2] then (D, cl), where $\mathrm{cl}(-) = \mathrm{acl}(-) \cap D$ has the equivalent geometry to $(\mathcal{M}^1, \mathrm{acl})$. Hence, it suffices to turn our attention to (D, cl) to comprehend the geometry of the strongly minimal structure. Indeed, for *any* SU-rank 1 strong type p (over $\emptyset$ after suppressing $\mathrm{dom}(p)$) with the solution set D, the same closure relation gives the geometry to D. Moreover, inside D, $\dim(-/A) = SU(-/A)$ and $\mathop{\smile\!\!\!\!\!\!\mid}$ = dimensional independence.

In this chapter, we study such that $D = (D, \mathrm{cl})$ in a simple theory. Understanding the geometry is the suitable generalization of the study of strong minimal structures. Furthermore, often, particularly infinite rank cases, information on the rank 1 types is a building block in analyzing the whole structure. In the spirit of Zilber's Principle for the strong minimality context, we come up with the following questions for simple theory context.

Question 8.0.3

(1) Relationship between D being 1-based (= modular in the sense of Definition 5.5.14) and its geometry being (locally) modular.[3]

(2) Assume (D, cl) is non-trivial locally modular or more generally D is non-trivial 1-based.

---

1. Here the dimension is that of the geometry.

2. Here $|D| = \bar{\kappa}$.

3. From now on, not to be confused with a *modular* set or theory as in Definition 5.5.14 or Section 6.1, when we refer to modularity in 8.0.1, we say either D has a *modular geometry* or (D, cl) *is modular*.

(a) Is the geometry of (D, cl) isomorphic to that of an affine or a projective space over some division ring?

(b) Can we recover a canonical infinite vector space from D?

(3) If D is not modular, then under which conditions are we able to get an infinite field and what kind of a field do we obtain?

Indeed, describing how much have been accomplished in solving the questions is the whole theme of the rest of this book. Answering the same questions for stable theories are also a part of major topics in geometric stability theory. An extensive exposition on geometric stability theory is made in [99]. Let us briefly summarize what has been achieved for *stable* theories in regard to the above questions: Fact 8.1.7 answers question (1) that the two notions are equivalent. As pointed out in the introduction of Chapter 6, Zilber solved question (2) affirmatively when D is ω-categorical. See also Fact 8.1.7. Later Hrushovski extended the result in *arbitrary* stable context (Fact 9.0.1) using the group configuration argument (Theorem 9.0.4). The theoretical and applicational evolvement of stability theory for question (3) is mentioned in the former paragraph.

For simple theories in general, some are generalized but not all. For (1), only one direction of the equivalence holds in simple context (Fact 8.3.5 and Remark 8.3.7). Similarly, there are counterexamples for (2)(a) (Section 8.2 and Remark 8.3.7). However, the positive answer to (2)(b) in stable context is reasonably generalized for simple theories (Corollary 8.3.13, Theorem 8.3.16, and Theorem 9.4.1). The answer for (3) is beyond the reach of our current knowledge. Even we do not know what kind of a field it should be if it exists. It is only conjectured that any supersimple field is pseudo-algebraically closed, and some partial results are obtained [90], [91].

The same questions can be imposed more generally in the context of regular types. More deeper arguments are developed for geometric stability theory by investigating problems around the questions. Little is known about generalization into simplicity context.

8.1 *SU*-Rank 1 Types

We begin this section by recalling facts on the (local) modularity of a pregeometry. The following is not hard to prove.

Fact 8.1.1 *The following are equivalent:*

(1) (S, cl) *is modular.*

(2) *For any subset* A *and* $b \in S$, *whenever* $c \in \mathrm{cl}(Ab)$, *then* $c \in \mathrm{cl}(ab)$ *for some* $a \in \mathrm{cl}(A)$.

(3) *If* $a, b \in S$, $A \subseteq S$, $\dim(a, b) = 2$, *and* $\dim(ab/A) = 1$, *then there is* $c \in (\mathrm{cl}(ab) \cap \mathrm{cl}(A)) - \mathrm{cl}(\emptyset)$.

Remark 8.1.2

(1) Suppose that (S, cl) has a non-modular triple (ca_1a_2). Then (S, cl) is not modular. Let u_i be as in 8.0.1(5). First, suppose $\mathrm{cl}(u_1) = \mathrm{cl}(u_2)$. Since $a_1, a_2 \in \mathrm{cl}(cu_1) \cup \mathrm{cl}(cu_2) = \mathrm{cl}(cu_1)$, we get $2 = \dim(cu_1) = \dim(ca_1a_2u_1) = 3$, a contradiction. Hence, $\dim(u_1u_2) = 2$. Now $a_1 \notin \mathrm{cl}(u_1u_2)$, as otherwise, since $c \in \mathrm{cl}(a_1u_1u_2) = \mathrm{cl}(u_1u_2)$ and $a_2 \in \mathrm{cl}(cu_2)$, $\dim(u_1u_2) = \dim(ca_1a_2u_1u_2) = 3$, a contradiction. Similarly, $a_2 \notin \mathrm{cl}(u_1u_2)$. Recall that $\mathrm{cl}(a_1a_2) = \mathrm{cl}(a_1) \cup \mathrm{cl}(a_2)$. Hence, $\mathrm{cl}(a_1a_2) \cap \mathrm{cl}(u_1u_2) = \mathrm{cl}(\emptyset)$. Thus, $4 = \dim(a_1a_2) + \dim(u_1u_2) \neq \dim(a_1a_2u_1u_2) + \dim(\mathrm{cl}(a_1a_2) \cap \mathrm{cl}(u_1u_2)) = 3$. This shows that S is not modular.

(2) If (S, cl) is a modular geometry and $\mathrm{cl}(cd) = \{c, d\}$ for any two distinct points $c, d \in S$ then S is trivial. By (4) of 4.4.3, it is sufficient to show that the closure of any finite set is itself. Suppose not; say there exists $d \notin A = \{a_1, \ldots, a_n\}$ such that $d \in \mathrm{cl}(A)$. Then $\mathrm{cl}(a_nd) = \{a_n, d\}$ and $\mathrm{cl}(a_1 \ldots a_{n-1}) = \{a_1, \ldots, a_{n-1}\}$ by the induction hypothesis. By modularity $n = \dim(a_1 \ldots a_n) = \dim(a_1 \ldots a_{n-1}) + \dim(a_nd) = n + 1$, a contradiction.

Let us now state two important classical results on the geometry that will be used later. Recall that the geometry of a vector space (or its corresponding projective space) over some division ring, as exemplified in 8.0.2, is called a *projective geometry over the division ring*. Similarly, the geometry of an affine space over a field is called an *affine geometry over the field.*

But a (*locally*, resp.) *projective* (*pre*)*geometry* simply means a non-trivial (locally, resp.) modular (pre)geometry.

Fact 8.1.3 [3] *A projective geometry of dimension ≥ 4, in which each closed set of dimension 2 contains at least 3 elements, is isomorphic to a projective geometry over some division ring.*

Fact 8.1.4 [34] *A locally projective locally finite geometry of dimension > 4, in which all closed sets of dimension 2 have the same size, is an affine or a projective geometry over a finite field.*

As promised, we shall now consider an SU-rank 1 Lascar strong type and its canonical pregeometry. As usual, we work with a hyperimaginary type in a simple theory T. Fix D, the solution set of a Lascar strong type of SU-rank 1 over a hyperimaginary $\mathrm{dom}(D)$. Throughout this chapter we work with (D, cl), where $\mathrm{cl}(-) = \mathrm{bdd}(- \cup \mathrm{dom}(D)) \cap D$. Clearly, the closure relation satisfies 4.4.3(1),(2), and (3), too, since $a \in \mathrm{cl}(Cb)$ iff $a \not\downarrow_{\mathrm{dom}(D)} Cb$. Moreover, by finite character in 4.2.11, the closure relation satisfies 4.4.3(4) as well. Therefore, (D, cl) forms a pregeometry. By the same reason, for $B \supseteq \mathrm{dom}(D)$, $D_B := (D, \mathrm{cl}_B)$ with $\mathrm{cl}_B(-) = \mathrm{bdd}(- \cup B) \cap D$ forms a pregeometry, too.

Definition 8.1.5 For $A \supseteq \mathrm{dom}(D)$, we say that D_A is *linear* if for any two points $a, b \in D$ and a parameter B with $SU(ab/AB) = 1$, $SU(e/A) \leq 1$, where $e = \mathrm{Cb}(ab/AB)$; if $SU(e/A) \leq k (< \omega)$ for any such a, b, AB, e, we call D_A *pseudolinear* (or *k-linear*).

From now on D' denotes the geometry of D. We shall use the same closure notation for the pregeometry and its geometry. We shall see that the two notions of linearity and 1-basedness for D coincide (Theorem 8.3.6).

In the rest, for notational simplicity we omit $\mathrm{dom}(D)$ to $\emptyset$.

Now we recall some of important facts on (D, cl) when T is stable and D consists of finite real tuples. The following obviously follows from 8.1.3 and 8.1.4 with stationarity.

Fact 8.1.6 *T is stable.*

(1) If (D', cl) is projective, then it is isomorphic to a projective geometry over a division ring.

(2) If (D', cl) is ω-categorical, non-modular but locally modular, then it is isomorphic to an affine geometry over some finite field.

If the stable D is ω-categorical, it is strongly minimal and 1-based, and D' lives in $D^{\mathrm{eq}} := \mathrm{dcl}(D) \cap \mathcal{M}^{\mathrm{eq}}$ via the definable equivalence relation $x \in \mathrm{cl}(y)$. Hence, if (D, cl) is non-trivial and modular (non-modular and locally modular, resp.), then the corresponding geometry (D', cl) is projective (affine, resp.) over some *finite* field F. Then in either case, by Zilber's results [126] on ω-categorical strongly minimal structures, a vector space $(V, +, \lambda)_{\lambda \in F}$ over F is definable in D^{eq}, possibly over some finite parameter.

Fact 8.1.7 *T is stable. The following are equivalent:*

(1) D is 1-based (i.e., modular).

(2) (D, cl) is locally modular.

(3) For some set A, (D, cl_A) is modular.

(4) D is linear.

The main purpose of this chapter is to generalize the above results to the context of simple theories (Theorem 8.3.6). The essential parts of the results are extended, but not all straightforwardly. For example, as said the equivalence of (1) and (4) are preserved, but (1)⇒(2) does not hold, and the local modularity of the geometry is no longer preserved when non-forking extensions are undertaken (Remark 8.3.7), while, of course, being so 1-basedness. This generalization will also give better understanding of the rank 1 types, and reveal that the stable case is only a particular side of the whole picture. In particular, the modular (rather than locally modular) geometry is the right one related to 1-basedness (8.3.6).

In the rest we return to the general simple D case, and begin our observation when D happens to have a modular pregeometry.

Lemma 8.1.8 *((D, cl) modular.) Suppose there exist $a, b \in D$ such that* $\dim\{a, b\} = 2$ *and* $\mathrm{cl}(ab) = \mathrm{cl}(a) \cup \mathrm{cl}(b)$. *Then D is trivial.*

Proof We can assume that D forms a geometry. Then the condition in the lemma simply says that $\mathrm{cl}(ab) = \{a, b\}$.

Now, by Remark 8.1.2, it is enough to show that $\mathrm{cl}(cd) = \{c, d\}$ for any two distinct points c, d: Suppose not; say, there is a third point $u' \in \mathrm{cl}(cd)$. Let $q(x, d) = \mathrm{tp}(c/d)$. As D is the solution set of a Lascar strong type, by the independence theorem for Lascar strong types, we can amalgamate $q(x, a)$ and $q(x, b)$. Let c' be a realization of a common non-forking extension of $q(x, a)$ and $q(x, b)$. Then $\dim(a, b, c') = 3$, and there are u, v such that $tp(cdu') = tp(c'bu) = tp(c'av)$. Hence, the tuple $(c'ab)$ forms a non-modular triple. Again by 8.1.2, this contradicts the modularity of D and finishes the proof. ■

By now, we have the following theorem due to 8.1.3 and the previous lemma.

Theorem 8.1.9 *If (D, cl) is modular, then either D is trivial or the geometry D' of D is projective over a division ring.*

Suppose now that D is ω-categorical. Let R_n be an $(n+1)$-ary relation on $D'(\subseteq D^{\mathrm{eq}})$, such that $(a_0, a_1, \ldots, a_n) \in R_n$ if and only if $a_0 \in \mathrm{cl}(a_1, \ldots, a_n)$. Then R_n is $\emptyset$-definable in D', and $(D', R_n)_{n\in\omega}$ is a (definitional) reduct of D'. Hence, if (D, cl) is non-trivial and modular, then by the previous theorem, $(D', R_n)_{n\in\omega}$ is stable, strongly minimal, and quantifier eliminable, having the same projective geometry (D', cl) over a finite field F. Hence, as mentioned previously, a vector space $(V, +, \lambda)_{\lambda\in F}$ over F is definable in D^{eq}, too, for simple D.

Now we investigate the locally modular case. We fix some notation. Choose $p \in D$, and let D_0 denote the pregeometry $D \setminus \mathrm{cl}(p)$ with closure operation cl_p. D_0' denotes the geometry of D_0. Recall that we use the same closure notation for the pregeometry and its geometry. Note that $x \in \mathrm{cl}_p(y)$ is an equivalence relation on D_0. We obtain the following results, 8.1.10 and 8.1.11, analogous to the previous 8.1.8 and 8.1.9. However, 8.1.10 and 8.1.11 are not automatic consequences of 8.1.8 and 8.1.9, since in the simple context D_0 need not be the solution set of a unique type over p.

In the rest of this section, (D, cl) is locally modular.

Lemma 8.1.10 *Suppose that there are distinct $a', b' \in D_0'$ such that $\mathrm{cl}_p(a'b') = \{a'b'\}$. Then $\mathrm{cl}_p(c'd') = \{c', d'\}$ for any two points c', d' in D_0'. Hence, D_0' is trivial.*

Proof Triviality will follow from the main assertion and 8.1.2. Fix points a, b in D from a', b', respectively. Due to the limitation on type amalgamation, we need more complicated arguments to prove the lemma. We shall calculate dimension inside D, unless stated otherwise.

Suppose that the lemma is not true. Then there is $u' \in D_0'$ such that $u' \in \mathrm{cl}_p(c'd') \setminus \{c', d'\}$. Now choose (in D) points c, d, u from c', d', u', respectively. Then clearly $\dim(pcdu) = 3$, and $\dim(pX) = 3$ for each 2-subset X of $\{c, d, u\}$ (*).

Case I $u \in \mathrm{cl}(cd)$: Then the closure of any two of the u, c, d contains the third point (**). Now, we claim the following.

Claim There are $v_i \in D$ with $v_i \in v_i' \in D_0'$ (for $i = 1, 2, 3$) such that $tp(cdu) = tp(v_1bv_2) = tp(v_1av_3)$ and $\dim(pabv_1) = 4$.

Proof of Claim Let $q(x, d) = tp(c/d)$. Then we can amalgamate $q(x, b)$ and $q(x, a)$. Let v_1 be a realization of a common non-forking extension. Then clearly there are v_2 and v_3 such that $tp(cdu) = tp(v_1 b v_2) = tp(v_1 a v_3)$. Moreover, we can assume that $v_1 \underset{ab}{\perp} p$, so that $\dim(pabv_1) = 4$. In particular, $p \notin \mathrm{cl}(v_1 b)$. Now by (**), $v_2 \in \mathrm{cl}(v_1 b)$. Hence, $\mathrm{cl}(p) \neq \mathrm{cl}(v_2)$, and $v_2' \in D_0$. Similarly, $v_3' \in D_0$. Therefore, the claim is proved.

Since $u \in \mathrm{cl}(cd) \setminus \{c, d\}$, we have $v_3 \in \mathrm{cl}(v, a) \setminus \{v, a\}$ and $v_2 \in \mathrm{cl}(v, b) \setminus \{v, b\}$. If $v_3 \in \mathrm{cl}(ap)$, then $p \in \mathrm{cl}(av_3) \subseteq \mathrm{cl}(av_1)$, contradicting $\dim(pabv_1) = 4$. Similarly, $v_3 \notin \mathrm{cl}(v_1 p)$ and $v_2 \notin \mathrm{cl}(bp) \cup \mathrm{cl}(v_1 p)$. Thus, $(v_1' a' b')$ forms a non-modular triple in D_0', contradicting modularity. Hence Case I does not happen. We consider the remaining case.

Case II $u \notin \mathrm{cl}(cd)$: Then by (*), each 3 subset of $\{p, c, d, u\}$ has dimension 3 (***). This time, let $q(x, cd) = tp(p/cd)$. There are u_2 and u_4 such that $tp(cd) = tp(au_2) = tp(bu_4)$ and $\dim(abu_2u_4) = 4$. Amalgamate $q(x, au_2), q(x, bu_4)$ and let $u_1 \in D$ be a realization of a common non-forking extension, independent of p_0. Then $\dim(pu_1abu_2u_4) = 6$. Now there are $u_3 \in u_3' \in D_0'$ and $u_5 \in u_5' \in D_0'$ such that $tp(pcdu) = tp(u_1 a u_2 u_3) = tp(u_1 b u_4 u_5)$. By use of (***), one can see that $\dim(pu_2u_3u_4u_5) = 5$ and $a', b' \notin \mathrm{cl}_p(u_2'u_3'u_4'u_5')$. Thus, $\mathrm{cl}_p(a'b') \cap \mathrm{cl}_p(u_2'u_3'u_4'u_5') = \emptyset$. This again violates the modularity of D_0'. Hence, Case II does not happen either. Therefore Lemma 8.1.10 is proved. ■

Then again due to 8.1.3, we have the following:

Theorem 8.1.11 *D_0' is either trivial or projective over a division ring.*

Moreover, triviality is preserved between D and D_0.

Lemma 8.1.12 *D is trivial if and only if D_0 is trivial.*

Proof We can safely assume that D forms a geometry. It is enough to show that triviality of D_0 implies that of D. Let D_0 be trivial.

Claim The closure of any two distinct points $\{p, q\}$ in D is itself.

Proof of Claim Suppose not; say $r \in \mathrm{cl}(pq) \setminus \{p, q\}$. Choose $s \notin \mathrm{cl}(pq) = \mathrm{cl}(pqr)$ (*). Since D_0 is trivial, $\mathrm{cl}(pqs) = \mathrm{cl}(ps) \cup \mathrm{cl}(pq)$ and $\dim(pqs) = 3$. Hence, $q, r \notin \mathrm{cl}(ps)$. Now, since the localized pregeometry of D at s is equivalent to D_0 and D_0 is trivial, $\mathrm{cl}(qs) \cup \mathrm{cl}(ps) = \mathrm{cl}(pqs)$. In particular, $r \in \mathrm{cl}(qs) \cup \mathrm{cl}(ps)$. Hence, as $r \notin \mathrm{cl}(ps)$, $r \in \mathrm{cl}(qs)$. As $\dim(qs) = 2$, $s \in \mathrm{cl}(qr)$, this contradicts (*). Thus, the claim is verified.

Now, by the claim, for $n + 1$ points $p, q_1, \ldots, q_n$, since D_0 is trivial, $\mathrm{cl}(pq_1 \ldots q_n) = \mathrm{cl}(pq_1) \cup \ldots \cup \mathrm{cl}(pq_n) = \{p, q_1, \ldots, q_n\}$. Hence, D is also trivial. ■

By the same argument described after 8.1.9, if D is locally modular ω-categorical, after naming a point p in D, a sort in $(D_p)^{eq}$ representing D'_0 has a generic (independence-preserving) stable reduct (over p). Hence, if D is non-trivial, a vector space over a finite field is again interpretable over a finite parameter. We shall, in fact, prove the interpretability of a vector space in the more general 1-based context (Theorem 8.3.16).

By now, one would naturally expect that, as in stable theories, a (non-modular) locally modular ω-categorical D would be affine. However, the example in the following section shows that this is *not* the case.

8.2 Non-affine Locally Modular Types

First, we fix an extended language $\mathcal{L}' = \mathcal{L}_A \cup \{P\}$ where P is a unary predicate and $\mathcal{L}_A$ is the language of the theory of affine space over a finite field F of cardinality at least 5. (We shall see below a number of reasons why this condition is imposed. In fact, one can do the same construction with a projective space over a finite field of cardinality ≥ 4.) Now, suppose V_1 is a finite affine space over F generated by independent elements $\{x_0, \ldots, x_n\}$. Let P color some of the elements in V_1. We call the coloring *good* if for every line l in V_1, we can find at least 2 points in l colored by P, and let $\mathrm{gc}(n) > 0$ be the finite number of good colorings, for $n \geq 1$, described by $\mathcal{L}'$-formulas $\mathrm{Col}_{r,n}(x_0, x_1, \ldots, x_n)$. Now, suppose $V_2 \supset V_1$ is an affine space with basis $\{x_0, \ldots, x_{n+1}\}$. Then, given a good coloring $\mathrm{Col}_{r,n}(x_0, \ldots, x_n)$ of V_1, there exists a good coloring extending it to V_2. (For example, color every point in $V_2 \setminus V_1$ with P while leaving the coloring on V_1. Since every line in V_2 either is already in V_1 or hits V_1 once, the condition $q \geq 5$ implies that this is a good coloring.) We denote the finite number of all possible such colorings by $\mathrm{gc}(r, n)$ and let them be described by $\mathcal{L}'$-formulas $\mathrm{Col}^{s,n+1}_{r,n}(x_0, x_1, \ldots, x_n, x_{n+1})$, for $1 \leq s \leq \mathrm{gc}(r, n)$. For $n = 0$, we interpret $\mathrm{Col}^{s,1}_{0,0}(x_0, x)$ and $\mathrm{Col}^{s,1}_{1,0}(x_0, x)$ to mean that P gives a good coloring on the line generated by $\{x_0, x\}$ extending the colorings $\neg P$ and P, respectively, on x_0, and we let $\mathrm{gc}(0, 0)$ and $\mathrm{gc}(1, 0)$ denote the number of such colorings.

Now let T_A denote the theory of affine space over F. Then the $\mathcal{L}_A$-formulas $x \in \mathrm{cl}(y_1, \ldots, y_n)$ and $\dim(y_1 \ldots y_n) = m(\leq n)$ make sense. (Remember that the (pre)geometry dimension notion we use here is one more than the usual affine subspace dimension.) The theory T_{LM} is given in the language $\mathcal{L}'$ by the axioms

$$T_{LM} \equiv T_A \cup \{LM\} \cup \{F_n : n \geq 0\},$$

where LM(local modularity) is the sentence in $\mathcal{L}'$,

$$\forall x \forall y\, (x \neq y \rightarrow \exists z_1 \exists z_2\, (z_1 \neq z_2 \wedge P(z_1) \wedge P(z_2) \wedge z_1, z_2 \in \mathrm{cl}(x, y)));$$

F_n(color extension), for $n \geq 1$, is the $\mathcal{L}'$-sentence

$$F_n \equiv \bigwedge_{1\le r\le \mathrm{gc}(n)} \forall y_0 \dots y_n \Bigg(\dim(y_0 \dots y_n) = n+1 \wedge \mathrm{Col}_{r,n}(y_0, y_1, \dots, y_n)$$
$$\to \bigwedge_{1\le s\le \mathrm{gc}(r,n)} \exists x \big(\dim(y_0 \dots y_n x) = n+2 \wedge \mathrm{Col}_{r,n}^{s,n+1}(y_0, y_1, \dots, y_n, x)\big) \Bigg);$$

and F_0 is the sentence

$$F_0 \equiv \forall y \Bigg(\neg P(y) \to \bigwedge_{1\le s\le \mathrm{gc}(0,0)} \exists x \big(\dim(yx) = 2 \wedge \mathrm{Col}_{0,0}^{s,1}(y,x)\big) \Bigg)$$
$$\wedge\ \forall y \Bigg(P(y) \to \bigwedge_{1\le s\le \mathrm{gc}(1,0)} \exists x \big(\dim \mathrm{cl}(yx) = 2 \wedge \mathrm{Col}_{1,0}^{s,1}(y,x)\big) \Bigg).$$

We first show that the theory T_{LM} is consistent. This can be done by a union of chains argument inside a model V of T_A. We first point out the following on *any* geometry.

Remark 8.2.1 Suppose that (S, cl) is a geometry. Let $\{A_i \in S | i \in I\}$ be independent over $B \subseteq S$; i.e., for finite $\bar{a}_i \in A_i$, $\dim(\bar{a}_i/B) = \dim(\bar{a}_i/B \cup \bigcup_{j\neq i} A_j)$.

(1) $\mathrm{cl}(B) = \mathrm{cl}(BA_{i_1} \dots A_{i_n}) \cap \mathrm{cl}(BA_{j_1} \dots A_{j_m})$ for $i_1, \dots, i_n, j_1, \dots, j_m$ distinct.

(2) Each line in $\mathrm{cl}(B \cup \bigcup_i A_i)$ either already lies in one of $\mathrm{cl}(BA_i)$ or hits $\bigcup_i \mathrm{cl}(BA_i)$ at most twice.

Proof The first statement clearly comes from the fact that

$$\dim\left(\bar{a}_{i_1} \dots \bar{a}_{i_n}/B\right) = \dim\left(\bar{a}_{i_1} \dots \bar{a}_{i_n}/B \cup \bigcup_{j\neq i_1,\dots,i_n} A_j\right).$$

For the second, assume that a line in $\mathrm{cl}(B \cup \bigcup_i A_i)$ hits $\bigcup_i \mathrm{cl}(BA_i)$ 3 times, but does not lie in any $\mathrm{cl}(BA_i)$. Let $\{x, y, z\}$ be the hit points. Then clearly $x \in \mathrm{cl}(BA_k)$, $y \in \mathrm{cl}(BA_m) \setminus \mathrm{cl}(BA_k)$, $z \in \mathrm{cl}(BA_n) \setminus (\mathrm{cl}(BA_m) \cup \mathrm{cl}(BA_k))$ for distinct k, m, n, whereas $z \in \mathrm{cl}(xy)$ and $z \in \mathrm{cl}(BA_n) \cap \mathrm{cl}(BA_k A_m)$; this contradicts (1). ■

Now we proceed in a series of steps.

Step 1 Choose a $y_0 \in V$ and label y_0 by P. Now add independent elements $\{x_1, \dots, x_{\mathrm{gc}(1,0)}\}$ over y_0 and color each $\mathrm{cl}(y_0 x_i)$ to witness F_0 for y_0. Let $W_1 = \mathrm{cl}(y_0, x_1, \dots, x_{\mathrm{gc}(1,0)})$. Now color all points in $W_1 \setminus \bigcup_i \mathrm{cl}(y_0 x_i)$ with P. Then W_1 satisfies LM; if a line in W_1 already lies in one of $\mathrm{cl}(y_0 x_i)$, then we are done by construction. Otherwise, by the above remark (2), the assumption $\mathrm{Card}(F) \geq 5$ ensures that we can pick up at least 3 points on the line labelled with P.

Step 2 We now want to satisfy F_0 in $W_1 \setminus \{y_0\}$, so we repeat the method of Step 1 for each point in $W_1 \setminus \{y_0\}$. Namely, for $a \in W_1 \setminus \{y_0\}$, find independent $\{z_1, \ldots, z_{\mathrm{gc}(k,0)}\}$ which are also independent from W_1 ($k = 0, 1$ depends on $a \in \neg P$ or $a \in P$). Then color all the points in $\mathrm{cl}(W_1 \cup \{z_i\}_i)$ with P, except $\bigcup_i \mathrm{cl}(az_i)$ and W_1. W_1 is already colored. Color each $\mathrm{cl}(az_i)$ to witness F_0 for a. Then for the same reason as in Step 1, with Remark 8.2.1, $\mathrm{cl}(W_1 \cup \{z_i\}_i)$ satisfies LM. Now for $b \in W_1 \setminus \{y_0, a\}$, do the same thing over $\mathrm{cl}(W_1 \cup \{z_i\}_i)$, and so on. We then obtain a subspace $U_0 \supset W_1$ such that U_0 satisfies LM and for all $x \in W_1$, F_0 is witnessed inside U_0.

Step 3 Now by modifying Steps 1 and 2, find $U_1 (\supset U_0)$ to make F_1 hold for W_1 inside U_1. Namely, given a distinct $x, y \in W_1$, choose points $\{w_k\}_k$ independent from U_0 to extend the good coloring on $\mathrm{cl}(xy)$ to all possible good colorings on distinct affine planes $\mathrm{cl}(xyw_k)$. (These colorings are compatible due to the fact that $\mathrm{cl}(xy) = U_0 \cap \mathrm{cl}(xyw_k) = \mathrm{cl}(xyw_i) \cap \mathrm{cl}(xyw_j)$ (for $i \neq j$ by 8.2.1).) Then color all the points in $\mathrm{cl}(U_0 \cup \{w_k\}_k)$ other than U_0 and $\bigcup_k \mathrm{cl}(xyw_k)$ with P. Again, as in Step 1, the fact that $\mathrm{Card}(F) \geq 5$ and Remark 8.2.1 guarantee that $\mathrm{cl}(U_0 \cup \{w_k\}_k)$ satisfies LM. For the next line from W_1 other than $\mathrm{cl}(xy)$, repeat the process over $\mathrm{cl}(U_0 \cup \{w_k\}_k)$. By iterating this argument, we obtain a chain of spaces $W_1 \subset U_0 \subset \ldots \subset U_n$ so that U_n witnesses $\{F_0, \ldots, F_n\}$ for W_1, and U_n satisfies LM. We can choose n large enough (depending on the dimension of W_1), so that in fact U_n witnesses all F_i (for $i \geq 0$) for W_1.

Rename U_n as W_2, and repeat the previous steps for W_2. Continuing in this way we obtain a chain of spaces $y_0 \subset W_1 \subset W_2 \subset \ldots W_i \subset \ldots$ such that each W_i satisfies LM (by 8.2.1), and all F_i ($i \geq 0$) are witnessed for W_i inside W_{i+1}. Let $W = \bigcup_i W_i$, then by construction $W \models T_A \cup \{LM\} \cup \{F_i : i \geq 0\}$. Hence, T_{LM} is consistent.

We now have the following theorem.

Theorem 8.2.2 *T_{LM} is ω-categorical and complete and has quantifier elimination. Moreover, T_{LM} is a model companion of $T_A \cup \{LM\}$.*

The axioms $\{F_n : n \geq 0\}$ of color extensions allow a partial isomorphism to be continued. Hence, quantifier elimination follows. In particular, for an independent tuple $(a_0, \ldots, a_n)$, $\mathrm{tp}(a_0, \ldots, a_n)$ is determined by the coloring on $\mathrm{cl}(a_0, a_1, \ldots, a_n)$. ω-categoricity and, hence, completeness also follow by a straightforward back and forth partial isomorphism argument. Now, given a model M of $T_A \cup \{LM\}$, by doing the above process starting with M, one can obtain an extension N of M which is a model of T_{LM}. Hence, T_{LM} is a model companion of $T_A \cup \{LM\}$.

We now want to show that T_{LM} is a simple theory. As we observed, the condition $\mathrm{Card}(F) \geq 4$ is enough to obtain the previous theorem. However, we shall see that we need $\mathrm{Card}(F) \geq 5$ to ensure T_{LM} is simple, particularly for the independence theorem. Let us first define a notion of independence on subsets A, B, C of a saturated model $\bar{V}$ of T_{LM} by $A \downarrow_C B$ iff A is *independent* from B over C in the sense of affine space (i.e., in $\mathcal{L}_A$). It follows easily that this notion is automorphism invariant and satisfies symmetry, transitivity, finite character, and local character. We need to check the extension axiom and the independence theorem. Below we keep using cl to denote affine closure.

The extension axiom For this, let $\bar{a}$, a tuple, and B, C, two small subsets of $\bar{V}$, be given. Then clearly there is $\bar{a}'$ such that $\bar{a}'$ is ($\mathcal{L}_A$-) independent from C over B. Hence, as in 8.2.1, $\text{cl}(\bar{a}'B) \cap \text{cl}(BC) = \text{cl}(B)$. Therefore, the coloring of $\text{cl}(\bar{a}'B)$ according to $\text{cl}(\bar{a}B)$ is compatible with the coloring on $\text{cl}(BC)$. Again, color all the points in $\text{cl}(\bar{a}'BC)$ with P except $\text{cl}(\bar{a}'B)$ and $\text{cl}(BC)$. By an argument similar to that in Remark 8.2.1, $\text{cl}(\bar{a}'BC)$ with the new coloring satisfies LM. Hence, by the axioms $\{F_n\}$ we can embed $\bar{a}'$, while fixing BC with the new coloring, into $\bar{V}$, and then get $\bar{a}''$ having the same $\mathcal{L}'$-type of $\bar{a}$ over B and $\bar{a}'' \downarrow_B C$.

The independence theorem for types We will verify this over any closed set B. Here, types are $\mathcal{L}'$-types. Let $\bar{a}$ and $\bar{b}$ be independent tuples over B, and $\bar{c}, \bar{d}$ tuples such that $\bar{c} \downarrow_B \bar{a}$, $\bar{d} \downarrow_B \bar{b}$ and $\text{tp}(\bar{c}/B) = \text{tp}(\bar{d}/B)$. By the extension property above, we can choose a tuple $\bar{e}$ independent from $\bar{b}$ over $B\bar{a}$ such that $\text{tp}(\bar{c}/B\bar{a}) = \text{tp}(\bar{e}/B\bar{a})$. Then $\{\bar{e}, \bar{a}, \bar{b}\}$ is independent over B. Consider $\text{cl}(B\bar{e}\bar{a}\bar{b})$. We now newly color $\text{cl}(B\bar{e}\bar{a}\bar{b})$ while leaving the (old) coloring on $\text{cl}(B\bar{e}\bar{a}) \cup \text{cl}(B\bar{a}\bar{b})$, as follows. First, color $\text{cl}(\bar{e}; B\bar{b})$ according to $\text{cl}(\bar{d}; B\bar{b})$. Since $\text{tp}(\bar{e}/B) = \text{tp}(\bar{d}/B)$, on $\text{cl}(B\bar{e}\bar{b}) \cap (\text{cl}(B\bar{a}\bar{b}) \cup \text{cl}(B\bar{a}\bar{e})) = \text{cl}(B\bar{b}) \cup \text{cl}(B\bar{e})$ (Remark 8.2.1), the two colorings are the same. Hence, the new coloring is compatible with the coloring on $\text{cl}(B\bar{e}\bar{a}) \cup \text{cl}(B\bar{a}\bar{b})$. Then color the rest of the points in $\text{cl}(B\bar{e}\bar{a}\bar{b}) \setminus (\text{cl}(B\bar{a}\bar{b}) \cup \text{cl}(B\bar{e}\bar{a}) \cup \text{cl}(B\bar{e}\bar{b}))$ with P. Now, since every line in $\text{cl}(B\bar{e}\bar{a}\bar{b})$ either already lies in $\text{cl}(B\bar{a}\bar{b}) \cup \text{cl}(B\bar{e}\bar{a}) \cup \text{cl}(B\bar{e}\bar{b})$ or hits the set at most 3 times, the condition $\text{Card}(F) \geq 5$ ensures that this new coloring on $\text{cl}(B\bar{e}\bar{a}\bar{b})$ is a good coloring. As we can embed $\bar{e}$ with the new coloring on $\text{cl}(B\bar{e}\bar{a}\bar{b})$ while fixing $B\bar{a}\bar{b}$, the independence theorem for types holds. (Hence, the notion of type coincides with that of Lascar strong type.)

So T_{LM} is simple. Moreover, in $\bar{V}(\models T_{LM})$, as $\mathcal{L}'$-independence is the same as $\mathcal{L}_A$-independence, T_{LM} is a SU-rank 1 theory.

Theorem 8.2.3 *T_{LM} is simple with SU-rank 1 having the same forking independence notion (and so algebraic relations) as T_A.*

Now we take the solution set $\bar{V}(P) = \{a \in \bar{V} | P(a)\}$. As we checked the independence theorem over any closed set, in particular over $\emptyset$, $\bar{V}(P)$ is a 1-(Lascar strong) type. We also have that $\bar{V}(P)$ is simple, rank-1, and ω-categorical. It remains to show that $\bar{V}(P)$ is locally modular but not affine.

Local modularity Choose $d \in \bar{V}(P)$. Let ef be a pair and B a set from $\bar{V}(P)$ such that $\dim_d(ef) = 2$ and $\dim_d(ef/B) = 1$ inside $\bar{V}(P)$. Working in $\bar{V}$, we also have $\dim_d(ef) = 2$, $\dim_d(ef/B) = 1$. As the geometry of $\bar{V}$ is affine, $\text{acl}(def) \cap \text{acl}(dB) = \text{acl}(dg)$, where $g \neq d$ is in $\bar{V}$. By the axiom LM for $\bar{V}$ we can assume that $g \in \bar{V}(P)$. Hence, $\dim_d(\text{cl}_1(def) \cap \text{cl}_1(dB)) = 1$, where $\text{cl}_1(-) = \text{acl}(-) \cap \bar{V}(P)$ is the closure inside $\bar{V}(P)$. So $\bar{V}(P)$ is a locally modular geometry.

Not affine For this it is sufficient to observe that a good coloring of a line in $\bar{V}$ can label r points, where $2 \leq r \leq \text{Card}(F)$.

So we have constructed a non-trivial, locally modular ω-categorical rank-1 structure/type whose geometry is neither affine nor projective over a finite field. Another example of this kind can be constructed, as mentioned previously, by repeating the same process on any projective space over a finite field of cardinality ≥ 4. (This cardinality requirement is necessary [32].) As said in the introduction, there is no such example in a stable structure, and also by 8.1.9 there is no example of a modular one.

8.3 1-Based Types

We begin this section by introducing a canonical extension or "completion" of (D, cl). Recall that for an e-invariant set X, we put $X^{\mathrm{heq}} := \mathrm{dcl}(Xe)$.

Definition 8.3.1 We use $G(D)$ to denote the collection of every countable hyperimaginary in D^{heq} having rank 1 over $\mathrm{dom}(D)$.

Clearly, $D \subseteq G(D)$, and for any small set A containing $\mathrm{dom}(D)$, $G(D)_A$ forms a pregeometry with the usual bounded closure relation over A. We can give the notion of linearity in 8.1.5 to $G(D)$ as well.

Definition 8.3.2 For a set A containing $\mathrm{dom}(D)$, we say that $G(D)_A$ is *linear* if for any two elements $a, b \in G(D)$ and a hyperimaginary B with $SU(ab/AB) = 1$, we have $SU(e/A) \leq 1$ where $e = \mathrm{Cb}(ab/AB)$.

Remark 8.3.3

(1) When defining linearity for D or $G(D)$, it suffices to assume that the parameter B is a subset of D or $G(D)$ since we can preserve the rank by taking a Morley sequence as a new parameter.

(2) Let p be a Lascar strong type of some tuple $\bar{c}$ from D^{heq} over a parameter containing $\mathrm{dom}(D)$. If the canonical base $e = \mathrm{Cb}(p)$ has rank 1 over $\mathrm{dom}(D)$, then we can find a singleton b in $G(D)$ that can be considered as a canonical base *up to interboundedness* over $\mathrm{dom}(D)$. By 4.3.9(5), e is in $\mathrm{dcl}(\bar{c}_i | i < \omega)$, where $\langle \bar{c}_i : i < \omega \rangle (\subset D^{\mathrm{heq}})$ is a Morley sequence of p. Hence, we can assume that e lives in D^{heq}. More precisely, as pointed out in 4.1.3, e is (interdefinable with) some sequence $I = \langle b_i \rangle_i$ of (countable) hyperimginaries $b_i \in D^{\mathrm{heq}}$. Thus, any $b \in I - \mathrm{bdd}(\mathrm{dom}(D))$ forks with e over $\mathrm{dom}(D)$. Since $SU(e/\,\mathrm{dom}(D)) = 1$, b is interbounded with e and has SU-rank 1, over $\mathrm{dom}(D)$ (hence, $b \in G(D)$).
From now on, for the case $SU(\mathrm{Cb}(p)/\,\mathrm{dom}(D)) = 1$, when we refer to a *canonical base* of p, we actually mean *any* such $b \in G(D)$. Since we are mainly interested in the geometry of D (or $G(D)$), as we shall see, this abuse of notation will not cause any problems.

We now point out easy but useful arguments used throughout occasionally.

Remark 8.3.4

(1) Suppose that

$$W_0, W_1 \subseteq \mathrm{bdd}(XB) \cap \mathrm{bdd}(YB)$$

and

$$X \underset{W_0B}{\perp} Y \text{ and } X \underset{W_1B}{\perp} Y.$$

Then $\mathrm{bdd}(W_0B) = \mathrm{bdd}(W_1B)$, since $W_1 \perp_{W_0B} Y$ and $W_0 \perp_{W_1B} Y$.

(2) Assume that $D \subseteq F \subseteq G(D)$, and let $(F, \mathrm{cl}(-) \cap F)$ form a modular pregeometry. Then for any $A \subseteq D^{\mathrm{heq}}$, there is $B \subseteq F$ such that A and B are interbounded over $\mathrm{dom}(D)$. Note that there is $A_0 \subseteq D$ such that $A \subseteq \mathrm{dcl}(A_0 \cup \mathrm{dom}(D))$. Let $A_1 (\subseteq D)$ realizing $\mathrm{tp}(A_0/A\,\mathrm{dom}(D))$ be independent from A_0 over $A\,\mathrm{dom}(D)$. Let $B = \mathrm{bdd}(A_0\,\mathrm{dom}(D)) \cap \mathrm{bdd}(A_1\,\mathrm{dom}(D)) \cap F$. Since F is modular, A_0 and A_1 are independent over $B\,\mathrm{dom}(D)$. Then by (1), $\mathrm{bdd}(B\,\mathrm{dom}(D)) = \mathrm{bdd}(A\,\mathrm{dom}(D))$.

Hence, if $a \in G(D)$, then there is a tuple $\bar{b}$ from F such that $a, \bar{b}$ are interbounded over $\mathrm{dom}(D)$. In particular, a component $b (\in F)$ of $\bar{b}$ is in $\mathrm{bdd}(a \cup \mathrm{dom}(D))$, so $a \not\perp_{\mathrm{dom}(D)} b$. Since $SU(a/\mathrm{dom}(D)) = 1$, we have $a \in \mathrm{bdd}(b)$; i.e., a, b are interbounded over $\mathrm{dom}(D)$. This says that the geometries of F and $G(D)$ are identical.

Now, we shall observe relationships between D, the solution set of a rank-1 Lascar strong type, being 1-basedness and having a modular pregeometry. We remind that $\mathrm{dom}(D)$ is omitted to $\emptyset$.

Fact 8.3.5 *If D has a locally modular pregeometry, then D is 1-based.*

Proof Let $p \in D$. Since (D, cl_p) is modular, by the similar argument in Remark 8.3.4(2), it easily follows that for any $A \subset D^{\mathrm{heq}}$ there is $A_0 \subset D$ containing p such that A and A_0 are interbounded over p (*).

Claim Given $A, B \subset D^{\mathrm{heq}}$, A and B are independent over $\mathrm{bdd}(Ap) \cap \mathrm{bdd}(Bp)$ (i.e., D is 1-based over p): By 6.1.2(3), it suffices to show $A \perp_U B$, where $U = \mathrm{bdd}(Ap) \cap \mathrm{bdd}(Bp) \cap D^{\mathrm{heq}}$. Now, due to (*), there is $U_0 \subset D$ such that U and U_0 are interbounded over p. Let $U_1 := \mathrm{bdd}(Ap) \cap \mathrm{bdd}(Bp) \cap D$. Then clearly $U_0 \subseteq U_1 \subseteq U$. Hence, U_0, U_1, and U are all interbounded over p. Moreover, since (D, cl_p) is modular, we have $A \perp_{U_1} B$, so as desired $A \perp_U B$. The claim is proved.

Now we claim that 6.1.2(6) holds in D, so D is 1-based. Let $I = \langle c_i | i \in \omega \rangle$ be an indiscernible sequence from D. By 2.1.6, we can clearly find $q \in D$ such that $q \perp I$ and I is indiscernible over q. Now by (*) with a proof similar to that of 6.1.2(4)$\Rightarrow$(6), $I \setminus \{c_0\}$ is a Morley sequence over qc_0. As q is independent from I, by transitivity $I \setminus \{c_0\}$ is Morley over c_0. ■

Contrary to the stable case 8.1.7, the converse of above 8.3.5 is false in general (Remark 8.3.7). Instead, we have the following equivalences.

Theorem 8.3.6 *The following are all equivalent:*

(1) *D (or D^{heq}) is 1-based.*

(2) *D is linear.*

(3) *$G(D)$ is linear.*

(4) *$(G(D), \mathrm{cl})$ has a modular pregeometry.*

(5) *$G(D)_A$ is linear, for any (some) small A.*

(6) *$(G(D)_A, \mathrm{cl}_A)$ is modular, for any (some) small A.*

Proof (1)⇒(2) Let D be 1-based. Suppose that $SU(ab/B) = 1$ ($a, b \in D$ and $B \subset D$). By 1-basedness, $C = \mathrm{Cb}(ab/B)$ is in $\mathrm{bdd}(ab)$. Now,

$$SU(abC) = SU(ab/C) + SU(C) = 1 + SU(C) = SU(C/ab) + SU(ab) \leq 2.$$

Hence, $SU(C) \leq 1$. Thus, D is linear.

(2)⇒(3) Assume that D is linear. Now let $x, y \in G(D)$, $B \subseteq G(D)$ and $SU(xy/B) = 1$ while $SU(xy) = 2$ (*). Let $C = \mathrm{Cb}(xy/B)$. Then

$$C \subseteq \mathrm{bdd}(B) \text{ and } SU(xy/C) = 1 \left(xy \underset{C}{\perp} B\right). \qquad (\star)$$

We will show that $C \subseteq \mathrm{bdd}(xy)$. Now $\{x, y\}$ is in the definable closure of a small subset of D. In fact, we can find an independent subset $F \cup \{a, b\}$ of D such that, $\{x, y\}$ and $\{a, b\}$ are interbounded over F, and F, $\{x, y\}$ are independent. Then there is $F' \subseteq D$ such that F and F' realize the same type over xy, and

$$F' \underset{xy}{\perp} abBF, \text{ hence } F' \underset{\emptyset}{\perp} xyabBF. \qquad (^{**})$$

Now, as F and F' realize the same type over xy, there is $\{a'b'\}$ such that abF and $a'b'F'$ realize the same type over xy. Note that

$$\mathrm{bdd}(xyF') = \mathrm{bdd}(a'b'F'). \qquad (^{***})$$

Claim 1 $SU(a'b'/F'B) = 1$: First, we note that $a'b' \not\perp_{F'} B$. (Otherwise, by (**) and (***), $\{x, y\}$ is independent from B, which contradicts (*).) Now if $SU(a'b'/F'B) = 0$, then by (***) $x, y \in \mathrm{bdd}(F'B)$, and then by (**) $x, y \in \mathrm{bdd}(B)$, again contradicting (*). Hence, $SU(a'b'/F'B) = 1$, and Claim 1 is proved.

Therefore by Claim 1 and the linearity of D, $SU(a'b'/\mathrm{bdd}(a'b') \cap \mathrm{bdd}(F'B)) = 1$. Now

$$\mathrm{bdd}(a'b') \cap \mathrm{bdd}(F'B) \subseteq \mathrm{bdd}(a'b'F') \cap \mathrm{bdd}(F'B) \subseteq \mathrm{bdd}(F'B).$$

Thus, by transitivity with Claim 1 and (***), we have $SU(a'b'/W) = 1$, where

$$W = \text{bdd}(a'b'F') \cap \text{bdd}(F'B) = \text{bdd}(xyF') \cap \text{bdd}(F'B). \qquad (\dagger)$$

Claim 2 $SU(xy/W) = SU(xy/WC) = 1(= SU(xy/C))$: Since $F' \subseteq W$, by (***) and ($\dagger$), it follows that $xy \not\perp_{F'} W$. Hence, $SU(xy/W) = 0$ or 1. But if $SU(xy/W) = 0$ then by (**) $xy \in \text{bdd}(B)$, which is not the case. Thus, $SU(xy/W) = 1$. Similarly, since $SU(xy/C) = 1$, if $SU(xy/WC) \neq 1$, then by ($\star$) and ($\dagger$), $xy \in \text{bdd}(WC) \subseteq \text{bdd}(CF'B) \subseteq \text{bdd}(F'B)$. Then by (**), $xy \in \text{bdd}(B)$. This contradicts (*) again. Hence, $SU(xy/WC) = 1$ and Claim 2 is proved.

Now since $\text{Ltp}(xy/C)$ is a canonical type, by Claim 2, $C \subseteq \text{bdd}(W)$. Therefore, by ($\dagger$), $C \subseteq \text{bdd}(xyF')$. Then $C \subseteq \text{bdd}(xy)$, since (**) with ($\star$) says that $C \perp_{xy} F'$. We have proved that $G(D)$ is linear.

(3)⇒(6),(4) Suppose that $G(D)$ is linear. Let $a, b \in G(D)$, and $SU(ab/A) = 2$, whereas $SU(ab/AB) = 1$ for $B \subseteq G(D)$. Now, there is a Morley sequence I of $\text{tp}(ab/AB)$ in $G(D)$ such that $SU(ab/AB) = SU(ab/ABI) = SU(ab/I) = 1$. By the linearity of $G(D)$, we have $c \in \text{bdd}(ab) \cap \text{bdd}(I) \cap \text{bdd}(AB)$, where $c = \text{Cb}(ab/I) = \text{Cb}(ab/AB)$, and $SU(c) = 1$. Hence, $c \in G(D)$ as in Remark 8.3.3(2). Moreover, by the rank calculation of $SU(cab/A)$, we have $SU(c/A) = 1$; i.e., $c \notin \text{bdd}(A)$. Thus, the pregeometry of $G(D)_A$ is modular by 8.1.1.

(4)⇒(1) This proof will be similar to that of Fact 8.3.5. Assume (4). Due to 6.1.2(3), it suffices to show for $A, B \subseteq D$ that $A \perp_U B$, where $U = \text{bdd}(A) \cap \text{bdd}(B) \cap D^{\text{heq}}$. Now, by Remark 8.3.4(2), there is $U_0 \subset G(D)$, which is interbounded with U. Let $U_1 := \text{bdd}(A) \cap \text{bdd}(B) \cap G(D)$. Then $U_0 \subseteq U_1 \subseteq U$ and U_1, U are interbounded. From (4), we have $A \perp_{U_1} B$, so $A \perp_U B$ follows as wanted.

(3)⇒(5) Suppose that $G(D)$ is linear. Let us show that $G(D)_A$ is linear. Let $a, b \in G(D)$, and $SU(ab/B) = 1$ for B containing A. Now there is a Morley sequence I of $\text{tp}(ab/B)$ in $G(D)$ such that $SU(ab/B) = SU(ab/BI) = SU(ab/I) = 1$. Then by the linearity of $G(D)$, we have $SU(c/A) \leq SU(c) \leq 1$ where $c = \text{Cb}(ab/I) = \text{Cb}(ab/B)$. Thus, $G(D)_A$ is linear.

(5) or (6)⇒(3) Let $a, b \in G(D)$, $B \subseteq G(D)$, and $SU(ab/B) = 1$. Let $c = \text{Cb}(ab/B)$. If $SU(ab) = 1$; i.e., $\{a, b\}$ is independent from B, then as $c \in \text{bdd}(\emptyset)$, there is nothing to prove. Thus, we can assume that $SU(ab) = 2$. Now we can further assume that abB is independent from A. Hence, $SU(ab/BA) = 1$, whereas $SU(ab/A) = 2$. Then by the modularity (or linearity) of $G(D)_A$, there is $d \in \text{bdd}(abA) \cap \text{bdd}(AB)$ and $SU(d/A) = 1$. Hence, d and ab are dependent over A. Thus $SU(ab/Ad) = 1 = SU(ab/AB)$. Therefore, $c = \text{Cb}(ab/AB) \in \text{bdd}(Ad)$, and as $SU(d/A) = 1$, by a rank calculation, we obtain $SU(c/A) \leq 1$. Since c and A are independent, $SU(c) \leq 1$. ■

Remark 8.3.7

(1) Similarly, it follows that D is linear iff D_A is linear for any (some) A. Note that the geometry of a non-forking extension of D over A is mere a subgeometry of D_A. But still linearity (1-basedness) is invariant under parallelism. Let q be a non-forking

extension of D over A. It suffices to check that D is linear when q is linear. The proof is the same as 8.3.6(5)⇒(3) with a trick that when an independent pair (a_0, a_1) is chosen from D, by amalgamation there is (b_0, b_1) such that $\text{tp}(a_0, a_1) = \text{tp}(b_0, b_1)$ and each b_i realizes q.

However, (D, cl) being locally modular is no longer invariant via parallelism in the simple theory context (while clearly 1-basedness is so). An affine or projective space equipped with a binary random graph relation as in Section 6.3.1 serves an example, by the similar reason described in (2) below.

(2) In general, even if D is 1-based, D_A need not be modular over a set A. Let V be a (saturated) affine space over a field, and let P be a generic predicate on V, as in Section 6.3.1. Then there are two 1-types over $\emptyset$ isolated by $P(x)$ and $\neg P(x)$, both of which have SU-rank 1. By the genericity of P, both 1-types are 1-based. V is already non-modular over $\emptyset$. Now let W be a non-empty small subspace of V and consider a line l, spanned by $\{a, b\}$ in V, independent from W, so $\dim(l/W) = 2$. Now let l', spanned by $\{a'b'\}$, be parallel to l such that $\dim(l/Wl') = 1$. Then there is $c \in V \setminus W$ such that $\text{acl}(cW) = \text{acl}(lW) \cap \text{acl}(l'W)$. However, as W is small, by genericity of P, we can find l, l' such that $\{a, b, a', b'\} \subseteq P$, whereas $P \cap \text{acl}(cW) = P \cap W$. Then clearly P cannot be modular over W and similarly for $\neg P$.

In conclusion, there are simple unstable examples of non-trivial, 1-based D that are either *non-locally modular* or *locally modular but neither affine nor projective* (Section 8.2). These say that Fact 8.1.7 (the stable case) is only a particular case of Theorem 8.3.6 (the simple case).

Now since triviality of a (pre)geometry implies modularity, we have the following lemma by Remark 8.3.4(2).

Lemma 8.3.8 *D is trivial iff $G(D)$ is trivial.*

Theorem 8.3.9 *Suppose D is 1-based. Then D is trivial or the geometry of $G(D)$ is a projective geometry over a division ring.*

Proof Assume that D is non-trivial. By the above lemma and Fact 8.1.3, it suffices to show that any independent pair $(a, b) \in G(D)$ (that is, $\dim(ab) = 2$) is non-trivial in $G(D)$ (i.e., there is $c \in \text{cl}(ab) \setminus (\text{cl}(a) \cup \text{cl}(b))$ where cl is the closure of $G(D)$).

Claim 1 Any independent pair $(c, d) \in D$ is non-trivial in $G(D)$: Since $G(D)$ is modular, if there is an independent pair from D trivial in $G(D)$, then any independent pair from D is trivial in $G(D)$. (Otherwise, by type amalgamation one can find a non-modular triple that violates modularity of $G(D)$. See the proof of 8.1.8.) Then any independent triple $(a, b, c) \in D$ is also trivial in $G(D)$ (otherwise, say $d \in \text{cl}(abc) \setminus (\text{cl}(a) \cup \text{cl}(b) \cup \text{cl}(c))$, then ab and cd violate modularity of $G(D)$). Iterating this argument shows that D is trivial, which is not the case.

Claim 2 Any independent pair $(a, b) \in G(D)$ with $a \in D$ is non-trivial in $G(D)$: Suppose not; say the independent pair $(a, b) \in G(D)$ with $a \in D$ is trivial. Now since $b \in G(D)$, there is an independent tuple $\bar{c} = (c_1, \ldots c_n) \in D$ such that $b \in \mathrm{cl}(c_1, \ldots c_n)$. By Claim 1, we can assume that $n > 1$. We can further assume that any $n - 1$ subtuple of $\bar{c}$ is independent from b, and $\{a, b, c_1, \ldots, c_{n-1}\}$ is independent. Hence, $\dim(a, b, \bar{c}) = n + 1$ (*). By Claim 1, there is $d \in \mathrm{cl}(ac_1) \setminus \mathrm{cl}(a) \cup \mathrm{cl}(c_1)$. Then it follows that $d \notin \mathrm{cl}(c_2, \ldots, c_n)$, (otherwise, $a \in \mathrm{cl}(\bar{c})$ violating (*)). Hence, the dimension of the set $B = \{d, c_2, \ldots, c_n\}$ is n, whereas by (*) and triviality of (a, b), it can easily be seen that $\mathrm{cl}(B) \cap \mathrm{cl}(ab) = \emptyset$. However, due to the modularity of $G(D)$, this contradicts (*). Hence, Claim 2 follows.

Claim 3 Any independent pair $(d, e) \in G(D)$ is non-trivial: Suppose not; say the independent pair $(d, e) \in G(D)$ is trivial. There is a point $c \in D$ that is independent from (d, e). By Claim 2, (c, d) and (c, e) are non-trivial. Again this violates modularity of $G(D)$, as (c, d, e) forms a non-modular triple. This finishes the proof of Claim 3 and the theorem. ∎

If the geometry of (D, cl) is modular, then it is isomorphic to that of $(G(D), \mathrm{cl})$ by 8.3.4(2). It is an easy exercise to show that the same equivalence holds between (D_p, cl_p) and $G(D)$, when (D, cl) is locally modular.

Now for rest of this section, we assume D **is 1-based.**

The (pre)geometry of $G(D)$ need not be definable even if D is a definable set. However, with the reducibility condition we shall discuss below, $G(D)$ can be dealt with as a definable object when D is ω-categorical.

Definition 8.3.10 We say $G(D)$ is *reducible into* (or has a reduction in) D^k if for any point $a \in G(D)$ there is a k-tuple $\bar{b} \in D$ such that $a \in \mathrm{bdd}(\bar{b})$.

Indeed, the following says that $G(D)$ is reducible into D^3 for 1-based D.

Theorem 8.3.11 *Let $u \in G(D)$.*

(1) *Suppose that (x, y) is a fixed independent pair from D. Then $u \in \mathrm{bdd}(x'y'z)$, where $x', y', z \in D$ and $\mathrm{tp}(xy) = \mathrm{tp}(x'y')$.*

(2) *Suppose that there is independent $\{x, y\}$ from D such that for any $z \in \mathrm{bdd}(xy) \cap G(D)$, we have $z \in \mathrm{bdd}(z')$ for some $z' \in \mathrm{bdd}(xy) \cap D$; i.e., a full line is already in D. Then $G(D)$ is reducible into D^2.*

(3) *For given $d_1 \in D$, there are $d_2, d_3 \in D$ such that $u \in \mathrm{bdd}(d_1 d_2 d_3)$.*

Proof We first prove (1) using induction on n where n, is the length of a sequence from D over which u is bounded. The case $n = 0$, that is, $u \in \mathrm{bdd}(\emptyset)$, is trivial. Now suppose that $u \in \mathrm{bdd}(a_1 \ldots a_{n+1})$ $(a_i \in D)$ with the induction hypothesis for n. We will verify (1) for $n + 1$. By choosing a basis, we can assume that $\{a_1, \ldots, a_{n+1}\}$ is independent. Now let $g = \mathrm{Cb}(ua_1/a_2 \ldots a_{n+1})$. Thus, $SU(g) = 1$ and $g \in G(D)$. Since $a_1 \notin \mathrm{bdd}(g) \subseteq \mathrm{bdd}(a_2 \ldots a_{n+1})$,

$$u \text{ is in the line generated by } \{g, a_1\}. \qquad (*)$$

Then, since $g \in \mathrm{bdd}(a_2 \ldots a_{n+1})$, by the induction hypothesis, $g \in \mathrm{bdd}(bcd)$ for some $c, d \in D$ such that $\mathrm{tp}(cd) = \mathrm{tp}(xy)$. If g is already in $\mathrm{bdd}(cd)$, then from (*), we have $u \in \mathrm{bdd}(ga_1) \subseteq \mathrm{bdd}(cda_1)$. Thus, in this case, the theorem holds. Therefore, we only need to consider the case when $g \notin \mathrm{bdd}(cd)$. We can also clearly assume that $g \notin \mathrm{bdd}(b)$ (otherwise, we are done). Then, since $G(D)$ has a modular geometry, we can find that $v \in \mathrm{bdd}(cd) \cap \mathrm{bdd}(bg) \cap G(D)$ such that $g \in \mathrm{bdd}(bv)$ (†), and $\dim(vg) = 2$. Also, at least one of $\{v, c\}$ or $\{v, d\}$ (say $\{v, d\}$) is independent (**). Now, since $d, a_1 \in D$, we have $\mathrm{Ltp}(d) = \mathrm{Ltp}(a_1)$. Hence, we can find v' such that $\mathrm{Ltp}(vd) = \mathrm{Ltp}(v'a_1)$ (⋆). In particular, $\mathrm{Ltp}(v) = \mathrm{Ltp}(v')$. Then by (*),(**),(⋆), we can amalgamate types $\mathrm{tp}(v'/a_1)$ and $\mathrm{tp}(v/g)$, so that we obtain $v'' \models \mathrm{tp}(v'/a_1) \cup \mathrm{tp}(v/g)$. Hence, there are b', c' such that $\mathrm{tp}(vgb) = \mathrm{tp}(v''gb')$, $\mathrm{tp}(vdc) = \mathrm{tp}(v''a_1c')$ (by (⋆)), then (†),

$$\mathrm{tp}(c'a_1) = \mathrm{tp}(xy),\ v'' \in \mathrm{bdd}(c'a_1),\ \text{and}\ g \in \mathrm{bdd}(b'v''). \tag{‡}$$

Thus, by (*),(‡),

$$u \in \mathrm{bdd}(ga_1) \subseteq \mathrm{bdd}(ga_1v'') \subseteq \mathrm{bdd}(a_1b'v'') \subseteq \mathrm{bdd}(c'a_1b').$$

Since $\mathrm{tp}(c'a_1) = \mathrm{tp}(xy)$, the $(n+1)$th induction hypothesis for (1) is deduced. Therefore, we have proved (1).

(2) comes from (1) and the modularity of $G(D)$.

For (3), use the proof of (1), replacing g by u, and a_1 by d_1. ■

Remark 8.3.12 From the theorem above, we can identify any element u in $G(D)$ as $\bar{d}/\bar{E} \in G(D)$ (i.e., $\mathrm{bdd}(u) = \mathrm{bdd}(\bar{d}/\bar{E})$) where $\bar{d} \in D^6$ and $\bar{E}$ is a type-definable equivalence relation on D^6 (over $\mathrm{dom}(D)$), as follows. As before we take $\mathrm{dom}(D) = \emptyset$. By 8.3.11, there are triples $(a, b, c), (a', b', c') \in D^3$ such that

$$u \in \mathrm{bdd}(abc),\ (a'b'c') \models \mathrm{tp}(abc/u),\ \text{and}\ abc \mathop{\perp}_{u} a'b'c'.$$

Then as in Remark 8.3.4(1), we have $\mathrm{bdd}(u) = \mathrm{bdd}(g)$, where $g = \mathrm{Cb}(abc/a'b'c')$. Now by 4.3.9(4), we have $g \in \mathrm{dcl}(abca'b'c')$. Hence, due to 4.1.5(3), we can identify

$$g = (abca'b'c')/\bar{E}\ (\in G(D)),$$

where $\bar{E}$ is a type-definable equivalence relation on D^6.

In particular, if D is ω-categorical, then as there are only finitely many types and equivalence relations over $\mathrm{dom}(D)$ on D^6, by the above remark, the geometry of $G(D)$ is isomorphic to that of some union of finitely many definable sets in finitely many sorts in $G(D)$. Since the union itself is ω-categorical even as a 1-sorted structure, we have the following results.

Corollary 8.3.13 *D ω-categorical. Then, either D is trivial or the geometry of G(D) is projective over a finite field and there is an ω-categorical strongly minimal reduction in D^{eq} preserving the same geometry as G(D).*

Now, we study the relationship between the whole theory and rank-1 types. As in the stable case, the notion of triviality can be imposed on the underlying theory.

Definition 8.3.14 We say that T is *trivial* if for hyperimaginaries a, b, c, A, whenever $\{a, b, c\}$ is pairwise independent over A, then $\{a, b, c\}$ is independent over A.

In the proof below, the following easy fact will be used. In any simple T, if a hyperimaginary a is not in $\operatorname{bdd}(\emptyset)$, then there is a complete extension p of $\operatorname{tp}(a)$ such that $SU(p) = 1$. Let $\operatorname{tp}(a) = p_0 \subseteq p_1 \ldots \subseteq p_i \ldots$ be a maximal non-bounded forking chain. (The length is bounded.) Then $p = \bigcup_i p_i$ is non-bounded as well and, by maximality, $SU(p) = 1$.

Lemma 8.3.15 *Suppose that T is 1-based. Then T is trivial if and only if all rank-1 types are trivial.*

Proof We only need to prove the right-to-left direction, as the other direction is obvious. Suppose that there exists a non-trivial triple (a, b, c) over A; i.e., (a, b, c) is pairwise independent over A, whereas $\{a, b, c\}$ is dependent over A. We will find a non-trivial rank-1 type.

Claim 1 We may assume that $a \in \operatorname{bdd}(bcA)$, $b \in \operatorname{bdd}(acA)$, and $c \in \operatorname{bdd}(baA)$ (but $\{a, b, c\}$ is still pairwise independent over A): If we choose

$$a_0 \in \operatorname{Cb}(a/bcA) - \operatorname{bdd}(A),\ b_0 \in \operatorname{Cb}(b/a_0cA) - \operatorname{bdd}(A),\ c_0 \in \operatorname{Cb}(c/a_0b_0A) - \operatorname{bdd}(A)$$

such that $\operatorname{Cb}(a/bcA) \subseteq \operatorname{bdd}(a_0)$ and similarly for b_0, c_0, then, by 1-basedness, these elements will work. Claim 1 is shown.

Now, for notational convenience, put $A = \emptyset$.

Claim 2 There is F such that $SU(a/F) = SU(b/F) = SU(c/F) = 1$ and (a, b, c) is still a non-trivial triple over F: As pointed out before 7.3.15, there is d such that $SU(a/d) = 1$, and we can assume that $d = \operatorname{Cb}(a/d)$. Then by 1-basedness, we have $d \in \operatorname{bdd}(a)$, and

$$bd \perp_F cd \text{ where } F = \operatorname{bdd}(bd) \cap \operatorname{bdd}(cd).$$

Then $b \perp_F c$. Now we have $SU(a/F) = 1$, since otherwise $a \in \operatorname{bdd}(bd)$, so $a \in \operatorname{bdd}(d)$ (as $ad \perp b$), contradicting that $SU(a/d) = 1$. Similarly, $a \perp_F b$ and $a \perp_F c$. Now by the rank calculation of $SU(ab/cF)$, it follows that $SU(b/cF) = SU(b/F) = 1$ since $SU(a/cF) = 1$. Similarly, by the rank calculation of $SU(ac/bF)$, we have $SU(c/bF) = SU(c/F) = 1$. Hence, Claim 2 is proved.

Claim 3 $\operatorname{tp}(a/F)$ is non-trivial. (This finishes the proof): Firstly, there are a_1, a_2, b', c' such that

$$\operatorname{tp}(a_1c'/Fb) = \operatorname{tp}(ac/Fb),\ c' \underset{F}{\perp} abc,\ \operatorname{tp}(a_2b'/Fc) = \operatorname{tp}(ab/Fc) \qquad (*)$$

and $b' \mathop{\smile\!\!\!\!\!\!\;\mid}_F abcc'$. Hence, $b'c' \mathop{\smile\!\!\!\!\!\!\;\mid}_F abc$ (**). Now we subclaim that a, a_1, a_2, all realizing the rank-1 type $\mathrm{tp}(a/F)$ form a non-trivial triple over $Fb'c'$. For pairwise independence, we only show $a_1 \mathop{\smile\!\!\!\!\!\!\;\mid}_{Fb'c'} a_2$; the other cases follow similarly. We have $b \mathop{\smile\!\!\!\!\!\!\;\mid}_F c$, hence $b \mathop{\smile\!\!\!\!\!\!\;\mid}_{Fb'c'} c$ by (**). Thus, $Fbc' \mathop{\smile\!\!\!\!\!\!\;\mid}_{Fb'c'} Fcb'$. Then it follows that $a_1 \mathop{\smile\!\!\!\!\!\!\;\mid}_{Fb'c'} a_2$ by Claim 2 and (*). Now, as $a \in \mathrm{bdd}(bc)$, again by (*) and Claim 2, each of a, a_1, a_2 is bounded over $Fb'c'$ together with the other two. Hence, the subclaim is verified.

Secondly, by 1-basedness, there is a tuple $a_3a_4a_5$ realizing $\mathrm{tp}(aa_1a_2/Fb'c')$ such that

$$aa_1a_2 \mathop{\smile\!\!\!\!\!\!\;\mid}_{a_3a_4a_5} Fb'c' \text{ and } aa_1a_2 \mathop{\smile\!\!\!\!\!\!\;\mid}_{Fb'c'} a_3a_4a_5.$$

Then, it can be easily seen that (a, a_1, a_2) forms a non-trivial triple over $Fa_3a_4a_5$. This means that the localized geometry of $\mathrm{tp}(a/F)$ at $\{a_3, a_4, a_5\}$ is non-trivial since all the a and a_i realize the rank-1 type $\mathrm{tp}(a/F)$. Hence, the original type $\mathrm{tp}(a/F)$ cannot be trivial either. We have proved Claim 3 and the lemma. ■

By previous results in this section, we obtain the following.

Theorem 8.3.16 *Let T be non-trivial, 1-based and ω-categorical. Then an infinite dimensional vector space over a finite field, in particular the infinite additive group, is definable in $\mathcal{M}^{\mathrm{eq}}$ over a finite parameter.*

8.4 Bibliographical Remarks

Most of the major results in this chapter are from [32]. Lemma 8.3.15 is obtained by mimicking the proof for the stable case [99, 4.2.5]. The equivalent of (1) and (2) in Theorem 8.3.6 was obtained first by Vassiliev [116] using his notion of a generic pair, which simply is a lovely pair (in Section 6.3.3) for the SU-rank 1 structure D. He also showed [117] that the generic pair of D only takes values of SU-rank 1 (if and only if D is trivial), 2 (if and only if D is non-trivial 1-based), or ω, generalizing the corresponding result when D is strongly minimal [10].

9

Generalized Amalgamation and the Group Configuration Theorem

In this introduction, we work in $\mathcal{M}^{\mathrm{eq}}$ of a *stable* theory, and tuples are finite. Let again D be the solution set of a strong type having SU-rank 1. As mentioned in Chapter 8, if D is ω-categorical non-trivial, then Zilber showed it is 1-based and a vector space over a finite field having the same geometry can definably be recovered. Later Hrushovski extended the result as follows [48].

Fact 9.0.1 *If the stable D is 1-based non-trivial, then there is a type-definable group $(G, +)$ such that $(V = G/G_0, +)$, where $G_0 = \mathrm{acl}(\mathrm{dom}(G)) \cap G$ forms a vector space over the division ring of the definable endomorphisms of V. Moreover, there $\mathop{\smile\!\!\!\!\!\mid}$ is linear independence (see Chapter 1).*

In other words, Hrushovski gives a positive answer to 8.0.3(2)(b) (in fact, (a) as well) for stable theories even in the non ω-categorical context. When he constructs the group, the following group configuration theorem plays a central role.

Definition 9.0.2 By *a group configuration* over a set A we mean a 6-tuple $C = (f_1, f_2, f_3, x_1, x_2, x_3)$ such that, for $\{i, j, k\} = \{1, 2, 3\}$,

(1) $f_i \in \mathrm{acl}(f_j, f_k; A)$,

(2) $x_i \in \mathrm{acl}(f_j, x_k; A)$, and

(3) All other triples from C are independent over A.

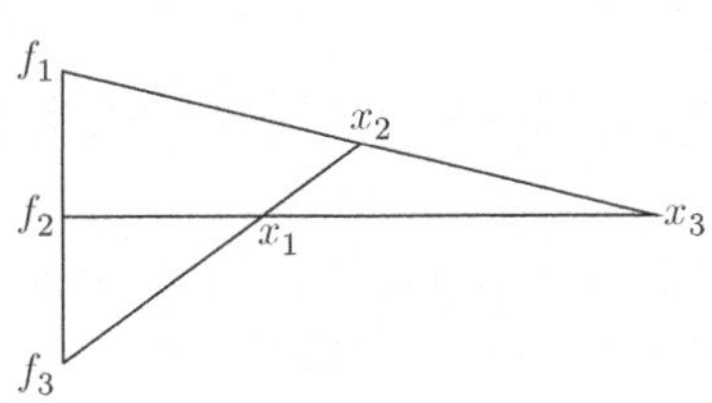

Figure 9.1

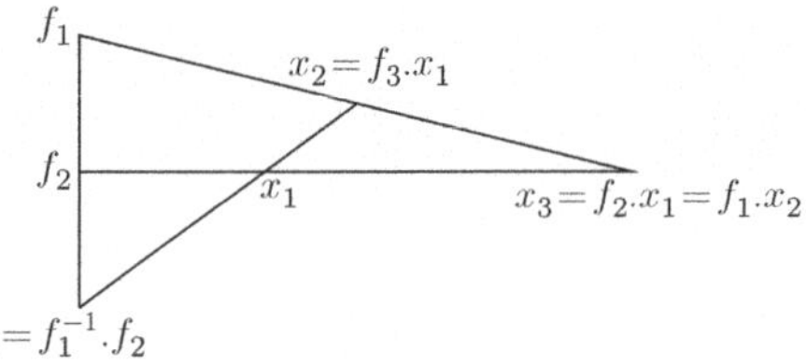

Figure 9.2

If there is $C' = (f_1', f_2', f_3', x_1', x_2', x_3')$ such that $\mathrm{acl}(f_iA) = \mathrm{acl}(f_i'A), \mathrm{acl}(x_iA) = \mathrm{acl}(x_i'A)$, then C' also forms a group configuration over A. In this case, we say C and C' are *equivalent* over A.

Definition 9.0.3 We say $C = (f_1, f_2, f_3, x_1, x_2, x_3)$ is a *group configuration from a type-definable homogeneous space* (G, X) over A if (G, X) and the group action of G on X are all type-definable over A, and $f_i \in G$, $x_i \in X$ are generic elements such that $\{f_1, f_2, x_1\}$ is A-independent; $f_1^{-1}.f_2 = f_3$; $x_2 = f_3.x_1$; and $x_3 = f_2.x_1$. (Of course, then C is a group configuration over A.)

Theorem 9.0.4 (Hrushovski [51]) (The group configuration theorem for stable theories) *Assume T is stable. A group configuration $C = (f_1, f_2, f_3, x_1, x_2, x_3)$ over $\emptyset$ is given. Then a type-definable homogeneous space (G, X) over $A(\mathop{\smile\!\!\!\!\!\!|}\,C)$ is constructed so that C is equivalent over A to some group configuration from (G, X).*

The group configuration theorem also plays a key role in producing deep results in *geometric stability theory*. (For a complete exposition of it, see [99]). For example, it is pivotal in the proof of the dichotomy theorem for Zariski' structures (see [59].) It is fair to say the group configuration theorem is one of the foundational theorems in geometric stability theory and its applications to algebraic geometry.

The whole topic of this chapter is how to generalize the group configuration theorem to the simple theory context. Surprisingly, there is a deep connection between this question and the so-called notion of generalized amalgamation. We will discuss the issue in the following section. In the rest of this chapter T **is simple** and we work with hyperimaginaries. We shall show that under 4-amalgamation, Fact 9.0.1 and Theorem 9.0.4 are fully generalized in the context of simple theories with hyperimaginaries (Theorem 9.4.1, Theorem 9.5.2); so it again answers 8.0.3(2)(b) positively.

9.1 Generalized Amalgamation

We introduce the concept of generalized amalgamation. This will be used to achieve the group configuration theorem for simple theories, but the notion itself is very intriguing in its own right.

Definition 9.1.1 Let $I = \mathcal{P}(n)^{-}(= \mathcal{P}(n) \setminus \{n\})$, ordered by inclusion. Let $(\{A_i\}_{i \in I}, \{\pi_j^i\}_{i \leq j \in I})$ be a directed family. Namely, each $\pi_j^i : A_i \to A_j$ is an elementary map; $\pi_i^i = \mathrm{id}_{A_i}$, and $\pi_k^j \circ \pi_j^i = \pi_k^i$ for $i \leq j \leq k \in I$. We say T has *n-amalgamation* if whenever

(1) $\{\pi_u^{\{i\}}(A_{\{i\}}) : i \in u\}$ is $\pi_u^{\emptyset}(A_{\emptyset})$-independent,
(2) $A_u = \mathrm{acl}(\bigcup_{i \in u} \pi_u^{\{i\}}(A_{\{i\}}))$,

for any $u \in I$, then we can extend the direct family to the one indexed by $\mathcal{P}(n)$ (by finding A_n and π_n^j) so that (1) and (2) hold for n, too. We say T has *n-amalgamation* over A, if $A_{\emptyset} = \mathrm{bdd}(A) = A$.

One can restate the definition in terms of functorial terminology. Recall that by a *category* $\mathcal{C} = (\mathrm{Ob}(\mathcal{C}), \mathrm{Mor}(\mathcal{C}))$, we mean a class $\mathrm{Ob}(\mathcal{C})$ of members called *objects* of the category, equipped with a class $\mathrm{Mor}(\mathcal{C}) = \{\mathrm{Mor}(a, b) | \ a, b \in \mathrm{Ob}(\mathcal{C})\}$, where $\mathrm{Mor}(a, b) = \mathrm{Mor}_{\mathcal{C}}(a, b)$ is the class of *morphisms* between objects a and b (we write $f : a \to b$ to denote $f \in \mathrm{Mor}(a, b)$, and a and b are said to be a source and a target of f, resp.) and composition maps

$$\circ : \mathrm{Mor}(a, b) \times \mathrm{Mor}(b, c) \to \mathrm{Mor}(a, c)$$

for each $a, b, c \in \mathrm{Ob}(\mathcal{C})$ such that:

(1) (Associativity) If $f : a \to b$, $g : b \to c$, and $h : c \to d$ then $h \circ (g \circ f) = (h \circ g) \circ f$ holds; and
(2) (Identity) For each object c, there exists a morphism $1_c : c \to c$ called the identity morphism for c, such that for $f : a \to b$, we have $1_b \circ f = f = f \circ 1_a$.

A *groupoid* is a category in which every morphism is invertible; i.e., for any $f \in \mathrm{Mor}(a, b)$, there is $g \in \mathrm{Mor}(b, a)$ such that $1_a = g \circ f$. So in a groupoid, $(\mathrm{Mor}(a, a), \circ)$ forms a group, called the *vertex group* of a. Vertex groups of objects a, b are isomorphic if they are connected (i.e., $\mathrm{Mor}(a, b) \neq \emptyset$).

Note that any ordered set $(P, \leq)$ is a category where objects are members of P, and $\mathrm{Mor}(a, b) = \{(a, b)\}$ if $a \leq b$; $= \emptyset$, otherwise. Given $\mathcal{M} \models T$, $\mathcal{C}_T$ (or $\mathcal{C}_{\mathcal{M}}$) denotes the category of boundedly closed sets; i.e.,

$$\mathrm{Ob}(\mathcal{C}_{\mathcal{M}}) = \left\{\mathrm{bdd}(A) | \ A \text{ a small subset of } \mathcal{M}\right\}$$

and morphisms are elementary maps.

Now we recall a functor F between two categories $\mathcal{C}$ and $\mathcal{D}$. The functor F sends an object $c \in \mathrm{Ob}(\mathcal{C})$ to $F(c) \in \mathrm{Ob}(\mathcal{D})$, and a morphism $f \in \mathrm{Mor}_{\mathcal{C}}(a, b)$ to $F(f) \in \mathrm{Mor}_{\mathcal{D}}(F(a), F(b))$ in such a way that

(1) (Associativity) $F(g \circ f) = F(g) \circ F(f)$ for $f : a \to b, g : b \to c$;
(2) (Identity) $F(1_c) = 1_{F(c)}$.

Definition 9.1.2

(1) A functor $a : W(\subseteq \mathcal{P}(n)) \to \mathcal{C}_T$ is given, where W is closed under subsets. For $u \subseteq w \in W$, we write $a(w) = a_w$, and a'_u (within a_w) means $a((u,w))(a_u)$. Now a is said to be *independence preserving* (*i.p.*) if for $w \in W$,

(a) $\{a'_{\{i\}} | i \in w\}$ is $a'_\emptyset$-independent (within a_w); and

(b) $a_w = \text{bdd}(\bigcup\{a'_{\{i\}} | i \in w\})$.

(2) We say T has *n-complete amalgamation* (n-CA) over $B = a_\emptyset = \text{bdd}(a_\emptyset)$ if any i.p. functor $a : W \to \mathcal{C}_T$ can be extended to i.p. $\hat{a} : \mathcal{P}(n) \to \mathcal{C}_T$.

Equivalently, we can define n-CA in terms of types.

Definition 9.1.3 We say T has *n-complete amalgamation* (n-CA) over a set $B = \text{bdd}(B)$ if the following holds: Let W be a collection of subsets of $\mathcal{P}(n)$, closed under subsets. For each $w \in W$, a complete type $r_w(x_w)$ over B is given where x_w is a set of variables for hyperimaginaries. Suppose that

(1) For $w \subseteq w'$, we have $x_w \subseteq x_{w'}$ and $r_w \subseteq r_{w'}$.

Moreover, for any $a_w \models r_w$,

(2) $\{a_{\{i\}} | i \in w\}$ is B-independent,

(3) a_w is as a set $\text{bdd}(\bigcup_{i \in w} a_{\{i\}}B)$ (and the map $a_w \to x_w$ is a bijection).

Then there is a complete type $r_n(x_n)$ over B such that (1),(2),(3) hold for all $w \in W \cup \{n\}$. We say T has n-CA if it has n-complete amalgamation over any boundedly closed set.

It is more or less a straightforward exercise to show that T has n-CA over B iff T has m-amalgamation over B for all $m \leq n$. Moreover, the usual type-amalgamation 4.2.17 is indeed equivalent to 3-amalgamation. Hence, any simple theory has 3-amalgamation.

One may wonder why such a bit complicated amalgamation is taken as the generalized version of type amalgamation. Recall that the usual type amalgamation (or the independence theorem) is as follows: For $A_0 \perp_A A_1$ with $A \subseteq A_0, A_1$, if $c_0 \equiv^L_A c_1$ and $c_i \perp_A A_i$ $(i = 0, 1)$, then there is $c \equiv^L_{A_i} c_i$ such that $c \perp_A A_0A_1$. Note that we can think of A_0, A_1 (after naming or omitting A) as two vertices of a base edge of a triangle and c a top vertex, and $\text{Ltp}(c_0/A_0), \text{Ltp}(c_1/A_1)$ are the two (edge) types to be amalgamated. One would expect generalized amalgamation to be a natural generalization of this amalgamation (= 3-amalgamation), using a tetrahedron and higher dimensional simplices instead of a triangle (*). For example, the candidate for 4-amalgamation would be the following: Let $\{A_0, A_1, A_2\}$ be independent over $A(\subseteq A_i = \text{bdd}(A_i))$. For $\{i, j, k\} = \{0, 1, 2\}$, let $c_i \equiv_{A_j} c_k$ and $c_i \perp_A A_jA_k$. Then there is $c \equiv^L_{A_jA_k} c_i$ such that $c \perp_A A_0A_1A_2$ (**).

However, the above candidate for the definition of 4-amalgamation does not work, even if it looks natural.

Example 9.1.4 Let M be the random graph in $\mathcal{L} = \{R\}$. Choose distinct $a_i, b_i, d_i \in M$ and imaginary elements $c_i = \{b_i, d_i\}$ $(i = 0, 1, 2)$. We can additionally assume that

$R(b_2, a_0) \wedge R(d_2, a_1) \wedge \neg R(b_2, a_1) \wedge \neg R(d_2, a_0)$ and $\text{tp}(b_2 d_2; a_0 a_1) = \text{tp}(b_1 d_1; a_0 a_2) = \text{tp}(b_0 d_0; a_1 a_2)$. Now it follows that $\text{Ltp}(c_1/a_0) = \text{Ltp}(c_2/a_0), \text{Ltp}(c_0/a_1) = \text{Ltp}(c_2/a_1)$, and $\text{Ltp}(c_0/a_2) = \text{Ltp}(c_1/a_2)$. However, $\text{Ltp}(c_0/a_1 a_2)$, and $\text{Ltp}(c_1/a_0 a_2)$, $\text{Ltp}(c_2/a_0 a_1)$ have no common realization.

In the above example, $\{c_0, c_1, c_2\}$ can be considered as a base triangle, and $\text{Ltp}(d_0/c_0 c_1)$, and $\text{Ltp}(d_1/c_1 c_2)$, and $\text{Ltp}(d_2/c_2 c_0)$ form other three triangles attached to the base triangle. The example shows that even if the edges of the three triangles are compatible over the base vertices, there is no common vertex joining the three triangles. On the other hand, due to the nature of the random graph if we only work in the home sort, then any desired three types attached on a base triangle with compatible edges will be realized. As we want the notion of generalized amalgamation to be preserved in interpreted theories, we cannot take the above (**) as 4-amalgamation. Moreover, when we use inductive arguments for example, we often must consider not only the algebraic closures of vertices of amalgamated types but also those of higher dimensional surfaces as well, since after naming parameters the surface dimension is increasing.

Conclusively, the accurate definition for 4-amalgamation is then the one defined in 9.1.1–3. But note that indeed the essential philosophy of n-amalgamation mentioned in (*) is the same. We simply need to apply extra care not only to compatible surfaces but to the (enumerations of) bounded closures of all the lower dimensional surfaces.

The correct 4-amalgamation can be written in the following manner. Let $\{A_0, A_1, A_2\}$ be independent over $A(\subseteq A_i = \text{bdd}(A_i))$. For $\{i, j, k\} = \{0, 1, 2\}$, let $\overrightarrow{\text{bdd}(c_i A_j)} \equiv_{A_j} \overrightarrow{\text{bdd}(c_k A_j)}$ and $c_i \perp_A A_j A_k$, where $\overrightarrow{\text{bdd}(c_i A_j)}$ is a subsequence of a fixed sequence $\overrightarrow{\text{bdd}(c_i A_j A_k)}$.

Then there are c and an enumeration $\overrightarrow{\text{bdd}(c A_0 A_1 A_2)}$ such that

$$\overrightarrow{\text{bdd}(c A_j A_k)} \equiv_{\overrightarrow{\text{bdd}(A_j A_k)}} \overrightarrow{\text{bdd}(c_i A_j A_k)}$$

and $c \perp_A A_0 A_1 A_2$.

However, even in stable theories, n-amalgamation need not hold over an algebraically closed set. There is an example below.

Example 9.1.5

(1) Consider $[A]^2 = \{\{a, b\} | a \neq b \in A\}$ where A infinite. Let $B = [A]^2 \times \{0, 1\}$, where $\{0, 1\} = \mathbb{Z}/2\mathbb{Z}$. Also let $E \subseteq A \times [A]^2$ be a membership relation, and let P be a subset of B^3 such that $((w_1, \delta_1)(w_2, \delta_2)(w_3, \delta_3)) \in P$ iff there are distinct $a_1, a_2, a_3 \in A$ such that for $\{i, j, k\} = \{1, 2, 3\}$, $w_i = \{a_j, a_k\}$, and $\delta_1 + \delta_2 + \delta_3 = 0$.

Let $M = (A, [A]^2, B;\ E, P;\ \text{Pr}_1 : B \to [A]^2)$. Then since M is a reduct of $(A, \mathbb{Z}/2\mathbb{Z})^{\text{eq}}$, M is stable. We work in M^{eq} and argue that the example does not have 4-amalgamation over $\emptyset$. Note first that $\text{dcl}(\emptyset) = \text{acl}(\emptyset)$, and for $a \in A$, $\text{dcl}(a) = \text{acl}(a)$. Now choose distinct $a_1, a_2, a_3, a_4 \in A$. For $\{i, j, k\} \subseteq \{1, 2, 3, 4\}$, fix an enumeration $\overline{a_i a_j} = (b_{ij}, \ldots)$ of $\text{acl}(a_i a_j)$, where $b_{ij} = (\{a_i, a_j\}, \delta) \in B = [A]^2 \times \{0, 1\}$. Let $r_{ij}(x_{ij}) = \text{tp}(\overline{a_i a_j})$, and let x_{ij}^1 be the variable for b_{ij}. Note that $b_{ij} = (\{a_i, a_j\}, \delta)$ and $b'_{ij} = (\{a_i, a_j\}, \delta + 1)$ have the same type over $a_i a_j$. Hence, there is $(\overline{a_i a_j})' = (b'_{ij}, \ldots)$ also realizing $r_{ij}(x_{ij})$. Therefore, we have complete types $r_{ijk}(x_{ijk}), r'_{ijk}(x'_{ijk})$ both extending $r_{ij}(x_{ij}) \cup r_{ik}(x_{ik}) \cup r_{jk}(x_{jk})$ realized by some enumerations of $\mathrm{acl}(a_i a_j a_k)$ such that $P(x^1_{ij} x^1_{ik} x^1_{jk}) \in r_{ijk}$, whereas $\neg P(x^1_{ij} x^1_{ik} x^1_{jk}) \in r'_{ijk}$. Then it is easy to see that $r_{123} \cup r_{124} \cup r_{134} \cup r'_{234}$ is inconsistent.

(2) There is a standard way to consider a category as a first-order structure. A category consists of two sorts (Ob, Mor) representing the sets of objects and morphisms, and is equipped with maps $s : \mathrm{Mor} \to \mathrm{Ob}$, $t : \mathrm{Mor} \to \mathrm{Ob}$, and $\mathrm{id} : \mathrm{Ob} \to \mathrm{Mor}$ representing, respectively, the source and target of a given morphism, and the identity morphisms. Also there is a partial composition map between morphisms satisfying associativity and the identity rule.

Now given a finite group G, T_G is the theory of a groupoid that is connected (i.e., $\mathrm{Mor}(a, b) \neq \emptyset$ for any objects a, b) having a (every) vertex group G. It is easy to see that T_G is a totally categorical theory. In [41], it is observed that T_G has 4-amalgamation over $\emptyset$ iff $Z(G)$ is trivial. So, in particular, if G is non-trivial abelian, then T_G does not have 4-amalgamation over $\emptyset$.

Hrushovski [57] remedies the deficiency of generalized amalgamation for stable theories by introducing the so-called generalized imaginaries. More precisely, if $\mathcal{M} \models T$ is stable, then he is able to construct $(\mathcal{M} \subseteq)\mathcal{M}^* \models T^*$ in some $\mathcal{L}^*(\supseteq \mathcal{L})$ such that $\mathcal{M}$ is *stably embedded* into $\mathcal{M}^*$ (i.e., every subset of $\mathcal{M}$ definable in $\mathcal{M}^*$ is already definable in $\mathcal{M}$ within $\mathcal{L}(\mathcal{M})$), and $\mathcal{M}^*$ has n-amalgamation for all n over *any* algebraically closed sets. As the new elements in $\mathcal{M}^*$ can be called *generalized imaginaries*, we may write $\mathcal{M}^*$ as $\mathcal{M}^{\mathrm{geq}}$. In short, like $\mathcal{M}^{\mathrm{eq}}$, without loss of generality, we can freely work in $\mathcal{M}^{\mathrm{geq}}$ and assume to have n-amalgamation when T is stable.

Question Can we construct such $\mathcal{M}^*$ for simple T? If yes, then possibly we can remove the assumption of 4-amalgamation in the group configuration theorem for simple theories (Theorems 9.3.5, 9.5.2).

On the other hand, all stable theories have n-amalgamation over a model for all n (9.1.7). But for each $n > 2$, there is a simple unstable theory having n-CA but not having $(n + 1)$-CA over *any set* [77]. The examples also show n-amalgamation need not imply k-amalgamation for $k < n$.

Many important simple structures also have n-CA for all n such as the random graph (9.1.7), every PAC-structure (over some parameter) [55], and ACFA [23].

Now we prove any stable theory has n-CA over a model. As usual for a stable theory, we work in $\mathcal{M}^{\mathrm{eq}}$.

Lemma 9.1.6 *Let T be stable.*

(1) *Suppose that for a set C, whenever $a \mathop{\smile\!\!\!\!\!\!\,|}_C b$ with $b = b_1 \cup \ldots \cup b_n$, then*

$$\mathrm{dcl}(\mathrm{acl}(ab_1C) \ldots \mathrm{acl}(ab_nC)) \cap \mathrm{acl}(bC) = \mathrm{dcl}(\mathrm{acl}(b_1C) \ldots \mathrm{acl}(b_nC)). \quad (\sharp)$$

Then the following are satisfied.

(a) $\mathrm{tp}(\mathrm{acl}(ab_1C)\ldots\mathrm{acl}(ab_nC)/\mathrm{acl}(b_1C)\ldots\mathrm{acl}(b_nC))$ *is stationary.*

(b) *Let* $A = \{a_1,\ldots a_n\}$, $B = \{c_1,\ldots,c_n\}$ *be C-independent, respectively. For* $1 \leq i \leq n$, *let* $v_i = \{1,\ldots,n\} \setminus \{i\}$. *Now given* $k \leq n$, *assume there is a bijective map*

$$h : \bigcup_{1\leq i\leq k} \mathrm{acl}(a_{v_i}C) \to \bigcup_{1\leq i\leq k} \mathrm{acl}(c_{v_i}C),$$

where $a_{v_i} = \{a_j | j \in v_i\}$ *such that* $h(a_i) = c_i$, $h\lceil C = id$ *and, for each* v_i, $h\lceil\mathrm{acl}(a_{v_i}C)$ *is elementary. Then h is C-elementary.*

(2) *In fact, the condition* ($\sharp$) *holds when the set C is a universe of a model M. (Hence, (1)(a),(b) also are true over a model.)*

Proof We can safely assume a, a_i, b_i, c_i are finite tuples from $\mathcal{M}^{eq} = \mathcal{M}$.

(1) (a) is immediate from ($\sharp$) using $\mathrm{Cb}(c/d) \subseteq \mathrm{dcl}(cd) \cap \mathrm{acl}(d)$.

(b) For $1 \leq i \leq k$, let $h_i = h\lceil\mathrm{acl}(a_{v_i}C)$. Then put $h^j = h_1 \cup \ldots \cup h_j$ (where $h^k = h$) and $D_a^j = \mathrm{dom}(h^j) = \cup_{i=1,\ldots,j}\mathrm{acl}(a_{v_i}C)$ and $D_c^j = \mathrm{ran}(h^j) = \cup_{i=1,\ldots,j}\mathrm{acl}(c_{v_i}C)$. For induction, assume h^{j-1} is elementary for $j > 1$. We shall show that h^j is elementary too. Now, for each $i < j$ let $w_i = v_j \cap v_i$. Then $a_{v_i} = \{a_j\} \cup a_{w_i}$. Now since h_j is elementary, there is an automorphism $\hat{h}_j$ extending h_j. Then by induction $h^{j-1} \circ \hat{h}^{-1}$ maps $\hat{h}(D_a^{j-1})$ to D_c^{j-1}, so they have the same type, in particular, over the set $\bigcup_{i=1,\ldots,j-1}\mathrm{acl}(c_{w_i}C)$ fixed by $h^{j-1} \circ \hat{h}^{-1}$. Note now that $c_j \perp_C c_{w_1}\ldots c_{w_{j-1}}$ and $c_{v_i} = \{c_j\} \cup c_{w_i}$. Hence, we can apply (1)(a) to conclude that $\hat{h}(D_a^{j-1})$ and $D_c^{j-1} = \bigcup_{i=1,\ldots,j-1}\mathrm{acl}(c_{v_i}C)$ also have the same type over $\mathrm{acl}(c_{v_j}C)$; i.e., there is an elementary map g sending $\hat{h}(D_a^{j-1})$ to D_c^{j-1} fixing $\mathrm{acl}(c_{v_j}C) = \mathrm{ran}(h_j)$. Therefore, it follows that $h^j (\subseteq g \circ \hat{h})$ is elementary.

(2) It suffices to show that $e \in \mathrm{dcl}(\mathrm{acl}(b_1M)\ldots\mathrm{acl}(b_nM))$ for $e \in \mathrm{dcl}(\mathrm{acl}(ab_1M)\ldots\mathrm{acl}(ab_nM)) \cap \mathrm{acl}(bM)$. Since $e \in \mathrm{dcl}(\mathrm{acl}(ab_1M)\ldots\mathrm{acl}(ab_nM))$, there are $e_1\ldots e_n$ and $\mathcal{L}(M)$-formulas $\varphi(x; y_1\ldots y_n)$, $\psi_i(y_i, zw_i)$ with $\models \varphi(e; e_1\ldots e_n)$, $\models \psi_i(e_i, ab_i)$ such that $\models \varphi(u; v)$ implies that u is definable over vM, and $\psi_i(u', v')$ implies that u' is algebraic over $v'M$. Therefore,

$$\models \exists y_1\ldots y_n \left(\varphi(e, y_1\ldots y_n) \wedge \bigwedge_i \psi_i(y_i, ab_i)\right).$$

Now since $e \in \mathrm{acl}(bM)$, $a \perp_M eb$ and so $\mathrm{tp}(a/Meb)$ is a coheir extension of $\mathrm{tp}(a/M)$. Thus, we have $m \in M$ such that

$$\models \exists y_1\ldots y_n \left(\varphi(e, y_1\ldots y_n) \wedge \bigwedge_i \psi_i(y_i, mb_i)\right).$$

Hence, $e \in \mathrm{dcl}(\mathrm{acl}(b_1M)\ldots\mathrm{acl}(b_kM))$. ■

Proposition 9.1.7

(1) *Let T be stable. If a set C satisfies (♯) in 9.1.6, then for each n, T has n-CA over C.*
(2) *All stable theories have n-CA over a model.*
(3) *The random graph has n-CA over any set.*

Proof (1) In a stable theory T we can work in $\mathcal{M}^{eq}$ and substitute algebraic closures for bounded closures. We use the notation in 9.1.3. It suffices to show the case $W = \mathcal{P}(u_n)^-$ with corresponding types $r_w(x_w)$ (for $w \in W$). Again, for $1 \leq i < k \leq n$, let $v_i = \{1, \ldots, n\} \setminus \{i\}$ and $w_i = v_k \cap v_i$. We shall show that $\bigcup_{1 \leq i \leq n} r_{v_i}$ is consistent and realized by $\bigcup_{1 \leq i \leq n} a_{v_i}$ such that $\{a_{\{1\}}, \ldots, a_{\{n\}}\}$ is B-independent. (Then the type of its algebraic closure over B extending $\cup_{1 \leq i \leq n} r_{v_i}$ is the desired $r_{u_n}(x_{u_n})$.) Now due to usual amalgamation there is $a_{v_1} a_{v_2} \models r_{v_1} \cup r_{v_2}$ such that $\{a_{\{1\}}, \ldots, a_{\{n\}}\}$ is B-independent. Then for induction, assume that there is $a_{v_1} \ldots a_{v_{k-1}} \models r_{v_1} \cup \ldots \cup r_{v_{k-1}}$ (for $2 < k$) such that $a_{v_1} \ldots a_{v_{k-1}}$ extends $a_{\{1\}}, \ldots, a_{\{n\}}$. Now let $b_{v_k} \models r_{v_k}$. Then there is a map $h : \bigcup_{1 \leq i < k} b_{w_i} \to \bigcup_{1 \leq i < k} a_{w_i}$ such that h sends b_{w_i} to a_{w_i}. Hence, h is elementary by 9.1.6, and extends to an automorphism $\hat{h}$. Then we have $a_{v_k} = \hat{h}(b_{v_k}) \models r_{v_k}$. Now, $a_{v_1} \ldots a_{v_k}$ realizes $r_{v_1} \cup \ldots \cup r_{v_k}$ if for $y = x_{v_k} \cap (x_{v_1} \cup \ldots \cup x_{v_{k-1}})$, $a_{v_1} \ldots a_{v_{k-1}} \lceil y = a_{v_k} \lceil y$. But this clearly holds since by independence $x_{v_k} \cap x_{v_i} = x_{w_i}$, whence $y = x_{w_1} \cup \ldots \cup x_{w_{k-1}}$. This finishes the proof of (1).

(2) It follows from 9.1.6 and (1) above.

(3) Note that for the random graph $\mathcal{M} = (\bar{M}, R)$ we can work in $\mathcal{M}^{eq}$ and substitute algebraic closures for bounded closures. Now, since the random graph has weak elimination of imaginaries, for any A there is A' in the home sort $\bar{M}$ such that $\mathrm{acl}(A) = \mathrm{dcl}(A')$. Hence, when we check n-CA, we can assume each r_w is a type of a set in $\bar{M}$. Then in $\bar{M}$, since $\mathrm{tp}(A/B)$ is determined by equality and R relations of pairs in $A \cup B$, due to randomness of R we clearly have the generalized amalgamation. ■

The category theoretical treatments of generalized amalgamation lead to a development of homology theory for types in simple theories. But the topic is beyond the coverage of this book [42], [43].

9.2 The Group Configuration

We restate the definition of a group configuration in regard to hyperimaginaries.

Definition 9.2.1 By *a group configuration* we mean a 6-tuple of hyperimaginaries $C = (f_1, f_2, f_3, x_1, x_2, x_3)$ over a hyperimaginary e such that, for $\{i, j, k\} = \{1, 2, 3\}$,

(1) $f_i \in \mathrm{bdd}(f_j, f_k; e)$,
(2) $x_i \in \mathrm{bdd}(f_j, x_k; e)$,
(3) All other triples and all pairs from C are independent over e.

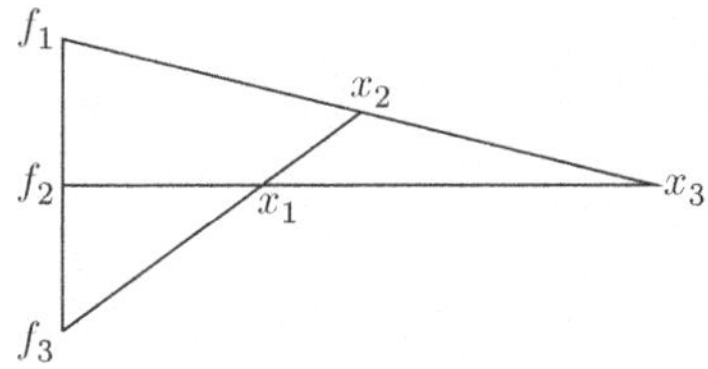

Figure 9.3

If the group configuration $C = (f_1, f_2, f_3, x_1, x_2, x_3)$ over e has the property that $\mathrm{bdd}(f_i; e) = \mathrm{bdd}(\mathrm{Cb}(x_j x_k / f_i e); e)$, we call such C *a bounded quadrangle.* If $f_i \in \mathrm{bdd}(x_j, x_k; e)$ then we call the group configuration *principal.*

Fact 9.2.2

(1) *If* $C = (f_1, f_2, f_3, x_1, x_2, x_3)$ *is a group configuration/bounded quadrangle over e and* $\mathrm{bdd}(f_i e) = \mathrm{bdd}(f_i' e), \mathrm{bdd}(x_i e) = \mathrm{bdd}(x_i' e)$, *then* $C' = (f_1', f_2', f_3', x_1', x_2', x_3')$ *is also a group configuration/bounded quadrangle over e. In this case, we say C and C′ are (boundedly) equivalent over e.*

(2) *For C, a group configuration/bounded quadrangle over e and* $e' \supseteq e$, *if* $C \mathrel{\raise0.2ex\hbox{$\smile$}\!\!\!\!\!\mid}_e e'$, *then C also is a group configuration/bounded quadrangle over* e'.

(3) *In a group configuration,* $(f_1, f_2, f_3, x_1, x_2, x_3)$ *over e even if we replace* f_i *by* $f_i' = \mathrm{Cb}(x_j x_k / e f_i)$ *for* $\{i, j, k\} = \{1, 2, 3\}$, $(f_1', f_2', f_3', x_1, x_2, x_3)$ *is still a group configuration (hence, a bounded quadrangle) over e.*

Proof We sketch the proof.

(1) Obvious for a group configuration. For a bounded quadrangle, note that, in general, $\mathrm{Cb}(a_1/a_2)$ and $\mathrm{Cb}(b_1/b_2)$ are interbounded as long as a_i and b_i are interbounded.

(2) Easy.

(3) Since $x_i x_j \mathrel{\raise0.2ex\hbox{$\smile$}\!\!\!\!\!\mid}_{f_k'} f_k e$ and $x_i \mathrel{\raise0.2ex\hbox{$\smile$}\!\!\!\!\!\mid}_{x_j f_k'} f_k e$, x_i and x_j are interbounded over f_k' (*). On the other hand, $x_j \mathrel{\raise0.2ex\hbox{$\smile$}\!\!\!\!\!\mid}_{f_k e} f_i$ implies that $x_i x_j \mathrel{\raise0.2ex\hbox{$\smile$}\!\!\!\!\!\mid}_{f_k e} f_i f_j$ (**), $x_i x_j \mathrel{\raise0.2ex\hbox{$\smile$}\!\!\!\!\!\mid}_{f_k f_i e} f_j$, and thus $x_i x_j x_k \mathrel{\raise0.2ex\hbox{$\smile$}\!\!\!\!\!\mid}_{f_k f_i e} f_j$, and $x_i x_j x_k \mathrel{\raise0.2ex\hbox{$\smile$}\!\!\!\!\!\mid}_{f_k f_i f_j f_j' e} f_j$. Then from (**), it follows that $x_i x_k \mathrel{\raise0.2ex\hbox{$\smile$}\!\!\!\!\!\mid}_{f_j'} f_i f_j f_k e$ and from (*) $x_i x_j \mathrel{\raise0.2ex\hbox{$\smile$}\!\!\!\!\!\mid}_{f_i' f_j'} f_i f_j e$. Therefore, $f_k' = \mathrm{Cb}(x_i x_j / f_k e) = \mathrm{Cb}(x_i x_j / f_i f_j e) \in \mathrm{bdd}(f_i' f_j')$. The other independences over e follow easily. ■

From now on, assume that a group configuration over $\hat{e} = A/\bar{E}$ is given. Our goal is, under 4-amalgamation, to produce a non-trivial canonical hyperdefinable group and its action from the configuration (Theorem 9.5.2). But at the moment we do not assume 4-amalgamation until we need it. Note now due to 9.2.2, we can replace the group configuration by a bounded quadrangle $C = (\hat{f}, \hat{g}, \hat{h}, \hat{a}, \hat{b}, \hat{c})$ over a model M containing A. After naming M, we freely assume that $\emptyset = \mathrm{bdd}(\emptyset)$. We further suppose $\hat{f}, \hat{g}, \hat{h}, \hat{a}, \hat{b}, \hat{c}$ are all boundedly closed (by extending each to its bounded closure, if necessary). Clearly, C still is a bounded quadrangle over $\emptyset$. We will work with this in the rest of this chapter.

Now let $p = \mathrm{tp}(\hat{f})(= \mathrm{Ltp}(\hat{f})), q = tp(\hat{g}), r = \mathrm{tp}(\hat{h})$ and let $\Gamma_q(uv) = q(u) \wedge q(v) \wedge u \perp v$. (Later, we shall omit q in Γ_q.) Now we can think of $\hat{h}$ as a multi-valued function such that $\mathrm{dom}(\hat{h}) = \mathrm{tp}(\hat{a}/\hat{h}) = \mathrm{Ltp}(\hat{a}/\hat{h})$ and $rag(\hat{h}) = \mathrm{Ltp}(\hat{b}/\hat{h})$. More precisely, $b \in k_r(a)$ means $k_r \models r$, $k_r ab \equiv \hat{h}\hat{a}\hat{b}$. Similarly, we write $a \in h_q(c)$, $b \in g_p(c)$ for $h_q ca \equiv \hat{g}\hat{a}\hat{c}$, $g_p cb \equiv \hat{f}\hat{b}\hat{c}$, respectively. In the same manner, $b \in \mathrm{dom}(f_p) \equiv \exists c(f_p bc \models \mathrm{tp}(\hat{f}\hat{b}\hat{c}))$, and so on.

We say a set A is n-independent if any subset of A having n elements is independent. Now, we define $R = R^q$ to be a symmetric type-definable relation over $\emptyset$ on the set of independent realizations of q such that

$$R(fg;f'g') \text{ iff } f,g,f'g' \models q,\ \{f,g,f',g'\} \text{ 3-independent,}$$
$$\text{and there are } b \text{ and } a \perp fgf'g' \text{ such that } f(a) \cap g(b) \neq \emptyset,$$
$$f'(a) \cap g'(b) \neq \emptyset.$$

It is easy to see that $a \perp fgf'g'$ above can be replaced with $b \perp fgf'g'$. Similarly, one can define R^p, R^{pq} by replacing $f,g,f'g' \models q$ by $f,g,f'g' \models p$ or $f,f' \models p,\ g,g' \models q$, respectively.

Lemma 9.2.3

(1) If $fg \models \Gamma_q$, and $c \in g(a) \cap f(b)$ with $c \perp fg$, then

(a) f,g are interbounded over $e := \mathrm{Cb}(ba/fg)$;

(b) $e \perp f, e \perp g$; and

(c) (f,g,e,a,b,c) forms a bounded quadrangle.

(2) (1) still holds when we replace $f,g \models q$ with $f,g \models p$ or $f \models p, g \models q$.

(3) If $(fg,f'g') \models R$ (or R^p, R^{pq}), then any element in $\{f,g,f',g'\}$ is in the bounded closure of the other three elements.

Proof (1)(a) Note that from $ab \perp_e fg$ and that a,b are interbounded over fg, it follows that a,b are interbounded over e, too (*). Now, from $c \perp_g f$, we have $ca \perp_g fe$. Moreover, from $c \perp_f g$, it follows that $cb \perp_{fe} g$, and then by (*), $ca \perp_{fe} g$. Hence, $g \in \mathrm{bdd}(\mathrm{Cb}(ca/g)) \subseteq \mathrm{bdd}(fe)$. By a similar argument, $f \in \mathrm{bdd}(ge)$ can be shown, too.

(1)(b) There are h_1u_1, h_2u_2 such that $h_1fbu_1c, h_2gau_2c \models \hat{f}\hat{g}\hat{a}\hat{b}\hat{c}$. Then since c is boundedly closed, by amalgamation we have

$$hu \models \mathrm{tp}(h_1u_1/cbf) \cup \mathrm{tp}(h_2u_2/cag),$$

such that $\{u,c,f,g\}$ is independent. Then we have k,k' such that $hgkauc, hfk'buc \models \hat{f}\hat{g}\hat{h}\hat{a}\hat{b}\hat{c}$. From $c \perp fgh$, we have $ba \perp_{fg} kk'$. From $b \perp fgh$ and $u,a \in \mathrm{bdd}(kk'b)$, it follows that $ba \perp_{kk'} fg$ and thus $e \in \mathrm{bdd}(kk')$ (+). On the other hand, $f \perp_h g$ implies that $k' \perp_h gk$ and $k' \perp gk$. Hence, $\{g,k,k'\}$ is independent. Similarly, $\{f,k,k'\}$ is independent. Then from (+), $e \perp f, e \perp g$.

(1)(c) This easily follows from (a),(b), and (*).

(2) Similar to (1).

(3) There are c, c', b and $a \perp fgf'g'$ such that $c \in f(a) \cap g(b)$, $c' \in f'(a) \cap g'(b)$. Hence, $ab \perp_{fg} f'g'$ and $ab \perp_{f'g'} fg$. Therefore, $e = \mathrm{Cb}(ab/fg)$ and $e' = \mathrm{Cb}(ab/f'g')$ are interbounded (**). From (1)(a), f, g are interbounded over e, and so are f', g' over e'. Hence, it follows from (**), $g' \in \mathrm{bdd}(f'e') = \mathrm{bdd}(f'e) \subseteq \mathrm{bdd}(f'fg)$ and similarly for the other relations. ■

Lemma 9.2.4

(1) *For $fg, f'g' \models \Gamma_q$, $R(fg, f'g')$ iff there are b and $a \perp fgf'g'$ such that $f(a) \cap g(b) \neq \emptyset, f'(a) \cap g'(b) \neq \emptyset$, and $fg \perp_e f'g'$ where $e = \mathrm{Cb}(ba/fg)$.*

(2) *Given independent $f, g \models q$, there are f', g' such that $R(fg, f'g')$.*

(3) *(1),(2) hold for R^p, R^{pq}, as well.*

Proof (1) ($\Rightarrow$) Note that since $a \perp fgf'g'$, $ba \perp_{fg} f'g'$, $ba \perp_{f'g'} fg$. Hence, $e, e' = \mathrm{Cb}(ba/f'g')$ are interbounded. Then, due to 9.2.3(1)(a), f', g' are interbounded over e. Now, from $fg \perp f'$, $fg \perp_e f'$, thus $fg \perp_e f'g'$.

($\Leftarrow$) Again since $a \perp fgf'g'$, e, e' are interbounded. Then from $fg \perp_e f'g'$, equivalently $fg \perp_{e'} f'g'$, and 9.2.3(1)(b), $\{f, g, f', g'\}$ is 3-independent.

(2) By amalgamation there is $c \in \mathrm{ran}(f) \cap \mathrm{ran}(g)$ such that $c \perp fg$. Choose $a \in g^{-1}(c), b \in f^{-1}(c)$. Then, by the extension axiom, we have $f'g'$ such that $\{ab, fg, f'g'\}$ is e-independent, where $e = \mathrm{Cb}(ba/fg)$ and $f'g' \equiv_{abe} fg$. Then the right-hand side of (1) follows easily.

(3) Clear. ■

The following lemma is crucial to our argument.

Lemma 9.2.5 *Let $R(fg, f'g')$ hold. Namely, $\{f, g, f', g'\}$ is 3-independent, and we can find c, c', d and $a \perp fgf'g'$ such that $c \in f(a) \cap g(d)$, $c' \in f'(a) \cap g'(d)$. Then there are $h, h' \models p$ and b such that $c \in h(b), c' \in h'(b)$, $b \perp hh'ff'gg'$, $\{f, h, f', h'\}$ is 3-independent, and $hh' \perp_{ff'} gg'$. It follows that $R^{pq}(hf, h'f')$ and $R^{pq}(hg, h'g')$.*

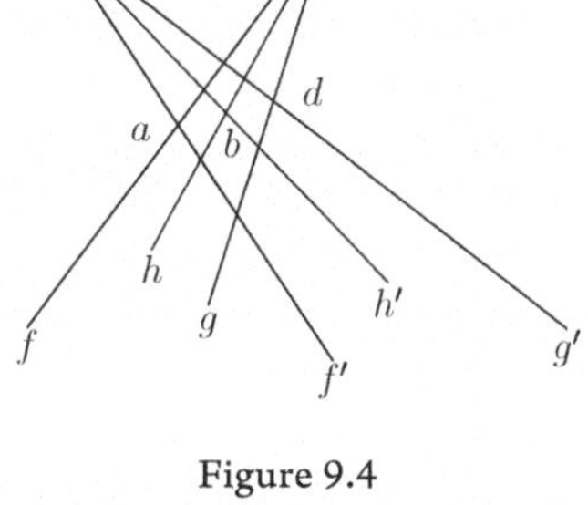

Figure 9.4

Proof Note that since $fac, f'ac' \models \hat{g}\hat{a}\hat{c}$, there are $k_0, b_0; k_1, b_1$ such that $f'k_0ab_0c'$, $fk_1ab_1c \models \hat{g}\hat{h}\hat{a}\hat{b}\hat{c}$. Then by amalgamation, we have kb, hh' such that

$$kb \models \mathrm{tp}(k_0b_0/af') \cup \mathrm{tp}(k_1b_1/af), kb \underset{a}{\perp} ff' \text{ and } h'f'kabc', hfkabc \models \hat{f}\hat{g}\hat{h}\hat{a}\hat{b}\hat{c}.$$

We can further assume that $hh' \perp_{ff'a} gg'$ (*). Now from $kb \perp_a ff'$ and $a \perp fgf'g'$, it follows that $ab \perp_k ff'$, $b \perp_k ff'hh'$, and $b \perp ff'hh'$. Then from (*), we have $a \perp_{hh'ff'} gg'$, $ab \perp_{hh'ff'} gg'$, and $b \perp ff'gg'hh'$. Moreover, from $f \perp_k f'$, we have that $hf \perp_k h'f'$, and thus $\{f, h, f', h'\}$ is 3-independent. Then since $hh' \perp_{ff'} gg'$, $\{g, h, g', h'\}$ is 3-independent as well. Therefore, $R^{pq}(hf, h'f')$ and $R^{pq}(hg, h'g')$. ■

Now define $R' = (R')^q$ by

$$\begin{aligned} &R'(fg; f'g') \text{ iff } f, g, f'g' \models q, \{f, g, f', g'\} \text{ 3-independent,}\\ &\text{and for any } a, b \text{ such that } f(a) \cap g(b) \neq \emptyset \text{ and } a \perp fg,\\ &\text{there are } a'b' \equiv^L_{fg} ab \text{ such that } a' \perp fgf'g', f'(a') \cap g'(b') \neq \emptyset. \end{aligned}$$

Again, we also define $(R')^p$, $(R')^{pq}$ by substituting $f, g, f'g' \models p$ or $f, f \models p, gg' \models q$, respectively.

Assumption In the rest of this chapter, we assume that T has 4-amalgamation. But as we shall point out in the proof of Theorem 9.2.6, our proof will fully cover the stable case.

Notation For bounded closed sequences $a_1, a_2, \ldots$, we use $\overline{a_1a_2\ldots}$ to denote some sequence of $\mathrm{bdd}(a_1a_2\ldots)$ extending the orderings of $a_1, a_2, \ldots$. More precisely, if $x_i \mapsto a_i$ is a bijective indexing of a_i by the variable x_i, then $\overline{a_1a_2\ldots}$ is an indexing $x(\supseteq x_1 \cup x_2 \cup \ldots) \mapsto \mathrm{bdd}(a_1a_2\ldots)$ extending each $x_i \mapsto a_i$.

We shall prove that R and R' are equivalent. Note that clearly R' implies R.

Theorem 9.2.6 *R and R' are equivalent.*

Proof We shall show that R implies R'. Let $R(fg, f'g')$ hold. We use Lemma 9.2.5 with the same notation. We then have cc', abd, hh' such that $c \in f(a) \cap h(b) \cap g(d)$; $c' \in f'(a) \cap h'(b) \cap g'(d)$; $b \perp fghf'g'h'$; $f' \in \mathrm{bdd}(fhh'), g' \in \mathrm{bdd}(hgh')$; and $\{f, g, h, h'\}$ is independent. Now let $\overline{hf}$, $\overline{hg}$, $\overline{fg}$, $\overline{hbc} = \mathrm{bdd}(hb) = \mathrm{bdd}(hc)$, $\overline{fac} = \mathrm{bdd}(fc) = \mathrm{bdd}(fa)$, and $\overline{gdc} = \mathrm{bdd}(gc) = \mathrm{bdd}(gd)$ be sequences of boundedly closed sets extending the boundedly closed sequences f, g, h, c, a, b, d (See Notation before 9.2.6). Since $hbc \equiv h'bc'$ we also have $\overline{h'bc'} \equiv \overline{hbc}$.

Now, to show $R'(fg, f'g')$, assume there are a_1, d_1, c_1 such that $c_1 \in f(a_1) \cap g(d_1)$ and $c_1 \perp fg$. We also have $\overline{fa_1c_1} \equiv \overline{fac} \equiv \overline{gd_1c_1} \equiv \overline{gdc}$. Then by 4-amalgamation (over a model which is named, so it covers the case of all stable theories), we can find c_2 such that

$$\mathrm{tp}\left(\overline{fgh}\right) \cup \mathrm{tp}\left(\overline{fhc}\right) \cup \mathrm{tp}\left(\overline{ghc}\right) \cup \mathrm{tp}\left(\overline{fgc_1}\right) \subseteq \mathrm{tp}\left(\overline{fghc_2}\right),$$

where $\overline{fa_1c_1}, \overline{gd_1c_1} \subset \overline{fgc_1}$, and $\overline{fac}, \overline{hbc} \subset \overline{fhc}$. Hence, there are b_2, a_2, d_2 such that

$$c_2 \in f(a_2) \cap g(d_2) \cap h(c_2).$$

In particular,

$$\overline{fa_2c_2}\ \overline{gd_2c_2}\ \overline{fg} \equiv \overline{fa_1c_1}\ \overline{gd_1c_1}\ \overline{fg};\ \overline{fh}\ \overline{hb_2c_2} \equiv \overline{fh}\ \overline{hbc};\ \text{and}\ \overline{gh}\ \overline{hb_2c_2} \equiv \overline{gh}\ \overline{hbc}\ (^*).$$

What we are going to amalgamate next are the strong types

$$\mathrm{Ltp}\left(\overline{hb_2c_2}/fg; h\right),\ \mathrm{Ltp}\left(\overline{hbc}/h'f; h\right),\ \text{and Ltp}\left(\overline{hbc}/h'g; h\right).$$

(Equivalently, $\mathrm{tp}(\overline{c_2fg; h})$, $\mathrm{tp}(\overline{ch'f; h})$, $\mathrm{tp}(\overline{ch'g; h})$, and $\mathrm{tp}(\overline{h'fg; h})$, where $h, b_2, c_2 \subset \overline{c_2h}$, and $h, b, c \subset \overline{ch}$.) The base parameter is h (here is the place where we need 4-CA over h. Since, indeed, the parameter is Mh and $\{f, g, h, h'\}$ is M-independent, our arguments will again cover all the stable case), and each realization is boundedly closed over the parameter. Each type does not fork over h. Additionally, due to (*), it can be seen that the three strong types are h-compatible. Hence, we have $\overline{hb_3c_3}$, a generic solution of the types.

Now we analyze consequences obtained from the generic solution. Note first that we have $\overline{fhabc} = \mathrm{bdd}(f, bc; h)$ extending $\overline{fac}, \overline{hbc}$. Also since $c' \in \mathrm{bdd}(h', bc; h)$, we have $\overline{hh'bcc'} = \mathrm{bdd}(h', bc; h)$ extending $\overline{hbc}, \overline{h'bc'}$. Similarly, there is $\overline{hgbdc} = \mathrm{bdd}(g, bc; h)$ extending $\overline{hbc}, \overline{gdc}$. Note here that the existence of the generic solution indeed says that there exist elementary maps $\tilde{h}_1, \tilde{h}_2, \tilde{h}_3$ with

$$\begin{aligned}
\mathrm{dom}(\tilde{h}_1) &= \mathrm{bdd}(f, bc; h)\,\mathrm{bdd}(h', bc; h)\,\mathrm{bdd}(fh'; h),\\
\mathrm{dom}(\tilde{h}_2) &= \mathrm{bdd}(f, b_2c_2; h)\,\mathrm{bdd}(g, b_2c_2; h)\,\mathrm{bdd}(fg; h),\\
\mathrm{dom}(\tilde{h}_3) &= \mathrm{bdd}(h', bc; h)\,\mathrm{bdd}(g, bc; h)\,\mathrm{bdd}(h'g; h),
\end{aligned}$$

fixing $\mathrm{bdd}(fh'; h)$, $\mathrm{bdd}(fg; h)$ $\mathrm{bdd}(gh'; h)$, respectively, such that $\tilde{h}_1(\overline{hbc}) = \tilde{h}_2(\overline{hb_2c_2}) = \tilde{h}_3(\overline{hbc}) = \overline{hb_3c_3}$, the generic solution. Moreover, they are compatible with elementary maps sending $\overline{fhabc} \to \overline{fha_2b_2c_2}$, $\overline{hgbdc} \to \overline{hgb_2d_2c_2}$ and $\overline{hh'bcc'} \xrightarrow{id} \overline{hh'bcc'}$. (In particular, maps $\tilde{h}_1 \lceil \mathrm{bdd}(h', bc; h) = \tilde{h}_3 \lceil \mathrm{bdd}(h', bc; h)$.) Hence, there are a_3, c'_3, d_3 such that $a_3 = \tilde{h}_1(a) = \tilde{h}_2(a_2)$, $c'_3 = \tilde{h}_1(c') = \tilde{h}_3(c')$, $d_3 = \tilde{h}_2(d_2) = \tilde{h}_3(d)$ and

(1) $\overline{fha_3b_3c_3}\,\overline{hh'b_3c_3c_3'} \equiv_{\mathrm{bdd}(fh';h)} \overline{fhabc}\,\overline{hh'bcc'}$;

(2) $\overline{fha_3b_3c_3}\,\overline{hgb_3d_3c_3} \equiv_{\mathrm{bdd}(fg;h)} \overline{fha_2b_2c_2}\,\overline{hgb_2d_2c_2}$;

(3) $\overline{hh'b_3c_3c_3'}\,\overline{hgb_3d_3c_3} \equiv_{\mathrm{bdd}(gh';h)} \overline{hh'bcc'}\,\overline{hgbdc}$.

Then by (*) and (2), $a_1d_1 \equiv^L_{fg} a_2d_2 \equiv^L_{fg} a_3d_3$. Also since $f' \in \mathrm{bdd}(fh'h)$, from (1), $c_3' \in f'(a_3) \cap h'(b_3)$ and $c_3 \in f(a_3)$. Note that $g' \in \mathrm{bdd}(gh'h)$, since $R(hg, h'g')$ holds. Then, similarly from (3), $c_3' \in g'(d_3)$ and $c_3 \in g(d_3)$. Therefore, $R'(fg, f'g')$. ■

9.3 The Generic Group Operation on $\Gamma/\tilde{R}$

We begin to show, using $R \equiv R'$ (Theorem 9.2.6), that the transitive closure of R is type-definable. The proof will use Proposition 4.3.6.

Let $\tilde{R}$ be the transitive closure R. We remark that if both $\{a, b, c, d\}$, $\{a', b', c, d\}$ are 3-independent and $ab \perp_{cd} a'b'$, then $\{a, b, a', b'\}$ is also 3-independent.

Lemma 9.3.1 *Suppose that $R(fg, hk)$, $R(hk, f'g')$ and $fg \perp_{hk} f'g'$. Then $R(fg, f'g')$ holds and $fg \perp_{f'g'} hk$.*

Proof By the previous remark, $\{f, g, f', g'\}$ is 3-independent (*). Now since $R(fg, hk)$, there are b and $a \perp fghk$ such that $f(a) \cap g(b) \neq \emptyset$, $h(a) \cap k(b) \neq \emptyset$ (so $ab \perp_{hk} fg$). Then since $R'(hk, f'g')$, there are $a'b' \equiv^L_{hk} ab$ such that $a' \perp hkf'g', f'(a') \cap g'(b') \neq \emptyset$. Hence, by amalgamation, we have

$$a''b'' \models \mathrm{Ltp}(ab/hk, fg) \cup \mathrm{Ltp}(a'b'/hk, f'g'), \text{ and } a''b'' \underset{hk}{\perp} fgf'g'.$$

It follows then that $a'' \perp fgf'g'$ and $f(a'') \cap g(b'') \neq \emptyset, f'(a'') \cap g'(b'') \neq \emptyset$. This with (*) says that $R(fg, f'g')$. It remains to show that $fg \perp_{f'g'} hk$. Since $g \perp hkf'g', g \perp_{f'g'} hk$. Now by 9.2.3(3), $f \in \mathrm{bdd}(gf'g')$, and, therefore, $fg \perp_{f'g'} hk$. The proof is finished. ■

Hence, due to 4.3.6 and 9.2.4(2), we obtain the following.

Theorem 9.3.2 *The following are equivalent.*

(1) $\tilde{R}(\bar{f}, \bar{g})$.

(2) For some $\bar{h}$, $R(\bar{h}, \bar{f})$ and $R(\bar{h}, \bar{g})$.

(3) For some $\bar{h}$ with $\bar{h} \perp_{\bar{f}} \bar{g}$ and $\bar{h} \perp_{\bar{g}} \bar{f}$, $R(\bar{h}, \bar{f})$ and $R(\bar{h}, \bar{g})$.

Recall that in Section 9.2, we define $\Gamma(xy) = \Gamma_q(xy) = q(x) \wedge q(y) \wedge x \perp y$, and now $\tilde{R}$ is an equivalence relation on Γ. By putting $(\tilde{R}(\bar{x}, \bar{y}) \wedge \Gamma(\bar{x}) \wedge \Gamma(\bar{y})) \vee \bar{x} = \bar{y}$, we can extend $\tilde{R}$ to a type-definable equivalence relation on the whole universe. We shall find the canonical hyperdefinable group from the hyperdefinable generic group operation on $\Gamma/\tilde{R}$. First, we state some more properties of R and $\tilde{R}$.

Lemma 9.3.3 *Let $\bar{f} = f_1 f_2, \bar{g} \models \Gamma$, and let $e = \bar{f}/\tilde{R}$.*

(1) $\tilde{R}(\bar{f},\bar{g})$ and $\bar{f} \perp_e \bar{g}$ iff $R(\bar{f},\bar{g})$.
(2) For a, b such that $f_1(a) \cap f_2(b) \neq \emptyset$ and $a \perp \bar{f}$, e is interbounded with $\mathrm{Cb}(ab/\bar{f})$.
(3) $f_i \perp e$, and f_1, f_2 are interbounded over e.

Proof (1) ($\Rightarrow$) By 9.3.2, there is $\bar{h}$ such that $\bar{h} \perp_{\bar{f}} \bar{g}$, $\bar{h} \perp_{\bar{g}} \bar{f}$, $R(\bar{h},\bar{f})$, and $R(\bar{h},\bar{g})$. Then since $\bar{f} \perp_e \bar{g}$, we have $\bar{f} \perp_{\bar{h}} \bar{g}$, and then by 9.3.1, $R(\bar{f},\bar{g})$.

($\Leftarrow$) $R(\bar{f},\bar{g})$ implies that $\tilde{R}(\bar{f},\bar{g})$. By the extension axiom, there is $\bar{g}' \models \mathrm{tp}(\bar{g}/e)$ such that $\bar{g}' \perp_e \bar{g}$ (*). Hence, $\tilde{R}(\bar{g},\bar{g}')$ and by the proof of ($\Rightarrow$), $R(\bar{g},\bar{g}')$. Again, by extension, we can assume that $\bar{f} \perp_{e\bar{g}} \bar{g}'$. Hence, $\bar{f} \perp_{\bar{g}} \bar{g}'$ and then by 9.3.1, $\bar{f} \perp_{\bar{g}'} \bar{g}$. Therefore, by (*), $\bar{f} \perp_e \bar{g}$.

(2) Let $e_1 = \mathrm{Cb}(ab/\bar{f})$. By extension, there is $\bar{f}' \models \mathrm{tp}(\bar{f}/abe_1)$ such that $\bar{f} \perp_{abe_1} \bar{f}'$. Hence, $\{ab, \bar{f}, \bar{f}'\}$ is e_1-independent and from 9.2.4(1), $R(\bar{f},\bar{f}')$ and $e_1 = \mathrm{Cb}(ab/\bar{f}')$. Then by (1), $\bar{f} \perp_e \bar{f}'$ (**). Let $e_2 = \mathrm{Cb}(\bar{f}/\bar{f}')$. Then due to (**), $e_2 \in \mathrm{bdd}(e)$. Moreover, since $\bar{f} \perp_{e_2} \bar{f}'$, $e \perp_{e_2} \bar{f}'$, and $e \in \mathrm{dcl}(\bar{f}')$, $e_2 \in \mathrm{bdd}(\bar{f}')$, we have $e \in \mathrm{bdd}(e_2)$. Thus, $\mathrm{bdd}(e) = \mathrm{bdd}(e_2)$. Similarly, since $\bar{f} \perp_{e_1} \bar{f}'$, it, too, can be seen that $\mathrm{bdd}(e_1) = \mathrm{bdd}(e_2)$. Therefore, $\mathrm{bdd}(e_1) = \mathrm{bdd}(e)$.

(3) follows from (2) and Lemma 9.2.3(1). ■

The proof of the following lemma uses 4-CA over a model.

Lemma 9.3.4 *Let $R(gh, vw)$. Then for any $c \in g(a), d \in v(a)$ with $a \perp gv$, there are $c'a'd' \equiv^L_{gv} cad$ and b' such that $a' \perp ghvw$, $c' \in h(b')$, $d' \in w(b')$.*

Proof By 9.2.5, there are a_0, b_0, c_0, d_0, t_0 and f, u such that $c_0 \in g(a_0) \cap f(t_0) \cap h(b_0)$, $d_0 \in v(a_0) \cap u(t_0) \cap w(b_0)$, $t_0 \perp fghuvw$, $\{f, g, u, v\}$ is 3-independent, and $fu \perp_{gv} hw$. It follows that $R^{pq}(fh, uw)$, $R^{pq}(fg, uv)$, and $gv \perp_{fu} hw$ (*). Let $e = \mathrm{bdd}(\mathrm{Cb}(t_0a_0/fg))$. Then from Lemma 9.2.4(3), $e = \mathrm{bdd}(\mathrm{Cb}(t_0a_0/uv))$ and $fg \perp_e uv$. Similarly, for $k = \mathrm{bdd}(\mathrm{Cb}(t_0b_0/fh)) = \mathrm{bdd}(\mathrm{Cb}(t_0b_0/uw))$, $fh \perp_k uw$. We now let $\overline{ea_0t_0} = \mathrm{bdd}(ea_0t_0) = \mathrm{bdd}(ea_0)$, $\overline{evu} = \mathrm{bdd}(ev)$, $\overline{egf} = \mathrm{bdd}(eg)$ $\overline{ga_0c_0} = \mathrm{bdd}(ga_0)$, $\overline{va_0d_0} = \mathrm{bdd}(va_0)$, $\overline{gv}$ be sequences of bounded closed sets (see Notation before 9.2.6). Note that there are sequences $\overline{gac}$, $\overline{vad}$ such that

$$\overline{ga_0c_0} \equiv \overline{gac} \text{ and } \overline{va_0d_0} \equiv \overline{vad}.$$

Now by 4-amalgamation over the named model, there is a_1 such that

$$\mathrm{tp}(\overline{egv}) \cup \mathrm{tp}(\overline{ega_0}) \cup \mathrm{tp}(\overline{eva_0}) \cup \mathrm{tp}(\overline{gva}) \subseteq \mathrm{tp}(\overline{egva_1}),$$

and $\{e, g, v, a_1\}$ is independent, where

$$\overline{efg},\ \overline{ga_0c_0},\ \overline{ea_0t_0} \subset \overline{ega_0};\ \overline{evu},\ \overline{va_0d_0},\ \overline{ea_0t_0} \subset \overline{eva_0};\ \text{and}\ \overline{gac},\ \overline{vad} \subset \overline{gva}.$$

In particular, there are $c_1(\subset \overline{ga_1})$, $d_1(\subset \overline{va_1})$, $t_1(\subset \overline{ea_1})$ (all are in $\overline{egva_1}$) such that

(1) $\overline{ga_1c_1}\ \overline{ea_1t_1} \equiv_{\overline{egf}} \overline{ga_0c_0}\ \overline{ea_0t_0}$;
(2) $\overline{va_1d_1}\ \overline{ea_1t_1} \equiv_{\overline{evu}} \overline{va_0d_0}\ \overline{ea_0t_0}$;
(3) $\overline{ga_1c_1}\ \overline{va_1d_1} \equiv_{\overline{gv}} \overline{gac}\ \overline{vad}$;

and, from 9.2.3(1), $\{t_1, f, u, e\}$ is independent (**). According to (1),(2), $ft_0c_0 \equiv ft_1c_1$ and $ut_0d_0 \equiv ut_1d_1$. Hence, there are enumerations such that

$$\overline{ft_0c_0} \equiv \overline{ft_1c_1} \text{ and } \overline{ut_0d_0} \equiv \overline{ut_1d_1}.$$

Similarly by 4-CA, we have t_2; c_2, d_2, b_2 such that

$$\text{tp}\left(\overline{kfu}\right) \cup \text{tp}\left(\overline{kft_0}\right) \cup \text{tp}\left(\overline{kut_0}\right) \cup \text{tp}\left(\overline{fut_1}\right) \subseteq \text{tp}\left(\overline{kfut_2}\right),$$

(4) $\overline{ft_2c_2}\ \overline{kb_2t_2} \equiv_{\overline{khf}} \overline{ft_0c_0}\ \overline{kb_0t_0}(\subset \overline{kft_0})$;
(5) $\overline{ut_2d_2}\ \overline{kb_2t_2} \equiv_{\overline{kwu}} \overline{ut_0d_0}\ \overline{kb_0t_0}(\subset \overline{kut_0})$;
(6) $\overline{ft_2c_2}\ \overline{ut_2d_2} \equiv_{\overline{fu}} \overline{ft_1c_1}\ \overline{ut_1d_1}(\subset \overline{fut_1})$;

and $\{t_2, f, u, k\}$ is independent. Hence, due to (*),(**), Lemma 9.2.3(1)(a), and (6), we can apply amalgamation to have

$$d'c't' \models \text{Ltp}(d_1c_1t_1/fu; gv) \cup \text{Ltp}(d_2c_2t_2/fu; hw)$$

such that $t' \perp_{fu} ghvw$ (†). Then there are the desired a', b' such that

(7) $d'c't'a' \equiv^L_{fguv} d_1c_1t_1a_1$, and $d'c't'b' \equiv^L_{fhuw} d_2c_2t_2b_2$.

Hence, it follows from (†), $a' \perp ghvw$. Moreover by (3),(7), $c'a'd' \equiv^L_{gv} cad$; by (4),(7), $c' \in h(b')$; and by (5),(7), $d' \in w(b')$. The proof is finished. ■

We are ready to define the promised generic operation on $\Gamma/\tilde{R}$. Let

$$\bullet(x_1y_1, x_2y_2; x_3y_3) := \exists xyz(\tilde{R}(x_1y_1; xy) \wedge \tilde{R}(x_2y_2; yz) \wedge \tilde{R}(xz; x_3y_3) \wedge \textstyle\bigwedge_{i=1,2,3} \Gamma(x_iy_i)$$
$$\wedge\ x_1y_1/\tilde{R} \perp x_2y_2/\tilde{R}\ \wedge \{x, y, z\} \text{ is independent}).$$

Note that for $x_1y_1, x_2y_2 \models \Gamma$, $x_1y_1/\tilde{R} \perp x_2y_2/\tilde{R}$ iff $\exists x'_1y'_1x'_2y'_2\tilde{R}(x_1y_1; x'_1y'_1) \wedge \tilde{R}(x_2y_2; x'_2y'_2) \wedge \{x'_1, y'_1, x'_2, y'_2\}$ independent. Hence, $\bullet$ is a partial type over $\emptyset$.

Claim 1 The relation $\bullet$ is a hyperdefinable partial type over $\emptyset$ such that, for any independent $e_1 = f_1g_1/\tilde{R}, e_2 = f_2g_2/\tilde{R} \in \Gamma/\tilde{R}$, there is $e_3 = fh/\tilde{R} \in \Gamma/\tilde{R}$ such that $(e_1, e_2; e_3)$ realizes

$\centerdot(x_1y_1, x_2y_2; x_3y_3)$: It suffices to show there exist f, g, h such that $e_1 = fg/\tilde{R}$, $e_2 = gh/\tilde{R}$ and $\{f, g, h\}$ is independent. Now by 9.3.3(3), $e_i \perp g_i$, $e_i \perp f_i$ and f_i, g_i are interbounded over e_i. Then, by amalgamation, there exists $g \models \mathrm{tp}(f_2/e_1) \cup \mathrm{tp}(g_1/e_2)$ and $\{g, e_1, e_2\}$ independent. We also have f, h such that $fg \equiv_{e_1} f_1g_1$, $gh \equiv_{e_1} f_2g_2$. Then, $\{f, g, h\}$ is independent too.

Claim 2 $e_1 \centerdot e_2 = e_3 = fh/\tilde{R}$ does not depend on the choice of f, g, h, i.e., $e_1 \centerdot e_2$ is unique: Suppose there are f', g', h' such that $e_1 = f'g'/\tilde{R}$, $e_2 = g'h'/\tilde{R}$ and $\{f', g', h'\}$ is independent. Then we can also find independent $\{u, v, w\}$ such that $u \perp_{e_1e_2} fghf'g'h'$ and $e_1 = uv/\tilde{R}$, $e_2 = vw/\tilde{R}$. Hence, from 9.3.3(2), $uvw \perp_{e_1e_2} fghf'g'h'$, and $\{f, u, e_1, e_2\}$ is independent (‡). We shall prove that $R(fh, uw)$. (Then the by the same proof, $R(f'h', uw)$ and thus $\tilde{R}(fh, f'h')$.) Note from (‡) and 9.3.3(1), $R(fg, uv)$ and $R(gh, vw)$. Hence, there are b, $a \perp fguv$ and $c \in g(a) \cap f(b)$, $d \in v(a) \cap u(b)$. Moreover, by 9.3.4, we have $a'b'c'd'$ such that $c'a'd' \equiv^L_{gv} cad$ and $a' \perp ghvw$, $c' \in g(a') \cap h(b')$, $d' \in v(a') \cap w(b')$. Now again due to (‡) and 9.2.3(1), we have $e_1 \perp_{gv} e_2$, $fu \perp_{gv} hw$ (∗), and $a \perp_{gv} fu$, $cad \perp_{gv} fu$, $c'a'd' \perp_{gv} hw$. Hence, by amalgamation, we have

$$c_1a_1d_1 \models \mathrm{Ltp}(cad/gv; fu) \cup \mathrm{Ltp}(c'a'd'/gv; hw),$$

such that $a_1 \perp_{gv} fhuw$ (∗∗). Then we have b_1, b'_1 such that

$$c_1a_1d_1b_1 \equiv_{fguv} cadb, c_1a_1d_1b'_1 \equiv_{ghvw} c'a'd'b'.$$

Thus, $c_1 \in f(b_1) \cap h(b'_1)$, $d_1 \in u(b_1) \cap w(b'_1)$ and from (∗∗), $b_1 \perp fhuw$. Moreover, from (∗) and the remark before 9.3.1, $\{f, h, u, w\}$ is 3-independent. Therefore, $R(fh, uw)$, as desired.

Claim 3 This generically given group satisfies the genericity properties Theorem 7.6.1(1),(2),(3): Note that since $\{f, g, h\}$ independent, it follows that, for $i = 1, 2$, $e_i \perp e_1 \centerdot e_2$. For generic associativity, let $\{k_1, k_2, k_3\}$ be independent realizations of $\Gamma/\tilde{R}$.

Sub Claim There exists independent $\{h_1, h_2, h_3, h_4\}$ such that $h_1h_2/\tilde{R} = k_1$, $h_2h_3/\tilde{R} = k_2$ and $h_3h_4/\tilde{R} = k_3$: As in the proof of Claim 1, we can find h_2, h_3, h_4 independent such that $h_2h_3/\tilde{R} = k_2$ and $h_3h_4/\tilde{R} = k_3$. Now, for $h'_1h'_2/\tilde{R} = k_1$, amalgamation of $\mathrm{Ltp}(h'_2/k_1)$ and $\mathrm{Ltp}(h_2/k_2k_3)$ gives the subclaim.

Now, by the subclaim, $k_1 \centerdot k_2 = h_1h_3/\tilde{R}$ and $k_2 \centerdot k_3 = h_2h_4/\tilde{R}$. Then $k_1 \centerdot (k_2 \centerdot k_3) = h_1h_4/\tilde{R}$ and $(k_1 \centerdot k_2) \centerdot k_3 = h_1h_4/\tilde{R}$ as well. Finally, it can be easily seen that $\centerdot$ is generically surjective. Hence, Claim 3 is verified.

Therefore due to Theorem 7.6.1, we have the following.

Theorem 9.3.5 *Given the group configuration, there exists a canonical hyperdefinable group and a hyperdefinable bijection mapping $\Gamma/\tilde{R}$ to the generic types of the group such that $\centerdot$ is mapped to the group multiplication generically.*

Actually, more are true. In Section 9.5 we also recover the hyperdefinable group action on the corresponding hyperdefinable set (Theorem 9.5.2).

9.4 An Application to 1-Based Theories

One application of Theorem 9.3.5 is the following result. This extends both Theorem 8.3.16 to the non-ω-categorical context and Fact 9.0.1 to the simple theory context.

Theorem 9.4.1 *Suppose that T is 1-based, non-trivial (having 4-CA). Then there is a hyperdefinable infinite bounded-by-abelian group V over a model M having SU-rank 1 generic types. Moreover, for the bounded subgroup $V_0 = V \cap \mathrm{bdd}(M)$, V/V_0 forms a vector space over the division ring R of $\mathrm{bdd}(M)$-endomorphisms of V such that for $b, a_1, \ldots, a_n \in V$, $b \in \mathrm{bdd}(a_1 \ldots a_n)$ iff $b + V_0 = \alpha_1(a_1 + V_0) + \cdots + \alpha_n(a_n + V_0)$ for some $\alpha_i \in R$.*

Proof By the proof of Lemma 8.3.15, there exists a non-trivial rank-1 Lstp p over some model M. For convenience, let $M = \emptyset$ after naming the model. As p is non-trivial, there exists $\{a, b, c\}$ realizing p such that b, c is independent and $a \in \mathrm{bdd}(b, c) \setminus \mathrm{bdd}(b) \cup \mathrm{bdd}(c)$. Let yx realize $\mathrm{tp}(ab/c)$ with $yx \mathop{\smile\!\!\!\!\!\!\,|}_c ab$. Then $SU(ay/bx) = 1$ as $y \in \mathrm{bdd}(abx)$ and $a \mathop{\smile\!\!\!\!\!\!\,|} bx$. Let $z = \mathrm{Cb}(\mathrm{Ltp}(ay/bx))$, then by 1-basedness, $z \in \mathrm{bdd}(ay) \cap \mathrm{bdd}(bx)$. Moreover, by a straightforward rank calculation, $SU(z) = 1$. This gives a bounded quadrangle (a, b, c, x, y, z). Now by Theorem 9.3.5, we obtain a hyperdefinable group G over $\emptyset$ such that the generic types all have SU-rank 1. The group G is 1-based since the underlying theory is 1-based. Now by Proposition 7.4.2, if we set $G^0 = V$, then V is the desired bounded-by-abelian hyperdefinable group. Note that by the above, V' is contained in the normal subgroup $V_0 = V \cap \mathrm{bdd}(\emptyset)$. Indeed, due to Proposition 7.4.7, the abelian group V/V_0 forms a vector space over a division ring R of $\mathrm{bdd}(\emptyset)$-endomorphisms of V, and dependence in V/V_0 is given by linear dependence of the vector space. ■

9.5 Recovering the Hyperdefinable Group Action

We restate Definition 9.0.3 in terms of hyperimaginaries in simple theories.

Definition 9.5.1 We say $(f_1, f_2, f_3, x_1, x_2, x_3)$ is a *group configuration obtained from* a hyperdefinable homogeneous space $((G, \circ), X, \cdot)$ over $\emptyset$ if $f_i \in G$, $x_i \in X$ generic elements such that $f_1 = f_2 \circ (f_3)^{-1}$, $x_2 = f_3 \cdot x_1$, $x_3 = f_2 \cdot x_1$, and $\{f_2, f_3, x_1\}$ is independent.

The following is the main theorem of this section. As assumed T has 4-CA, and a model is named so that $\emptyset = \mathrm{bdd}(\emptyset)$. We keep working with the bounded quadrangle $C = (\hat{f}, \hat{g}, \hat{h}, \hat{a}, \hat{b}, \hat{c})$.

Theorem 9.5.2 *Given the induced bounded quadrangle C from a group configuration over $\emptyset$, we can construct a hyperdefinable homogeneous space over $\emptyset$ having the transitive*

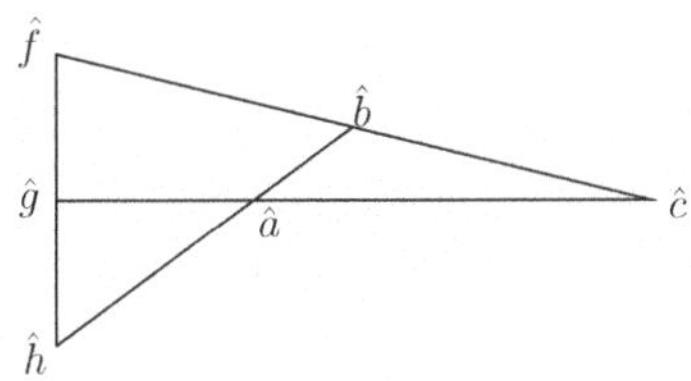

Figure 9.5

action, such that C and a bounded quadrangle obtained from the space are equivalent over some $A(\mathop{\smile\!\!\!\!\!\!|}\, C)$.

We summarize what we have done in Sections 9.2 and 9.3. A bounded quadrangle (induced from a group configuration) $C = (\hat{f}, \hat{g}, \hat{h}, \hat{a}, \hat{b}, \hat{c})$ over $\emptyset = \mathrm{bdd}(\emptyset)$ (after naming a model M) is fixed where $\hat{f}, \hat{g}, \hat{h}, \hat{a}, \hat{b}, \hat{c}$ are all boundedly closed.

Let $q = \mathrm{tp}(\hat{g})$ and let $\Gamma_q(uv)(= \Gamma(uv)) := q(u) \wedge q(v) \wedge u \mathop{\smile\!\!\!\!\!\!|}\, v$. Recall that $a \in h_q(c)$ means $h_q ca \models \mathrm{tp}(\hat{g}\hat{a}\hat{c})$. Similarly, we write that $b \in \mathrm{dom}(f) \equiv \exists c(fbc \models \mathrm{tp}(\hat{g}\hat{a}\hat{c}))$, and so on.

We define a binary relation R on Γ. Recall that a set is said to be *n-independent* if every subset having n elements is independent (over $\emptyset$):

$R(fg; f'g')$ iff $f, g, f'g' \models q$, $\{f, g, f', g'\}$ 3-independent, and there are $a \mathop{\smile\!\!\!\!\!\!|}\, fgf'g'$ and b such that $f(a) \cap g(b) \neq \emptyset, f'(a) \cap g'(b) \neq \emptyset$.

We have shown that $\tilde{R}$, the transitive closure of R, is a type-definable equivalence relation (over $\emptyset$). Moreover, we can define the hyperdefinable generic operation $\centerdot$ on $\Gamma/\tilde{R} := H$:

$$\centerdot(x_1y_1, x_2y_2; x_3y_3) := \exists xyz(\tilde{R}(x_1y_1; xy) \wedge \tilde{R}(x_2y_2; yz) \wedge \tilde{R}(xz; x_3y_3) \wedge \textstyle\bigwedge_{i=1,2,3} \Gamma(x_iy_i) \wedge x_1y_1/\tilde{R} \mathop{\smile\!\!\!\!\!\!|}\, x_2y_2/\tilde{R} \wedge \{x, y, z\} \text{ is independent}).$$

The partial operation $\centerdot$ satisfies the genericity conditions in Theorem 7.6.1. Hence, there is a unique (up to hyperdefinable isomorphisms) hyperdefinable group $(G, \centerdot)$ over $\emptyset$ such that $(H, \centerdot)$ is $\emptyset$-hyperdefinably and generically embedded into $(G, \centerdot)$. We identify H and its image in G. For $fg \models \Gamma$, let $[f, g]$ denote $fg/\tilde{R}$ (or its *interdefinable* image in G). Then $[f, g]^{-1} = [g, f]$. Recall that $H(u) \wedge H(v) \wedge u \mathop{\smile\!\!\!\!\!\!|}\, v$ is hyperdefinable.

Now we start to construct a hyperdefinable action of G on a hyperdefinable set equivalent to $\mathrm{tp}(\hat{a}) := \pi$. *For the rest of this section, we shall only use the usual type-amalgamation.* Let us define some hyperdefinable relations.

- $[fg, xy]$ iff $x \mathop{\smile\!\!\!\!\!\!|}\, fg(\models \Gamma)$ and $f(x) \cap g(y) \neq \emptyset$;
- $\Gamma_1(u; x) := H(u) \wedge \pi(x) \wedge u \mathop{\smile\!\!\!\!\!\!|}\, x$;
- $R_1(\sigma c; \sigma' c')$ iff $\sigma c, \sigma' c' \models \Gamma_1$, $\{\sigma, c, \sigma'\}$ independent, and there are $hh' \models \Gamma$ such that $[hh', cc']$ with $[h, h'] = \sigma^{-1} \centerdot \sigma'$.

Lemma 9.5.3

(1) *If* $[fg, ab]$ *and* $c \in f(a) \cap g(b)$, *then* $[f, g]$ *is interbounded with* $\mathrm{Cb}(ab/fg)$, *and* $(f, g, [f, g], b, a, c)$ *forms a bounded quadrangle.*

(2) *If* $R_1(\sigma a, \sigma' a')$, *then* a, a' *are interbounded over* $\sigma^{-1} \cdot \sigma'$, *and* $\{a, a', \sigma, \sigma'\}$ *is 3-independent.*

(3) *For* $\sigma a \models \Gamma_1$ *and* $a_1 \mathop{\smile\!\!\!\!|}\, \sigma a$, *there is* $\sigma_1 (\mathop{\smile\!\!\!\!|}\, a_1)$ *such that* $R_1(\sigma a, \sigma_1 a_1)$.

Proof (1) comes from Lemma 9.2.3.

(2) If $R_1(\sigma a, \sigma' a')$, then by (1), a, a' are interbounded over $\sigma^{-1} \cdot \sigma'$. Now since $\{\sigma, a, \sigma'\}$ independent, it follows that $a \mathop{\smile\!\!\!\!|}_{\sigma^{-1} \cdot \sigma'} \sigma\sigma'$ and $aa' \mathop{\smile\!\!\!\!|}_{\sigma^{-1} \cdot \sigma'} \sigma\sigma'$. Then 3-independence of $\{a, a', \sigma, \sigma'\}$ follows from the 2-independence of the sets $\{a, a', \sigma^{-1} \cdot \sigma'\}$ and $\{\sigma, \sigma', \sigma^{-1} \cdot \sigma'\}$.

(3) Since $a \mathop{\smile\!\!\!\!|}\, a_1$, by amalgamation and extension we have $f \mathop{\smile\!\!\!\!|}\, g (\mathop{\smile\!\!\!\!|}\, a$ and $\mathop{\smile\!\!\!\!|}\, a_1)$ such that $[fg, aa_1]$ and $fg \mathop{\smile\!\!\!\!|}_{aa_1} \sigma, fgaa_1 \mathop{\smile\!\!\!\!|}\, \sigma$. Let $[f, g] = \sigma'$ and let $\sigma_1 = \sigma \cdot \sigma'$. Then $\{\sigma', a, \sigma\}$ is independent and $a_1 \mathop{\smile\!\!\!\!|}\, \sigma_1$. Thus, $a \mathop{\smile\!\!\!\!|}_{\sigma} \sigma_1$, and $\{a, \sigma, \sigma_1\}$ is independent. Hence, $R_1(\sigma a, \sigma_1 a_1)$ follows. ■

Note that from 9.5.3, R_1 is symmetric. Let $\tilde{R}_1$ be the transitive closure of R_1. We shall show that $\tilde{R}_1$ is type-definable.

Lemma 9.5.4 *Suppose that* $R_1(\sigma a, \sigma' b)$, $R_1(\sigma' b, \sigma'' c)$, *and* $\sigma a \mathop{\smile\!\!\!\!|}_{\sigma' b} \sigma'' c$. *Then* $R_1(\sigma a, \sigma'' c)$, *and* $\sigma a \mathop{\smile\!\!\!\!|}_{\sigma'' c} \sigma' b$, .

Proof There are hh', kk' such that $[hh', ab], [kk'bc]$, $[h, h'] = \sigma^{-1} \cdot \sigma' := \alpha$, $[k, k'] = (\sigma')^{-1} \cdot \sigma'' := \alpha'$, and both $\{\sigma, a, \sigma'\}$ and $\{\sigma', b, \sigma''\}$ independent. Then since $\sigma a \mathop{\smile\!\!\!\!|}_{\sigma' b} \sigma'' c$, we have $\sigma a \sigma' \mathop{\smile\!\!\!\!|}\, \sigma''$ and that $\{a, \sigma, \sigma', \sigma''\}$ is independent. Now the following claim says that $R_1(\sigma a, \sigma'' c)$.

Claim There is $h_0 k_0'$ such that $[h_0 k_0', ac]$ and $[h_0, k_0'] = \sigma^{-1} \cdot \sigma''$: Since $[hh', ab], [kk'bc]$, we have that $b \mathop{\smile\!\!\!\!|}\, hh'$, $b \mathop{\smile\!\!\!\!|}\, kk'$ and there are d, d' such that $d \in h(a) \cap h'(b)$, $d' \in k(b) \cap k'(c)$ (*). Now again since $\sigma a \mathop{\smile\!\!\!\!|}_{\sigma' b} \sigma'' c$, $\{b, \sigma, \sigma', \sigma''\}$ is independent. Thus, it easily follows that $\{b, \alpha, \alpha'\}$ is independent (**). Hence, $\alpha \mathop{\smile\!\!\!\!|}_b \alpha'$ and then, from 9.5.3, $\alpha a \mathop{\smile\!\!\!\!|}_b \alpha' c$. Moreover, by (*), both $\{\alpha, h', b\}$ and $\{\alpha', k, b\}$ are independent and, hence, $h' \mathop{\smile\!\!\!\!|}_b \alpha$, $k \mathop{\smile\!\!\!\!|}_b \alpha'$ and $h'd \mathop{\smile\!\!\!\!|}_b a\alpha$, $kd' \mathop{\smile\!\!\!\!|}_b c\alpha'$ hold. Therefore, by amalgamation there is

$$h_0' d'' \models \mathrm{Ltp}(h'd/ab\alpha) \cup \mathrm{Ltp}(kd'/bc\alpha'),$$

with $h_0' d'' \mathop{\smile\!\!\!\!|}_b ac\alpha\alpha'$ (***). Hence, there are h_0, k_0' such that

$$hh'd \equiv^L_{ab\alpha} h_0 h_0' d'' \text{ and } kk'd' \equiv^L_{bc\alpha'} h_0' k_0' d''.$$

Then from (**),(***), it can easily be seen that $\{b, h_0', k_0', h_0\}$ is independent. Thus, from (*), $d'' \in h_0(a) \cap k_0'(c)$ and $\{a, h_0, k_0'\}$ independent; i.e., $[h_0 k_0', ac]$ holds.

Moreover, $[h_0, k'_0] = [h_0, h'_0] \cdot [h'_0, k'_0] = \alpha \cdot \alpha' = \sigma^{-1} \cdot \sigma''$. Therefore, the claim is proved.

Now from 9.5.3, we have $\sigma \perp \sigma' b \sigma'' c$ and $\sigma \perp_{\sigma'' c} \sigma' b$. According to $R_1(\sigma a, \sigma'' c)$ again with 9.5.3, $\sigma a \perp_{\sigma'' c} \sigma' b$ holds. ■

Therefore due to Proposition 4.3.6 and Lemma 9.5.3(3), we obtain the following.

Proposition 9.5.5 *The following are equivalent:*

(1) $\tilde{R}_1(\alpha x, \alpha' x')$.

(2) *For some* σz, $R_1(\sigma z, \alpha x)$ *and* $R_1(\sigma z, \alpha' x')$.

(3) *For some* σz *with* $\sigma z \perp_{\alpha' x'} \alpha x$ *(and* $\sigma z \perp_{\alpha x} \alpha' x'$*),* $R_1(\sigma z, \alpha x)$ *and* $R_1(\sigma z, \alpha' x')$.

Hence $\tilde{R}_1$ is a type-definable equivalence relation. Now let $[\sigma, a]$ denote a hyperimaginary $(\sigma, a)/\tilde{R}_1$ in $\Gamma_1/\tilde{R}_1 := X$. For $fg \models \Gamma$, $[fg, a] := [[f, g], a]$.

Lemma 9.5.6

(1) $\sigma \perp [\sigma, a]$, *and* $a \perp [\sigma, a]$.

(2) $[\sigma, a] = [\sigma', a']$ *and* $\sigma a \perp_{[\sigma,a]} \sigma' a'$ *iff* $R_1(\sigma a, \sigma' a')$.

(3) $[\sigma, a]$ *and* a *are interbounded over* σ.

Proof (1) As in 9.5.3(3), there is $\sigma' a' \models \Gamma_1$ such that $R_1(\sigma a, \sigma' a')$. Then due to 9.5.3(2), each of $\{\sigma, a', \sigma'\}, \{a, a', \sigma'\}$ is independent. Hence, $\sigma \perp [\sigma', a'] = [\sigma, a]$, and $a \perp [\sigma, a]$.

(2) ($\Rightarrow$) By 9.5.5, there is αc such that $\alpha c \perp_{\sigma a} \sigma' a'$, $R_1(\alpha c, \sigma a)$ and $R_1(\alpha c, \sigma' a')$. Then since $\sigma a \perp_{[\sigma,a]} \sigma' a'$, it follows $\sigma a \perp_{\alpha c} \sigma' a'$. Thus, by 9.5.4, $R_1(\sigma a, \sigma' a')$ holds.

($\Leftarrow$) By the extension axiom, there is $\sigma'' a'' \models tp(\sigma' a'/[\sigma, a])$ such that $\sigma'' a'' \perp_{[\sigma,a]} \sigma' a'$ (*). Hence, by the proof of ($\Rightarrow$), $R_1(\sigma'' a'', \sigma' a')$. Again, by extension, we can assume that $\sigma a \perp_{[\sigma,a]\sigma' a'} \sigma'' a''$. Hence, $\sigma a \perp_{\sigma' a'} \sigma'' a''$ and then by 9.5.4, $\sigma a \perp_{\sigma'' a''} \sigma' a'$. Therefore, by (*), $\sigma a \perp_{[\sigma,a]} \sigma' a'$.

(3) Given $\sigma \perp a$, it suffices to show $a \in \mathrm{bdd}(\sigma, [\sigma, a])$. There are f, g, g', h, a' such that $[f, g] = \sigma$ and $[g' h, a a']$. By amalgamation and extension, we can assume that $g = g'$ and $\{\sigma, g = g', a, h\}$ independent. Then it follows that $\{f, g, h\}$ independent, $[g, h] = \sigma^{-1} \cdot \sigma'$, where $\sigma' = [f, h]$, and $R_1(\sigma a, \sigma' a')$. Hence, from (2), $\sigma a \perp_{[\sigma,a]} \sigma' a'$ holds. Thus, $a \perp_{[\sigma,a]\sigma} \sigma^{-1} \cdot \sigma' a'$, and due to 9.5.3, it follows that $a \in \mathrm{bdd}(\sigma, [\sigma, a])$. ■

Lemma 9.5.7

(1) *Suppose that* $\sigma, x \models \Gamma_1$ *and there is independent* $\{f, g, x'\}$ *such that* $[fg, x' x]$ *and* $[f, g] = \sigma$. *Then* x' *and* $[\sigma, x]$ *are interbounded.*

(2) *For* $\alpha (\perp x)(\perp y)$, *if* $[\alpha, x] = [\alpha, y]$, *then* x, y *are interbounded.*

Proof (1) The proof will be similar to that of 9.5.6(3). Let $z \in f(x') \cap g(x)$. Then by extension, there is $x_0 g_0$ such that $g_0 x_0 \perp_{fx'z} gx$ and $g_0 x_0 \equiv^L_{fx'z} gx$ (*). Thus, $\{x', f, g, g_0\}$ independent, $[gg_0, xx_0]$, and $R_1(\sigma x, \sigma_0 x_0)$, where $\sigma_0 = [f, g_0]$. Note that x (x_0, resp.) and x' are interbounded over σ (σ_0, resp.). Now due to 9.5.6, $\sigma x \perp_{[\sigma,x]} \sigma_0 x_0$, $\sigma x x' \perp_{[\sigma,x]} \sigma_0 x_0 x'$, and $\sigma \perp_{[\sigma,x]} x'$, $\sigma \perp_{x'} [\sigma, x]$, $\sigma x \perp_{x'} [\sigma, x]$. Therefore, $x', [\sigma, x]$ are interbounded.

(2) By extension, there are β, z such that $\alpha xy \perp_{[\alpha x]} \beta z$ and $[\beta, z] = [\alpha, x]$. By 9.5.6, $R_1(\alpha x, \beta z)$ and $R_1(\alpha y, \beta z)$. Hence, there are $fg, f'g'$ such that $[fg, xz], [f'g', yz]$ and $[f, g] = [f', g'] = \alpha^{-1} \centerdot \beta$. Then from (1), each of x, y is interbounded with $[fg, z]$. Thus, x, y are interbounded. ■

Lemma 9.5.8 *Let $(\alpha, x), (\beta, y) \in \Gamma_1$ and $\sigma \in H$ such that $[\alpha, x] = [\beta, y]$ and both $\{\sigma, \alpha, x\}, \{\sigma, \beta, y\}$ independent. Then $[\sigma \centerdot \alpha, x] = [\sigma \centerdot \beta, y]$.*

Proof As we can apply the extension axiom, it suffices to show the case $\alpha x \perp_{\sigma[\alpha,x]} \beta y$ (*). Now since $\{\alpha, x, \sigma\}$ independent, it follows from 9.5.6, $\{[\alpha, x], \alpha, \sigma\}$ is independent (**). Then due to (*), $\alpha \perp_{[\alpha,x]} \beta y$. Again from 9.5.6, $\alpha x \perp_{[\alpha,x]} \beta y$ and $R_1(\alpha x, \beta y)$. Hence, there is fg such that $[fg, xy]$ with $[f, g] = \alpha^{-1} \centerdot \beta = (\sigma \centerdot \alpha)^{-1} \centerdot (\sigma \centerdot \beta)$. Moreover, similarly to (**), $\{[\alpha, x], \beta, \sigma\}$ is independent. Thus, by (*), $\alpha x \sigma \perp \beta$ and, hence, both $\{\alpha, x, \sigma, \beta\}$ and $\{x, \sigma \centerdot \alpha, \sigma \centerdot \beta\}$ are independent and $R_1(\sigma \centerdot \alpha x, \sigma \centerdot \beta y)$. The lemma is proved.

For $\sigma \in H$ and $[\alpha, x] \in X$ with $\sigma \perp [\alpha, x]$, define a hyperdefinable partial action $\centerdot$ of H on X as follows: $\sigma \centerdot [\alpha, x] = [\sigma \centerdot \beta, x']$, where $[\beta, x'] = [\alpha, x]$ and $\{\sigma, \beta, x'\}$ independent. We shall show that this is well defined and satisfies the genericity conditions in Theorem 7.6.2.

Claim 1 The action is well defined: For $\sigma \perp [\alpha, x]$, there are β, x' such that $[\alpha x] = [\beta x']$ and $\sigma \perp_{[\alpha,x]} \beta, x'$. Then $\{\sigma, \beta, x'\}$ is independent, and $x' \perp \sigma \centerdot \beta \in H$. Hence, the action is a non-void relation. Now, assume there is $\gamma x''$ such that $\{\sigma, \gamma, x''\}$ is independent and $[\gamma, x''] = [\alpha, x](= [\beta, x'])$. Then due to 9.5.8, $[\sigma \centerdot \beta, x'] = [\sigma \centerdot \gamma, x'']$ holds, and Claim 1 is proved.

Claim 2 The partial action satisfies generic independence; i.e., $\sigma \perp \sigma \centerdot [\alpha, x]$ and $[\alpha, x] \perp \sigma \centerdot [\alpha, x]$: As we can assume $\{\sigma, \alpha, x\}$ is independent (and $\sigma \centerdot [\alpha, x] = [\sigma \centerdot \alpha, x]$), so is each of $\{\alpha, \sigma \centerdot \alpha, x\}$ and $\{\sigma, \sigma \centerdot \alpha, x\}$. Thus, $\sigma \perp \sigma \centerdot [\alpha, x]$. Moreover, $\alpha \perp_x \sigma \centerdot \alpha$, $[\alpha, x] \perp_x [\sigma \centerdot \alpha, x]$ and, from 9.5.6, $[\alpha, x] \perp [\sigma \centerdot \alpha, x]$.

Claim 3 The generic action is generically associative. Given independent $\{\beta, \sigma, [\alpha, x]\}$, we can assume that $\{\beta, \sigma, \alpha, x\}$ is independent. Then clearly, $(\beta \centerdot \sigma) \centerdot [\alpha, x] = [\beta \centerdot \sigma \centerdot \alpha, x] = \beta \centerdot (\sigma \centerdot [\alpha, x])$.

Claim 4 The generic action is generically surjective. Given $\alpha \perp [\beta, y]$, we can assume that $\{\alpha, \beta, y\}$ is independent. Then there is $\gamma \in H$ such that $\beta = \alpha \centerdot \gamma$ and $\{\alpha, \gamma, y\}$ is independent. Thus, $\alpha \perp [\gamma \centerdot y]$ and $\alpha \centerdot [\gamma, y] = [\beta, y]$.

Claim 5 The partial action is generically transitive. Given $[\alpha, x] \downarrow [\beta, y]$, by extension we can assume that $\alpha x \downarrow_{[\alpha,x]} [\beta, y]$, $[\alpha, x] \downarrow_{[\beta,y]} \beta y$, and $\alpha x \downarrow_{[\alpha,x][\beta,y]} \beta y$. Then it easily follows that $\{\alpha, x, \beta, y\}$ is independent. Thus, from 9.5.3, there is $\alpha'(\downarrow x)$ such that $[\alpha', x] = [\beta, y]$ and again by extension we assume that $\alpha' \downarrow_{x[\beta,y]} \alpha$. Hence, $\{\alpha, \alpha', x\}$ is independent. Therefore, we have $\gamma(\downarrow \alpha, x)$ such that $\gamma \centerdot \alpha = \alpha'$ and $\gamma \centerdot [\alpha, x] = [\beta, y]$. Thus, $\centerdot$ is generically transitive.

Hence now we can apply Theorem 7.6.2 and conclude that there are a hyperdefinable set W and a hyperdefinable transitive group action $*$ of G generically extending that $\centerdot$ on X. To finish the proof of Theorem 9.5.2, it remains to show the equivalence of the original bounded quadrangle $C = (\hat{f}, \hat{g}, \hat{h}, \hat{a}, \hat{b}, \hat{c})$ and a bounded quadrangle obtained from the homogeneous space $(G, W, *)$. As before we identify $H \subseteq G$ and $X \subseteq W$.

Claim 6 Choose $f(\downarrow C)$ and b such that $\hat{c} \in f(b)$. In particular, $[f\hat{g}, b\hat{a}]$ holds. Then the bounded quadrangle $C' = (f, \hat{g}, [f, \hat{g}], \hat{a}, b, \hat{c})$ and C are equivalent over $e = \mathrm{Cb}(b\hat{b}/f\hat{f})$. Choose next $h(\equiv \hat{g})$ independent from all other hyperimaginaries and b' such that $\hat{c} \in h(b')$. Then a bounded quadrangle $C'' = ([h,f], [h,\hat{g}], [f,\hat{g}], [\hat{g}h, b'], [f\hat{g}, \hat{a}], [h\hat{g}, \hat{a}])$ obtained from (G, W) is equivalent to C' over h. Therefore, C and C'' are equivalent over eh. We sketch the proof. That $e \downarrow C$, $e \downarrow C'$ and the equivalence of C and C' over e come from the same proof of Lemma 9.2.3. Now since $h \downarrow C'$, it easily follows that $h \downarrow C''$. According to 9.5.7, b' and $[h\hat{g}, \hat{a}]$ are interbounded. Thus, $\hat{c}$ and $[h\hat{g}, \hat{a}]$ are interbounded over h, because $\hat{c} \in h(b')$. Then C' and C'' are equivalent over h, as the interboundedness of other pairs over h follows as well from 9.5.3 and 9.5.7. Note that since $\hat{c} \in \hat{g}(\hat{a}) \cap f(b) \cap h(b')$, $R_1([\hat{g}h]b'; [\hat{g}f]b)$, $R_1([f\hat{g}]\hat{a}; [fh]b')$, and $R_1([hf]b; [h\hat{g}]\hat{a})$ hold. Hence, C'' is a bounded quadrangle obtained from (G, W).

We have proved the main theorem 9.5.2. ■

Remark 9.5.9 We point out that there is a canonical almost injective map from the set of realizations of $\pi(x)$ to X. Define $j_\pi : x(\models \pi) \mapsto \alpha^{-1} \centerdot [\alpha, x] \in X$ for any $\alpha(\in H) \downarrow x$. It is a well-defined hyperdefinable map such that $j_\pi(x) = j_\pi(y)$ implies that $\mathrm{bdd}(x) = \mathrm{bdd}(y)$, and for $\alpha \in H(\downarrow x)$, $\alpha \centerdot j_\pi(x) = [\alpha, x]$. To show well-definedness, assume there are $\alpha, \beta \in H$, each independent with x. By extension we may additionally assume that $\beta \downarrow \alpha x$. Then $(\beta \centerdot \alpha^{-1}) \centerdot [\alpha, x] = [\beta, x]$ and since $\{\beta^{-1}, \beta \centerdot \alpha^{-1}, [\alpha, x]\}$ independent, from generic associativity, $\alpha^{-1} \centerdot [\alpha, x] = \beta^{-1} \centerdot [\beta, x]$ holds. Thus, clearly j_π is hyperdefinable as well. Now choose $\alpha(\in H) \downarrow xy$. Suppose that $\alpha^{-1} \centerdot [\alpha, x] = \alpha^{-1} \centerdot [\alpha, y]$. Then $[\alpha, x] = [\alpha, y]$ and from 9.5.7(2), x, y are interbounded. Moreover, $\alpha \centerdot j_\pi(x) = \alpha \centerdot (\alpha^{-1} \centerdot [\alpha, x]) = [\alpha, x]$.

We finish this section by stating the case when the group configuration is principal. (This is the case in the proof of Theorem 9.4.1 where T is non-trivial 1-based.)

Theorem 9.5.10 *Assume additionally the original bounded quadrangle $C = (\hat{f}, \hat{g}, \hat{h}, \hat{a}, \hat{b}, \hat{c})$ is principal. Then there is a bounded quadrangle C' over e equivalent to C such that the hyperdefinable homogeneous space constructed from C' is principal (i.e., the group action is regular).*

Proof First, we point out the following.

Claim 1 If $ac \equiv_{\hat{g}} \hat{a}\hat{c}$ and $c \downarrow \hat{g}\hat{a}\hat{c}$, then $(c\hat{c}k, \hat{a}a\hat{g})$ is a principal group configuration, where $k := \mathrm{Cb}(a\hat{a}/c\hat{c})$: The proof will be similar to that of 9.2.3(1). We first check that $(c\hat{c}k, \hat{a}a\hat{g})$ forms a group configuration. Since $k = \mathrm{Cb}(a\hat{a}/c\hat{c})$, $a \downarrow_{\hat{a}k} c\hat{c}$ holds. Then since C is principal, $a \in \mathrm{bdd}(\hat{a}\hat{c}c)$ and so $a \in \mathrm{bdd}(\hat{a}k)$. Similarly, $\hat{a} \in \mathrm{bdd}(ak)$, and $c, \hat{c}$ are interbounded over k as well. It remains to see $c \downarrow k$ and $\hat{c} \downarrow k$ (the rest follow easily). Note there are $f'h'$ such that $f'h'ac \equiv_{\hat{g}} \hat{f}\hat{h}\hat{a}\hat{c}$. Then by amalgamation there are $fgdb$ such that

$$fh \models \mathrm{Ltp}(f'h'/\hat{g}ac) \cup \mathrm{Ltp}(\hat{f}\hat{h}/\hat{g}\hat{a}\hat{c}),$$

$\{f, \hat{g}, \hat{c}, c\}$ is independent, and $f\hat{g}h\hat{a}b\hat{c} \equiv C \equiv f\hat{g}hadc$ (*). It follows then that $a \downarrow c\hat{c}f, a \downarrow c\hat{c}fdb$. By C being principal and (*), $\hat{a} \in \mathrm{bdd}(ac\hat{c}) \cap \mathrm{bdd}(adb)$. Hence, $a\hat{a} \downarrow_{c\hat{c}} db$ $a\hat{a} \downarrow_{db} c\hat{c}$ hold. Therefore, $k \in \mathrm{bdd}(db)$ and since $c \downarrow db, \hat{c} \downarrow db$, we have $c \downarrow k$, $\hat{c} \downarrow k$. Now in order to show $(c\hat{c}k, \hat{a}a\hat{g})$ is principal, it is enough to see that $k \in \mathrm{bdd}(a\hat{a})$. Note that $\{d, a, \hat{a}, c\}$ is independent. In particular, $d \downarrow_{a\hat{a}} c$. Then again $db \downarrow_{a\hat{a}} c\hat{c}$ and $k \downarrow_{a\hat{a}} k$. Therefore, $k \in \mathrm{bdd}(a\hat{a})$. We have proved Claim 1.

We resume the proof of 9.5.10. Choose $\hat{a}_0\hat{c}_0 \equiv_{\hat{g}} \hat{a}\hat{c}$ with $\hat{c}_0 \downarrow C$. Then by Claim 1, $C' = (\hat{f}, \hat{g}, \hat{h}, \hat{a}_0\hat{a}, \hat{b}, \hat{c}_0\hat{c})$ is a principal-bounded quadrangle over $e' := \mathrm{Cb}(\hat{a}_0\hat{a}/\hat{c}_0\hat{c})(\downarrow C)$ equivalent to C. Now if $k' \equiv_{e'} \hat{c}_0\hat{c}$, then we write $k' = k_0k$, where $k_0 \equiv_{e'} \hat{c}_0$ and $k \equiv_{e'} \hat{c}$. Similarly, for $x' \equiv_{e'} \hat{a}_0\hat{a}$, write $x' = x_0x$, where $x_0, x \models \pi$. Also for C', we shall use the same but primed notations which are defined in the process of the group action construction. For example, $[fg, a'b']'$ where $f, g \models \mathrm{tp}(\hat{g}/e')$, $a', b' \models \pi' = \mathrm{tp}(\hat{a}_0\hat{a}/e')$, means that there is $c' = c_0c$ such that $f, a', c' \equiv_{e'} g, b', c' \equiv_{e'} \hat{g}, \hat{a}_0\hat{a}, \hat{c}_0\hat{c}$ and $\{f, g, a'\}$ is e'-independent. In particular, as in Claim 1, (c_0ce', aa_0f) and (c_0ce', aa_0f) are principal group configurations, and $\{f, g, c_0, c\}$ is independent (**).

Claim 2 Suppose that $[fg, a'b']'$. Then

(i) $e := \mathrm{Cb}(ab/a_0b_0) \in \mathrm{bdd}(fg)$ and (a_0b_0e, bae') forms a principal group configuration; and

(ii) $\mathrm{Cb}(a'b'/fge')$ and e are interbounded over e'.

We have $c' = c_0c$ as in (**). Now the proof of (i) is the same as that of Claim 1. Namely, from that $a \downarrow a_0b_0c_0$, $a \downarrow a_0b_0c_0fg$, $ab \downarrow_{a_0b_0} fg$, and $ab \downarrow_{fg} a_0b_0$, we have that $a_0 \downarrow \mathrm{bdd}(fg) \ni e$ and $a \downarrow e$. Also $e \in \mathrm{bdd}(ab)$ comes from that $f \downarrow_{ab} a_0, fg \downarrow_{ab} a_0b_0$, and $e \downarrow_{ab} e$. The rest is easy. For (ii), let $k := \mathrm{Cb}(a'b'/fge')$. Thus, $a'b' \downarrow_k fge'$, and from (i), $e \downarrow_k e$. Hence, $e \in \mathrm{bdd}(k)$. Moreover, from (**), $fg \downarrow_{c_0} c$, $fga_0 \downarrow_{c_0} e'$, so $\{f, g, a_0, e'\}$ is independent and $a_0 \downarrow fge'$. Thus, $a_0 \downarrow_{ee'} fge'$, and then from (i), $a'b' \downarrow_{ee'} fge'$. Therefore, $k \in \mathrm{bdd}(ee')$. Claim 2 is proved.

We now possibly take another model N such that $C' \downarrow_{e'} N$ and $e' \in \mathrm{dcl}(N)$ (so that $C \downarrow N$) and may replace each component in C' by its bounded closure with

N. But for convenience call this new one, again, C' over N, which clearly is equivalent to C. We keep using the primed notations (defined over N) for C', and again if $a' \equiv \mathrm{bdd}(\hat{a}_0\hat{a}N) \in C'$, then a_0a is a subsequence of a' equivalent to $\hat{a}_0\hat{a}$.

Claim 3 If $[fg, a'b']', [hk, a'b']'$ hold, then $[f, g]' = [h, k]'$: By the extension axiom, it clearly suffices to show that $R'(fg, hk)$ holds when $fg \downarrow_{a'b'} hk$ (***). Note that 3-independence (over N) of $\{f, g, h, k\}$ follows from that of $\{f, g, a', b'\}$ and $\{h, k, a', b'\}$. It remains to show that $a' \downarrow_N fghk$. Let $e_1 := \mathrm{Cb}(a'b'/fg)$ and $e_2 := \mathrm{Cb}(a'b'/hk)$ and let $e = \mathrm{Cb}(ab/a_0b_0)$ be the same as in Claim 2. Now, according to the choice of N, it is easy to see that Claim 2(ii) with e' being replaced by N holds. Namely, $\mathrm{bdd}(e_1N) = \mathrm{bdd}(eN) = \mathrm{bdd}(e_2N)$. Thus, from (***), $fga'b' \downarrow_{e_2N} hk$, $fga'b' \downarrow_{e_1N} hk$, and $a'b' \downarrow_{e_1N} fghk$. Now since $[fg, a'b']'$ (over N), $a' \downarrow_N e_1$ holds. Therefore, as desired, $a' \downarrow_N fghk$.

Claim 4 If $[\alpha, a]' = [\beta, a]'$, then $\alpha = \beta$: By Lemma 9.5.6 and extension, there are γ, b such that $R_1'(\alpha a, \gamma b), R_1'(\beta a, \gamma b)$. Hence, there are fg, hk such that $[fg, ab]', [hk, ab]'$. Thus, from Claim 3, $\alpha^{-1} \cdot' \gamma = [f, g]' = [h, k]' = \beta^{-1} \cdot' \gamma$, and hence $\alpha = \beta$.

Due to Theorem 7.6.2 it suffices to show the following.

Claim 5 The partial action $\cdot'$ of H' on X' is generically regular: We already know the partial action is transitive. Now suppose for $\beta_i \downarrow [\alpha, x]'$, $\beta_1 \cdot [\alpha, x]' = \beta_2 \cdot [\alpha, x]'$ holds. We can clearly assume that $\{\beta_i, \alpha, x\}$ N-independent so that $\beta_i \cdot [\alpha, x]' = [\beta_i \cdot \alpha, x]'$. Then due to Claim 4, $\beta_1 \cdot \alpha = \beta_2 \cdot \alpha$ and $\beta_1 = \beta_2$ hold.

We have proved Theorem 9.5.10. ■

9.6 Bibliographical Remarks

Example 9.1.5(1) is due to Hrushovski. The topics related to *n-simplicity* are not covered [77], [72]. Sections 9.2–9.4 are from [33], and 9.5 is from [70]. Ben-Yaacov, Tomasic, and Wagner [8] also obtain the group configuration theorem for unconditional simple theories but the group they obtained is not hyperdefinable. Using their group configuration, the equivalence of pseudolinearity and linearity is proved for any simple ω-categorical SU-rank 1 structure [115]. The equivalence is known to be held in any stable rank 1 structure [10].

REFERENCES

[1] Hans Adler, *Explanations of Independence*, PhD Thesis, University of Freiburg (2005).

[2] Hans Adler, "Around the Strong Order Property and Mock Simplicity," unpublished note (2008).

[3] Emil Artin, *Geometric Algebra* (Interscience Publishers, New York, 1957).

[4] James Ax, "The Elementary Theory of Finite Fields," *Annals of Math.* 88 (1968) 239–271.

[5] John T. Baldwin, *Fundamentals of Stability Theory* (Springer-Verlag, New York, 1987).

[6] Andreas Baudisch, "Neostability-Properties of Fraïssé Limits of 2-Nilpotent Groups of Exponent $p > 2$," preprint (2013).

[7] Itaï Ben-Yaacov, Anand Pillay, and Evgueni Vassiliev, "Lovely pairs of models," *Annals of Pure and Applied Logic* 122 (2003), 235–61.

[8] Itaï Ben-Yaacov, Ivan Tomasic, and Frank O. Wagner, "Constructing an Almost Hyper-definable Group," *Journal of Math. Logic* 4 (2004), 181–212.

[9] Itaï Ben-Yaacov and Frank O. Wagner, "On Almost Orthogonality in Simple Theories," *Journal of Symbolic Logic* 69 (2004), 398–408.

[10] Steven Buechler, "Pseudoprojective Strongly Minimal Sets Are Locally Projective," *Journal of Symbolic Logic* 56 (1991), 1184–94.

[11] Steven Buechler, *Essential Stability Theory* (Springer-Verlag, New York, 1996).

[12] Steven Buechler, "Lascar Strong Types in Some Simple Theories," *Journal of Symbolic Logic* 64 (1999), 817–24.

[13] Steven Buechler, A. Pillay, and F. O. Wagner, "Supersimple Theories," *Journal of American Math. Society* 14 (2001), 109–24.

[14] Enrique Casanovas, "The Number of Types in Simple Theories," *Annals of Pure and Applied Logic* 98 (1999), 69–86.

[15] Enrique Casanovas, *Simple Theories and Hyperimaginaries* (Cambridge University Press, Cambridge, 2011).

[16] Enrique Casanovas and Byunghan Kim, "A Supersimple Nonlow Theory," *Notre Dame Journal of Formal Logic* 39 (1998), 507–18.

[17] Enrique Casanovas, Daniel Lascar, Anand Pillay, and Martin Ziegler, "Galois Groups of First Order Theories," *Journal of Math. Logic* 1 (2001), 305–19.

[18] Enrique Casanovas and Joris Potier, "Normal Hyperimaginaries," preprint, arXiv:1112.2049v2.

[19] Enrique Casanovas and Frank O. Wagner, "Local Supersimplicity and Related Concepts," *Journal of Symbolic Logic* 67 (2002), 744–58.

[20] C. C. Chang and H. Jerome Keisler, *Model Theory*, 3rd edn (North-Holland, Amsterdam, 1990).

[21] Zoe Chatzidakis, "Simplicity and Independence for Pseudo-algebraically Closed Fields," in *Models and Computability*, London Math. Society Lecture Notes Series 259 (Cambridge University Press, Cambridge, 1999), 41–61.

[22] Zoe Chatzidakis, "Generic Automorphisms of Separably Closed Fields," *Illinois Journal of Math.* 45 (2001), 693–733.

[23] Zoe Chatzidakis and Ehud Hrushovski, "Model Theory of Difference Fields," *Transactions of American Math. Society* 351 (1999), 2997–3071.

[24] Zoe Chatzidakis and Ehud Hrushovski, "Model Theory of Endomorphisms of Separably Closed Fields," *Journal of Algebra* 281 (2004), 567–603.

[25] Zoe Chatzidakis, Ehud Hrushovski, and Y. Peterzil, "Model Theory of Difference Fields II," *Proceedings of London Math. Society* 85 (2002), 257–311.

[26] Zoe Chatzidakis, Angus MacIntyre, and Lou van den Dries, "Definable Sets in Finite Fields," *Journal für die reine und angewandte Mathematik* 427 (1992), 107–35.

[27] Zoe Chatzidakis and Anand Pillay, "Generic Structures and Simple Theories," *Annals of Pure and Applied Logic* 95 (1998), 71–92.

[28] Gregory Cherlin, "Stable Division Rings," in *Logic Colloquium 1977* (North Holland, 1978), 99–111.

[29] Gregory Cherlin, Leo Harrington, and Alistair H. Lachlan, "ω-Categorical, ω-Stable Structures," *Annals of Pure and Applied Logic* 28 (1985), 103–35.

[30] Gregory Cherlin and Ehud Hrushovski, *Finite Structures with Few Types* (Princeton University Press, Princeton, 2003).

[31] Gregory Cherlin and Saharon Shelah, "Superstable Fields and Groups," *Annals of Math. Logic* 18 (1980), 227–70.

[32] Tristram de Piro and Byunghan Kim, "The Geometry of 1-Based Minimal Types," *Transactions of American Math. Society* 355 (2003), 4241–63.

[33] Tristram de Piro, Byunghan Kim, and Jessica Millar, "Constructing the Hyperdefinable Group from the Group Configuration," *Journal of Math. Logic* 6 (2006), 121–39.

[34] Jean Doyen and Xavier Hubaut, "Finite Regular Locally Projective Spaces," *Mathematische Zeitschrift* 119 (1971), 83–8.

[35] J. L. Duret, "Les corps faiblement algébriquement clos non séparablement clos ont la propriété d'indépendance," in *Model Theory of Algebra and Arithmetic*, Springer Lecture Note 834 (1980), 135–57.

[36] Clifton Ealy and Alf Onshuus, "Characterizing Rosy Theories," *Journal of Symbolic Logic* 72 (2007), 919–40.

[37] David M. Evans, "Ample dividing," *Journal of Symbolic Logic* 68 (2003), 1385–402.

[38] David M. Evans, "Expansions of Fields by Angular Functions," *Journal of the Institute of Math. of Jussieu* 7 (2008), 735–51.

[39] David Evans and Frank O. Wagner, "Supersimple ω-Categorical Groups," *Journal of Symbolic Logic* 65 (2000), 767–76.

[40] Jakub Gismatullin and Ludomir Newelski, "G-Compactness and Groups," *Archive for Math. Logic* 47 (2008), 479–501.

[41] John Goodrick and Alexei Kolesnikov, "Groupoids, Covers, and 3-Uniqueness in Stable Theories," *Journal of Symbolic Logic* 75 (2010), 905–29.

[42] John Goodrick, Byunghan Kim, and Alexei Kolesnikov, "Homology Groups of Types in Model Theory and the Computation of $H_2(p)$, " Journal of Symbolic Logic, forthcoming.

[43] John Goodrick, Byunghan Kim, and Alexei Kolesnikov, "Amalgamation Functors and Homology Groups in Model Theory," Preprint (2013).

[44] Victor Harnik and Leo Harrington, "Fundamentals of Forking," *Annals of Pure and Applied Logic* 26 (1984), 245–86.

[45] Bradd Hart, Byunghan Kim, and Anand Pillay, "Coordinatization and Canonical Bases in Simple Theories," *Journal of Symbolic Logic* 65 (2000), 293–309.

[46] Bradd Hart and Ziv Shami, "On the Type-Definability of the Binding Group in Simple Theories," *Journal of Symbolic Logic* 70 (2005), 379–88.

[47] Wilflid Hodges, *Model theory* (Cambridge University Press, Cambridge, 1993).

[48] Ehud Hrushovski, "Locally Modular Regular Types," in *Classification Theory*, Lecture notes in mathematics 1292 (Springer-Verlag, Berlin, 1987), 132–64.

[49] Ehud Hrushovski, "Finitely Based Stable Theories," *Journal of Symbolic Logic* 54 (1989), 221–25.

[50] Ehud Hrushovski, "Unidimensional Theories Are Superstable," *Annals of Pure and Applied Logic* 50 (1990), 117–38.

[51] Ehud Hrushovski, "Unimodular Minimal Theories," *Journal of London Math. Society* 46 (1992), 385–96.

[52] Ehud Hrushovski, "A New Strongly Minimal Set," *Annals of Pure and Applied Logic* 62 (1993), 147–66.

[53] Ehud Hrushovski, "The Mordell-Lang Conjecture for Function Fields," *Journal of American Math. Society* 9 (1996), 667–90.

[54] Ehud Hrushovski, "The Manin-Mumford Conjecture and the Model Theory of Difference Fields," *Annals of Pure and Applied Logic* 112 (2001), 43–115.

[55] Ehud Hrushovski, "Pseudo-Finite Fields and Related Structures," *Quaderni di Matematica* 11 (2002), 151–212.

[56] Ehud Hrushovski, "Simplicity and the Lascar Group," unpublished note (1998).

[57] Ehud Hrushovski, "Groupoids, Imaginaries and Internal Covers," Turkish Journal of Mathematics 36 (2012), 173–198.

[58] Ehud Hrushovski and Anand Pillay, "Groups Definable in Local Fields and Pseudo-Finite Fields," *Israel Journal of Math.* 85 (1994), 203–62.

[59] Ehud Hrushovski and Boris Zilber, "Zariski Geometries," *Journal of American Math. Society* 9 (1996), 1–56.

[60] Tapani Hyttinen, "Remarks on Structure Theorems for ω_1-Saturated Models," *Notre Dame Journal of Formal Logic* 36 (1995), 269–78.

[61] Thomas Jech, *Set Theory*, 3rd edn (Springer, New York, 2002).

[62] William M. Kantor, Martin W. Liebeck, and H. Dugald Macpherson, "$\aleph_0$-Categorical Structures Smoothly Approximated by Finite Structures," *Proceedings of London Math. Society* 59 (1989), 439–63.

[63] Itay Kaplan, Thomas Scanlon, and Frank O. Wagner, "Artin–Schreier Extensions in NIP and Simple Fields," *Israel Journal of Math.* 185 (2011), 141–54.

[64] Hirotaka Kikyo and Saharon Shelah, "The Strict Order Property and Generic Automorphisms," *Journal of Symbolic Logic* 67 (2002), 214–16.

[65] Byunghan Kim, *Simple First Order Theories*, PhD thesis, University of Notre Dame (1996).

[66] Byunghan Kim, "Forking in Simple Unstable Theories," *Journal of London Math. Society* 57 (1998), 257–67.

[67] Byunghan Kim, "A Note on Lascar Strong Types in Simple Theories," *Journal of Symbolic Logic* 63 (1998), 926–36.

[68] Byunghan Kim, "On the Number of Countable Models of a Countable Supersimple Theory," *Journal of London Math. Society* 60 (1999), 641–5.

[69] Byunghan Kim, "Simplicity, and Stability in There," *Journal of Symbolic Logic* 66 (2001), 822–36.

[70] Byunghan Kim, "Recovering the Hyperdefinable Group Action in the Group Configuration Theorem," *Journal of Symbolic Logic* 75 (2010), 12–24.

[71] Byunghan Kim, "The Lascar Group, and the Strong Types of Hyperimaginaries," *Notre Dame Journal of Formal Logic*, forthcoming.

[72] Byunghan Kim, Alexei Kolesnikov, and Akito Tsuboi, "Generalized Amalgamation and n-Simplicity," *Annals of Pure and Applied Logic* 155 (2008), 97–114.

[73] Byunghan Kim and Rahim Moosa, "Stable Definability and Generic Predicates," *Journal of Symbolic Logic* 72 (2007), 1163–76.

[74] Byunghan Kim and Anand Pillay, "Simple Theories," *Annals of Pure and Applied Logic* 88 (1997), 149–64.

[75] Byunghan Kim and Anand Pillay, "From Stability to Simplicity," *Bulletin of Symbolic Logic* 4 (1998), 17–36.

[76] Byunghan Kim and Anand Pillay, "Around Stable Forking," *Fundamenta Mathematicae* 170 (2001), 107–18.

[77] Alexei Kolesnikov, "n-Simple Theories," *Annals of Pure and Applied Logic* 131 (2005), 227–61.

[78] Krzysztof Krupiński and Anand Pillay, "On Stable Fields and Weight," *Journal of the Institute of Math. of Jussieu* 10 (2011), 349–58.

[79] Alistair H. Lachlan, "On the Number of Countable Models of a Countable Superstable Theory," in *Logic, Methodology, Philosophy of Science IV* (North Holland, Amsterdam, 1973), 45–56.

[80] Daniel Lascar, "Rank and Definability in Superstable Theories," *Israel Journal of Math.* 23 (1976), 53–87.

[81] Daniel Lascar, "On the Category of Models of a Complete Theory," *Journal of Symbolic Logic* 47 (1982), 249–66.

[82] Daniel Lascar, "Ordre de Rudin-Keisler et poids dans les theories stables," *Math. Logic Quarterly* 28 (1982), 413–30.

[83] Daniel Lascar, "The Group of Automorphisms of a Relational Saturated Structure," in *Proceedings, Banff, 1991*, NATO ASI Series C: Math. Phys. Sciences 411 (Kluwer Academic, Dordrecht, 1993), 225–36.

[84] Daniel Lascar, "Recovering the Action of an Automorphism Group," in *Logic: From Foundations to Applications*, Oxford Science Publications (Oxford University Press, New York, 1996), 313–26.

[85] Daniel Lascar and Anand Pillay, "Hyperimaginaries and Automorphism Groups," *Journal of Symbolic Logic* 66 (2001), 127–43.

[86] Daniel Lascar and Bruno Poizat, "An Introduction to Forking," *Journal of Symbolic Logic* 44 (1979), 330–50.

[87] Angus MacIntyre, "On ω_1-Categorical Theories of Fields," *Fundamenta Mathematicae* 71 (1971), 1–25.

[88] David Marker, *Model Theory: An Introduction* (Springer, New York, 2002).

[89] H. Dugald Macpherson, "Absolutely Ubiquitous Structures and $\aleph_0$-Categorical Structures," *Quarterly Journal of Math.* 39 (1988), 483–500.

[90] Amador Martin-Pizarro and Anand Pillay, "Elliptic and Hyperelliptic Curves over Supersimple Fields," *Journal of London Math. Society* 69 (2004), 1–13.

[91] Amador Martin-Pizarro and Frank O. Wagner, "Supersimplicity and Quadratic Extensions," *Archive for Math. Logic* 48 (2009), 55–61.

[92] Ronald B. Medina, "Model Theory of Differentially Closed Fields with a Generic Automorphism," preprint, arXiv:0907.3959.

[93] Michael Morley, "Categoricity in Power," *Transactions of American Math. Society* 114 (1965), 514–38.

[94] Ludomir Newelski. "The diameter of a Lascar Strong Type," *Fundamenta Mathematicae* 176 (2003), 157–70.

[95] Alf Onshuus, "Properties and Consequences of Thorn-Independence," *Journal of Symbolic Logic* 71 (2006), 121.

[96] Ya'acov Peterzil and Sergei Starchenko, "A Trichotomy Theorem for o-Minimal Structures," *Proceedings of London Math. Society* (3) 77 (1998), 481–523.

[97] Anand Pillay, *An Introduction to Stability Theory* (Clarendon Press, Oxford, 1983).

[98] Anand Pillay, "Definability and Definable Groups in Simple Theories," *Journal of Symbolic Logic* 63 (1988), 788–796.

[99] Anand Pillay, *Geometrical Stability Theory* (Oxford University Press, Oxford, 1996).

[100] Anand Pillay, "A Note on CM-Triviality and the Geometry of Forking," *Journal of Symbolic Logic* 65 (2000), 474–80.

[101] Anand Pillay, "On Countable Simple Unidimensional Theories," *Journal of Symbolic Logic* 68 (2003), 1377–84.

[102] Anand Pillay and Bruno Poizat, "Pas d'imaginaires dans l'infini!," *Journal of Symbolic Logic* 52 (1987), 400–3.

[103] Anand Pillay and Bruno Poizat, "Corps et chirurgie," *Journal of Symbolic Logic* 60 (1995), 528–33.

[104] Anand Pillay, Thomas Scanlon, and Frank O. Wagner, "Supersimple Fields and Division Rings," *Math. Research Letters* 5 (1998), 473–83.

[105] Anand Pillay and Evegueni Vassiliev, "Imaginaries in Beautiful Pairs," *Illinois Journal of Math.* 48 (2004), 759–68.

[106] Bruno Poizat, "Paires de structures stables," *Journal of Symbolic Logic* 48 (1983), 239–49.

[107] Bruno Poizat, *Stable Groups*, translated (American Math. Society, New York, 2001).

[108] G. Schlichting, "Operationen mit periodischen Stabilisatoren," *Archiv der Mathematik* 34 (1980), 97–9.

[109] Ziv Shami, "Definability in Low Simple Theories," *Journal of Symbolic Logic* 65 (2000), 1481–90.

[110] Ziv Shami and Frank O. Wagner, "The Binding Group in Simple Theories," *Journal of Symbolic Logic* 67 (2002), 1016–24.

[111] Ziv Shami, "Countable Hypersimple Unidimensional Theories," *Journal of LMS* 83 (2011), 309–32.

[112] Saharon Shelah, "Simple Unstable Theories," *Annals of Math. Logic* 19 (1980), 177–203.

[113] Saharn Shelah, *Classification Theory and the Number of Non-isomorphic Models*, revised (North-Holland, Amsterdam, 1990).

[114] Katrin Tent and Martin Ziegler, *A Course in Model Theory* (Cambridge University Press, Cambridge, 2012).

[115] Ivan Tomasic and Frank O.Wagner, "Applications of the Group Configuration Theorem in Simple Theories," *Journal of Math. Logic* 3 (2003), 239–55.

[116] Evegueni Vassiliev, "Generic Pairs of *SU*-Rank 1 Structures," *Annals of Pure and Applied Logic* 120 (2003), 103–49.

[117] Evegueni Vassiliev, "On Pseudolinearity and Generic Pairs," *Math. Logic Quarterly* 56 (2010), 35–41.

[118] Frank O. Wagner, "Relational Structures and Dimensions," in *Automorphism Groups of First-Order Structures* (Oxford University Press, Oxford, 1994), 153–80.

[119] Frank O. Wagner, *Stable Groups* (Cambridge University Press, Cambridge, 1997).

[120] Frank O. Wagner, *Simple Theories* (Kluwer Academic, Dordrecht, 2000).

[121] Frank O. Wagner, "Hyperdefinable Groups in Simple Theories," *Journal of Mathematical Logic* 1 (2001), 152–72.

[122] Frank O. Wagner, "Groups in Simple Theories," in *Logic Colloquium 2001*, Lecture Notes in Logic 20 (A. K. Peters, Wellesley, MA, 2005), 440–7.

[123] Carol Wood, "Notes on the Stability of Separably Closed Fields," *Journal of Symbolic Logic* 44 (1979), 412–16.

[124] Ikuo Yoneda, "Some Remarks on CM-Triviality," *Journal of Math. Society of Japan* 61 (2009), 379–91.

[125] Martin Ziegler, "Introduction to the Lascar Group," in *Tits Buildings and the Model Theory of Groups*, London Math. Lecture Note Series 291 (Cambridge University Press, 2002), 279–98.

[126] Boris I. Zilber, *Uncountably Categorical Theories*, American Math. Society Translations of Math. Monographs, Vol. 117 (American Math. Society, New York, 1993).

INDEX